Biotechnology of Natural Products

Wilfried Schwab
Bernd Markus Lange • Matthias Wüst
Editors

Biotechnology of Natural Products

Springer

Editors
Wilfried Schwab
Biotechnology of Natural Products
Technische Universität München
Freising, Germany

Matthias Wüst
Bioanalytics
Institute of Nutritional and Food Sciences (IEL)
University of Bonn
Bonn, Germany

Bernd Markus Lange
Institute of Biological Chemistry and
M.J. Murdock Metabolomics Laboratory
Washington State University
Pullman, WA, USA

ISBN 978-3-319-67902-0 ISBN 978-3-319-67903-7 (eBook)
https://doi.org/10.1007/978-3-319-67903-7

Library of Congress Control Number: 2017957822

Printed on acid-free paper

This Springer imprint is published by Springer Nature
The registered company is Springer International Publishing AG
The registered company address is: Gewerbestrasse 11, 6330 Cham, Switzerland

Preface

The world is losing farmland at a rapid rate to erosion and chemical depletion of soils, while the world's population is increasing tremendously. Plant scientists are challenged with the development of sustainable production systems for food, feed, fuel, and value-added bioproducts. The natural sources for bioactive natural products, which support industries from nutrition to flavors and health care, are subject to unfavorable weather conditions, insect infestations, poor quality of soil, and sociopolitical instabilities. Therefore, this volume assesses the current status of biotechnological processes for the production of valuable plant natural products.

What are the advantages attributed to a biotechnological production of natural products that justify such an effort?

(i) Legal status of "natural" as opposed to "artificial" or "synthetic."
(ii) If defined product is desired, enzymes with exquisite catalytic specificity can be employed.
(iii) If chemical diversity is desired, promiscuous enzymes can be employed.
(iv) Optimized conditions lead to high reliability and productivity.
(v) Multistep reactions, which are difficult or impossible to achieve in aqueous solutions by chemical means, proceed under mild conditions.

The enormous chemical diversity among plant natural products is reflected in distinct chemical properties and biological activities. For this book, we have assembled a cast of authors that cover a rather broad spectrum of research to illustrate the potential for harnessing the diversity of natural products through biotechnological approaches. Chapters on isoprenoids, polyphenols, alkaloids, and phenylpropanoids are included, with a special emphasis on economically important targets such as vanillin, caffeine, and morphine. Furthermore, two chapters discuss the legal framework for natural products and their quality control, with authenticity and quality assurance in the European Union serving as an example.

The legislation that determines the correct usage of the term "natural" in labeling consumer goods is quite complex, with several pitfalls, and has already led to several lawsuits in member countries of the European Union, especially in Germany.

We hope this book will be a resource for students, teachers, and researchers with interests in the fascinating field of natural product biotechnology.

The editors would like to express their sincere thanks to the esteemed colleagues who contributed to this compilation of eleven chapters and to the publisher for administrative support and patience during the writing and review process.

Contents

Contributors

Leanne Bourgeois Department of Biology, Centre of Applied Synthetic Biology, Centre for Structural and Functional Genomics, Concordia University, Montréal, QC, Canada

Alex Campbell Department of Biology, Centre of Applied Synthetic Biology, Centre for Structural and Functional Genomics, Concordia University, Montréal, QC, Canada

Meghan E. Davies Department of Biology, Centre of Applied Synthetic Biology, Centre for Structural and Functional Genomics, Concordia University, Montréal, QC, Canada

Jan C.R. Demyttenaere European Flavour Association (EFFA), Brussels, Belgium

Elena Fossati Department of Biology, Centre of Applied Synthetic Biology, Centre for Structural and Functional Genomics, Concordia University, Montréal, QC, Canada

Nethaji J. Gallage Plant Biochemistry Laboratory, Department of Plant and Environmental Sciences, University of Copenhagen, Center for Synthetic Biology, VILLUM Research Center for "Plant Plasticity", Copenhagen, Denmark

Cecilie I.C. Hansen Plant Biochemistry Laboratory, Department of Plant and Environmental Sciences, University of Copenhagen, Copenhagen, Denmark

Katja Härtl Biotechnology of Natural Products, Technische Universität München, Freising, Germany

Stefan Jennewein Department of Industrial Biotechnology, Fraunhofer Institute for Molecular Biology and Applied Ecology (IME), Aachen, Germany

Misako Kato Graduate School of Humanities and Sciences, Ochanomizu University, Tokyo, Japan

Institute for Human Life Innovation, Ochanomizu University, Tokyo, Japan

Mattheos A.G. Koffas Department of Chemical and Biological Engineering, Rensselaer Polytechnic Institute, Center for Biotechnology and Interdisciplinary Studies, Troy, NY, USA

Department of Biological Sciences, Rensselaer Polytechnic Institute, Center for Biotechnology and Interdisciplinary Studies, Troy, NY, USA

Bernd Markus Lange Institute of Biological Chemistry and M.J. Murdock Metabolomics Laboratory, Washington State University, Pullman, WA, USA

Vincent J.J. Martin Department of Biology, Centre of Applied Synthetic Biology, Centre for Structural and Functional Genomics, Concordia University, Montréal, QC, Canada

Christopher McElroy Institute for Molecular Biotechnology, RWTH Aachen University, Aachen, Germany

Kate McGraphery Biotechnology of Natural Products, Technische Universität München, Freising, Germany

Birger Lindberg Møller Plant Biochemistry Laboratory, Department of Plant and Environmental Sciences, University of Copenhagen, Center for Synthetic Biology, VILLUM Research Center for "Plant Plasticity", Copenhagen, Denmark

Fumiyo Nakayama Graduate School of Humanities and Sciences, Ochanomizu University, Tokyo, Japan

Lauren Narcross Department of Biology, Centre of Applied Synthetic Biology, Centre for Structural and Functional Genomics, Concordia University, Montréal, QC, Canada

Maike Petersen Institut für Pharmazeutische Biologie und Biotechnologie, Philipps-Universität Marburg, Marburg, Germany

Soheil Pezeshki Institut für Pharmazeutische Biologie und Biotechnologie, Philipps-Universität Marburg, Marburg, Germany

Michael E. Pyne Department of Biology, Centre of Applied Synthetic Biology, Centre for Structural and Functional Genomics, Concordia University, Montréal, QC, Canada

Julian Rüdiger Biotechnology of Natural Products, Technische Universität München, Freising, Germany

Wilfried Schwab Biotechnology of Natural Products, Technische Universität München, Freising, Germany

Jian Wang College of Engineering, University of Georgia, Athens, GA, USA

Matthias Wüst Bioanalytics, Institute of Nutritional and Food Sciences (IEL), University of Bonn, Bonn, Germany

Yajun Yan BioChemical Engineering Program, College of Engineering, University of Georgia, Athens, GA, USA

Yaping Yang College of Engineering, University of Georgia, Athens, GA, USA

Jian Zha Department of Chemical and Biological Engineering, Rensselaer Polytechnic Institute, Center for Biotechnology and Interdisciplinary Studies, Troy, NY, USA

Part I

Biotechnological Production of Selected Natural Products

Vanilla: The Most Popular Flavour

1

Nethaji J. Gallage and Birger Lindberg Møller

Abbreviations

°C	Celsius
cm	Centimetre
DNA	Deoxyribonucleic acid
ER	Endoplasmic reticulum
EU	European Union
g	Gram
h	Hour
kg	Kilogram
l	Liter
m	Meter
mM	Millimolar
sp.	Species
UDP	Uridine diphosphate glucose
US$	United States dollar

1.1 Introduction

Vanilla is a universally appreciated global delicacy and probably the most popular plant-derived flavour in the world. Vanilla flavour is obtained from the seedpods of the cultivated orchid, *Vanilla planifolia* and from several other vanilla species.

N.J. Gallage • B.L. Møller (✉)
Plant Biochemistry Laboratory, Department of Plant and Environmental Sciences, University of Copenhagen, Center for Synthetic Biology, VILLUM Research Center for "Plant Plasticity", Thorvaldsensvej 40, DK-1871 Frederiksberg C, Copenhagen, Denmark
e-mail: blm@plen.ku.dk

W. Schwab et al. (eds.), *Biotechnology of Natural Products*,
https://doi.org/10.1007/978-3-319-67903-7_1

Vanilla pods, vanilla extracts (the isolated extract from vanilla pods) and vanillin (the main flavour of the vanilla extract) constitute a multimillion-dollar industry.

The history of human utilization of vanilla flavour began with the Aztecs, in current day Mexico, during the 1300s. The Aztecs utilized the seedpods from vanilla orchids for flavour and fragrance. The earliest documented consumption of vanilla pods dates back to 1520, when the Spanish arrived in Mexico. The colonists were exposed to the Aztecs' use of vanilla to flavour a drink, which is considered the ancestor of hot chocolate. The Spanish were impressed by the flavour of vanilla and transported the pods back to Spain. Subsequently, the demand for vanilla pods in Spain, France and the rest of Europe increased, and this demand led to domestication of the vanilla orchid, *V. planifolia* [1].

Despite the popularity of the flavour, there is still a great deal of uncertainty regarding the biology of vanilla. Natural pollinators of the vanilla orchid species are not well investigated [2, 3]. Symbiosis with fungi is required for seed germination and growth, but the interactions are poorly understood [4]. The question of why some *Vanilla* species produce aromatic fleshy pods while other *Vanilla* species do not is unclear, and it is unknown why some *Vanilla* species rarely flower in cultivation or even in their native habitat [5]. Moreover, little is known of pod and seed dispersal mechanisms, and the taxonomy and systematics of the *Vanilla* genus is in a state of ambiguity [6]. The biosynthesis pathway of vanillin (4-hydroxy-3-methoxybenzaldehyde), the main flavour component of vanilla extract [7], has remained elusive for decades in spite of dedicated efforts. In 2014, an enzyme designated *Vp*VAN was isolated from *V. planifolia* and shown to efficiently convert ferulic acid and ferulic acid glucoside into vanillin and vanillin glucoside, respectively [8] (**see also Chap. 9 of this book**).

In contrary to vanilla orchid biology, bioengineering approaches to biosynthesis of vanillin in microorganisms and its status as a 'natural' ingredient are in the spotlight. In this book chapter, we review our current understanding of the **vanilla plants, vanilla flavour**, and **vanillin biosynthesis in the vanilla orchid** and summarize **emerging bioengineering possibilities of vanillin biosynthesis in microorganisms**.

1.2 Vanilla Orchids, Vanilla Flowers and the Pod

Vanilla plants are climbing orchids [6, 9]. They belong to the *Orchidaceae* family, one of the largest families of flowering plants. However, more detailed classification of vanilla plants into tribes, subfamilies and genera has proven to be challenging. The increased use of DNA-based systematic sequence studies has governed the latest classification of the vanilla species. According to these studies, the vanilla orchid belongs to the genus *Vanilla*, tribe Vanilleae, subfamily Vanilloideae and family Orchidaceae [6, 10–12]. In the genus of *Vanilla* there are about 110 vanilla species, among which there are three species known to be important for commercial cultivation and for local low scale vanilla farmers: *V. planifolia*, *V. tahitensis* and *V. pompona*. *V. planifolia* originates in current day Mexico and is the most valued of these three species because

Fig. 1.1 *V. planifolia* climbing orchids that produce beautiful yellow flowers when the vine is about 4–5 m long. Pods in this picture are only a few weeks old (Photo was taken at CIRAD shade houses, at La Reunion. Photo credit: Nethaji J. Gallage)

of its vanilla flavour quality. *V. planifolia* is widely cultivated and provides 95% of the world vanilla production [9]. *V. planifolia* differs very little from its ancestors in the wild. The vanilla plants that are grown in La Reunion, Madagascar, Mauritius and Seychelles are derived from a single cutting of *V. planifolia*. Stem cutting propagation results in a lack of genetic variation in vanilla plants [9].

The orchid, *V. planifolia* is a climbing perennial vine with a large, green, fleshy and succulent stem that is photosynthetic. *V. planifolia* has oblong, smooth, bright green leaves and adventitious aerial roots that grow opposite of each leaf, aiding lateral support. The roots are associated with endotrophic mycorrhiza [13]. This symbiosis provides the fungus access to the sugars produced by *V. planifolia* through photosynthesis and in return, vanilla is able to get water and minerals through the fungal mycelium. This particular association involves 'endomycorrhizae'- fungi whose hyphae actually enter the plant's root's cell membrane [4].

When the *V. planifolia* vine is approximately 4–5 m long, the plant starts flowering. The vanilla flowers are yellow, bisexual and usually sprout towards the top of the plant (Fig. 1.1). Once opened, these flowers only survive for 24 h. Although flowers are bisexual, they are not able to self-pollinate. Pollination requires outside action either by transferring pollen from an anther to the stigma using a soft tiny paintbrush or by lifting a thin-membrane that prevents self-fertilization, and subsequently pressing the flower's anther towards the stigma [14]. The lack of specific flower pollinators was a key issue encountered when introducing *V. planifolia* to the rest of the world from Mexico. Vanilla's pollen is also largely inaccessible to most pollinating insects due to the shape of the flower [14]. In nature, flowers are pollinated by small bees and hummingbirds that are capable of penetrating the tough

Fig. 1.2 Transverse picture of vanilla pod disc, photo is taken with Canon EF 200 mm f/2.8 L macro lens (Photo credit: Nethaji J Gallage)

membrane that separates the plant's pistil and stamen, although very little scientific information is available on this subject. In 1841, Edmond Albius had developed an artificial pollination technique of vanilla flowers by using bamboo sticks. Additionally, there are records by Charles Morren, in 1836, of the artificial pollination of vanilla flowers. Even today, hand pollination techniques are used for the flower pollination in vanilla production, and these techniques have not developed much further since 1841 [14, 15].

Botanically, the vanilla pod is a seed capsule, but is generally referred to as a vanilla bean or vanilla pod. The pod reaches its full size 10–15 weeks after pollination. Fully matured pods are about 15 cm long and are pale green to yellow in color. A transverse section of a mature green vanilla pod is triangular, containing a central cavity comprising numerous black seeds. From the outer part to the inner cavity, the pod consists of the following tissues: epicarp, mesocarp and endocarp. The mesocarp fills the majority of the fruit volume and consists of parenchyma cells. The cavity of the pod contains black seeds that are attached to a long narrow funicle [16] (Fig. 1.2).

Vanilla pods are harvested when they are 8–9 months old, before the pods begin to split from the end and become yellow in color. The immature green vanilla pods are almost odorless as the key flavour components are stored as glucosides. Freshly harvested pods are processed by curing to stop the natural vegetative process. This induces enzymes responsible for the formation of aromatic flavour constituents and prevents microbial growth, thus enabling long-term preservation of vanilla pods [9]. Curing methods can be different according to the country in which the plants are cultivated. As a result, the curing process has a major influence on the variety, quality and aromatic profile of the pods that are traded [17].

1.3 When the Green Becomes Black – Vanilla Pod Curing

In general, the curing process includes wilting, sweating, drying and conditioning of the pod. The main purpose of wilting is to stop the vegetative growth of the pod after harvest and to disrupt the cell structures. This process is also referred to as kilning, as it ceases the respiratory function of the plant tissues and promotes disruption of cell membranes thus creating better conditions for contact between certain metabolites and enzymes that release aroma compounds typically by catalyzing hydrolysis of a glycosidic linkage. Methods used to initiate kilning are sun drying, freezing, hot water, and ethylene gas treatments [18].

Sweating is the process by which the pod temperature is raised; moisture is initially allowed to escape to prevent harmful fermentation by microbial spoilage, but enough moisture is retained to promote enzymatic reactions. This is documented to be the most crucial step of the curing process, as most enzymes that are responsible for flavour and aroma development are active at this stage and are determinants of the quality of the cured pods. The sweating process is usually carried out for seven to ten days, and at the end of the process pods obtain a characteristic brown chocolate color due to the oxidation of polyphenolic compounds [13]. The pods are then dried at room temperature to reduce their moisture content so microbial spoilage can be avoided. The lower moisture content also reduces undesired enzymatic activities [9].

Conditioning/packaging is the last step of the curing process. Pods are placed in closed boxes for one to several months to initiate various biochemical reactions such as esterification, etherification and oxidative degradation etc., which produce the final high quality aromatic composition. After the curing process, the pod in general consists of sugars, proteins, free amino acids, fibers, cellulose, organic acids, oil, wax, resin, gum, pigments, minerals, volatile aromatics and essential oils. As mentioned, the most abundant aromatic compound in the vanilla extract is vanillin, followed by *p*-hydroxybenzadehyde, *p*-hydroxybenzoic acid, vanillic acid, *p*-hydroxybenzyl alcohol and vanillyl alcohol [18–20]. The ratio between these main flavour compounds in the pod is known to determine the quality and the final flavour of the vanilla pod and the vanilla extract. In addition to the curing process, the chemical constituents of the vanilla pod and its flavour are also determined by various other factors such as plant species, growing conditions, soil nutrition and maturity level at harvest [19].

Madagascar is the largest vanilla pod producer in the world. Madagascar, together with La Réunion islands, account for nearly 75% of the vanilla pod market. Vanilla pods from Madagascar and other islands east of Africa (La Réunion, Mauritius, the Comoros and the Seychelles) have the status 'Bourbon', and are considered the best vanilla pods in the current market. The vanilla growers in this region also produce the majority of the world's natural vanilla extract, which is an alcohol extract from matured and cured vanilla pods. The vanilla extract isolated from the vanilla pods includes more than 200 aromatic compounds [21].

1.4 Vanillin Is the Key Flavour Compound of the Complex Vanilla Extract

Often the terms 'vanilla' and 'vanillin' are confused in non-scientific communications. The term 'vanilla' refers to the complete extract of the vanilla pod and, as stated above, is known to include more than 200 different flavour compounds. However, the term vanilla is commonly used also to describe vanilla plants, vanilla pods and vanilla flavour. Vanillin is the main flavour compound in vanilla extract. The chemical structure of vanillin is quite simple and known to give the characteristic flavour and aroma that we associate with vanilla extracts [7].

Vanillin (4-hydroxy-3-methoxybenzaldehyde) (Fig. 1.3) is the most abundant compound in the cured vanilla pod and corresponds to 2.5–4.5% of the dry weight [7]. The French chemist Theodore Nicolas Gobley was able to isolate vanillin as the main flavour constituent of vanilla pod extracts in 1858. He also elucidated the chemical structure of vanillin [22]. Recent studies have demonstrated that vanillin starts to accumulate in a free or glycosylated form from the 15th week and continues to accumulate until the 30th week after pollination [23]. Vanillin is an aldehyde. The aldehyde group is very reactive, forming Schiff bases with, for example, free amino groups present on the side of the protein-bound lysine residues. This is the reason it is toxic to living organisms in high concentrations [24]. Vanilla plants store vanillin almost entirely in the glucose-conjugated form, vanillin-β-D-glucoside (commonly referred to as vanillin glucoside or glucovanillin) (**see also Chap. 9 of this book**). Vanillin glucoside can account for up to 15% of the vanilla pod dry weight. The highest concentration of vanillin glucoside is reached in the inner part of the pod, including mesocarp and placenta, 6 months after pollination. Small amounts of vanillin glucoside can also be found in the papillae of the pod. There is no vanillin found in the black vanilla seeds and interestingly, black vanilla seeds are free of aroma [16] (Fig. 1.4).

Currently, the annual worldwide consumption of vanilla pods, vanilla extract and vanillin is over 18,000,000 kg [25]. However, due to the slow growth of the orchids and the low concentration of vanillin in the vanilla pods (about 2% of the dry weight of cured vanilla beans), only about 0.25% of consumed vanillin originates from vanilla pods [26]. The production of vanilla beans and the isolation of vanillin from vanilla pods is a laborious and costly process [7]. Production of 1 kg

Fig. 1.3 Vanillin (4-hydroxy-3-methoxybenzaldehyde)

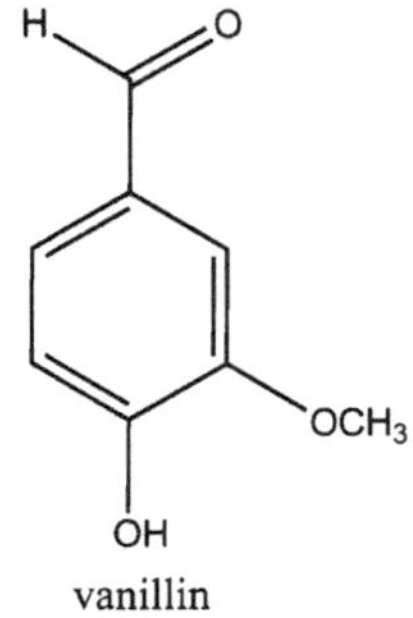

vanillin

Fig. 1.4 The most popular flavour (Photo credit: Nethaji J. Gallage)

of vanillin requires approximately 500 kg of vanilla pods, corresponding to the pollination of approximately 40,000 vanilla orchid flowers. The market cost of natural vanillin derived from vanilla pods is therefore high and fluctuates because of the unpredictable availability of vanilla pods. Crop yield is tightly associated with weather conditions, the incidence of diseases, as well as local and international political and economic issues. Vanillin extracted from vanilla pods has a market price varying from around US\$ 1200 kg^{-1} to more than US\$ 4000 kg^{-1} [27]. Thus, the increasing global demand for natural vanilla flavour appeals other sources of vanillin [28].

Currently the main source of vanillin is chemical synthesis, while, as mentioned, less than 1% is derived from the vanilla pod industry. Less than 20 years after its initial isolation, synthetically produced vanillin was marketed. Nowadays, guaiacol and lignin are favoured starting materials for synthetic vanillin. Synthetic vanillin is able to meet the global market demands, and it is also rather cheap with a market price below US\$ 15 kg^{-1} [1]. Chemical synthesis of vanillin suffers from serious drawbacks *e.g.* the use of hazardous chemicals. Chemical synthesis of vanillin via lignin has been calculated to require safe removal of 160 kg of waste per 1 kg of vanillin obtained. As a consequence, concerns are increasing regarding the negative environmental impact caused by chemical synthesis of vanillin [29]. Nevertheless, at present a substantial amount of synthetic vanillin is still derived from lignin [30]. Recent advances in biotechnology have allowed an alternative method to challenge the chemical synthesis of vanillin, namely bioengineering of natural vanillin.

1.5 How Does the Vanilla Plant Form Vanillin?

As the main constituent of the vanilla extract and the world's most popular flavour, vanillin is a compound of major interest to the flavour and fragrance industry. Although there are many research efforts and resources committed to engineering various microorganisms for vanillin biosynthesis, limited attention has been given to understanding the most efficient vanillin synthesizing machinery that is found in nature, namely the vanilla orchid. The vanilla orchid produces vanillin in the pods in such high concentrations that it cannot be compared to any other known biological system in nature.

This simple molecule, 4-hydroxy-3-methoxy benzaldehyde, was speculated to be formed through the operation of a multitude of pathways in the vanilla orchid. It was clear that vanillin biosynthesis in the vanilla pod proceeds from the amino acid phenylalanine and includes phenylpropanoid intermediates [31]. Vanillin glucoside and *p*-hydroxybenzaldehyde glucoside are the two most abundant components that produce aroma-active compounds upon hydrolysis in mature vanilla pods (**see also Chap. 9 of this book**). These compounds are structurally similar. Accordingly, a biosynthetic relationship between the formation of these two compounds had been hypothesized in the early literature, with *p*-hydroxybenzaldehyde as a putative precursor for vanillin glucoside biosynthesis [32–34]. Recently, the involvement of *p*-hydroxybenzaldehyde in vanillin biosynthesis has been ruled out by Gallage et al., who demonstrated that incubation with [^{14}C]-*p*-hydroxybenzaldehyde did not result in [^{14}C]-vanillin glucoside formation in the 6 month old vanilla pods after pollination [8]. This work also established a route to vanillin biosynthesis in the pod, namely via C3 side chain shorting of ferulic acid or its glucoside (Fig. 1.5). This argument was first brought forward by Zenk in 1965, who carried out radioactive precursor studies using [^{14}C]-ferulic acid and observed efficient conversion of [^{14}C]-ferulic acid to [^{14}C]-vanillin glucoside [31]. The conclusions of Zenk were confirmed by Negishi et al., who carried out a similar study employing radioactive precursors [35].

Ferulic acid and ferulic acid glucoside are ubiquitous phenylpropanoids that are derived from cinnamic acid. Ferulic acid is present as a constituent of the plant cell wall polymers. It is a component of lignocelluloses, where it confers rigidity to the cell wall by making the crosslink between polysaccharides and lignin. Ferulic acid is highly reactive and is often linked to a variety of metabolites including sugars as glycosidic conjugates, different esters and amides, thus forming a broad range of natural products [36]. Ferulic acid is formed by *O*-methylation of caffeic acid, and caffeic acid is formed from phenylalanine in approximately six enzyme catalysed steps. When produced from phenylalanine, the first intermediate is cinnamic acid, and the reaction is catalyzed by phenylalanine ammonia lyase (PAL) [37]. Subsequently, cinnamic acid 4-hydroxylase (C4H) [38] catalyzes the hydroxylation of cinnamic acid at the 4-position, resulting in the formation of *p*-coumaric acid. *p*-Coumaric acid-3-hydroxylase (C3H) [39] catalyzes hydroxylation of *p*-coumaric acid at the 3 position, resulting in the formation of caffeic acid. C3-hydroxylation step is shown to be proceeding via, *e.g.* quinate or shikimate esters. 4-Hydroxycinnamoyl-CoA ligase (4CL) and hydroxycinnamoyltransferase (HCT) are involved in quinate and shikimate ester formation [39]. Caffeic acid could, in principle, be *O*-methylated by an *O*-methyltransferase (OMT) [40] to afford ferulic acid.

The key enzyme that is involved in catalyzing C3 side chain shorting of ferulic acid or its glucoside in the pods of *V. planifolia* was recently identified by Gallage et al., [8]. A single enzyme named vanillin synthase (*Vp*VAN) was characterized to catalyze the double carbon bond cleavage of ferulic acid and ferulic acid glucoside to vanillin and vanillin glucoside, respectively. *Vp*VAN was isolated from *V. planifolia* and functionally characterized *in vitro*, in yeast and *in planta*. A route to vanillin biosynthesis mediated by *Vp*VAN is illustrated in Fig. 1.5. *Vp*VAN belongs to the enzyme family of cysteine proteases [8]. In a recent paper including Hailian Yang, Daphna Havkin-Frenkel and Richard A. Dixon as authors, the role of *Vp*VAN in vanillin biosynthesis was questioned [41].

Cysteine proteases are known to possess versatile physiological functions and do not have well-defined substrate specificities. In general, cysteine proteases are expressed as a pre-protein, with a N-terminal ER-targeting signal peptide being part of a pro-peptide domain comprising 130–160 residues [42]. To form the mature cysteine-proteinases, the pro-peptide sequence is removed either by a processing enzyme or by auto-catalytical processing [43]. The *Vp*VAN protein has not shown any evidence of autocatalytic processing. This indicates that the removal of the pro-peptide requires the action of a separate processing enzyme [8].

As vanillin is almost entirely stored as vanillin glucoside, it is apparent that the glycosylation step in the vanilla pod is highly efficient (**see also Chap. 9 of this book**). The glycosylation step has not been explored in detail, and it is not known at which step glucose incorporation occurs in the course of vanillin glucoside biosynthesis. Gallage et al., demonstrated, that *Vp*UGT72E1 possesses vanillin-specific glycosyltransferase activity. However, the glycosylation step in the vanillin biosynthesis machinery needs further study [8]. The enzyme that catalyses the reverse reaction hydrolysing vanillin glucoside to vanillin, vanillin-β-glucosidase, has previously been characterized [44].

1.6 Flavour Synthesis by Brewing – Bioengineering of Vanillin Biosynthesis

The increasing global demand for natural vanilla flavour can no longer be met with pods of the vanilla orchid as the sole source. This is why the market for vanillin is increasing [1]. The major source of marketed vanillin originates from chemical synthesis. In recent years, demand from consumers for natural products has increased. Though the approved use of the attribute 'natural' is not well defined and not evident to most consumers [45], many consumers equate the term "natural" with food quality and food safety and maybe also with enhanced environmental friendliness [46] (see also Chap. 11 of this book). Accordingly, and guided by novel technologies, research on bioengineering of microorganisms for flavour production is rapidly growing.

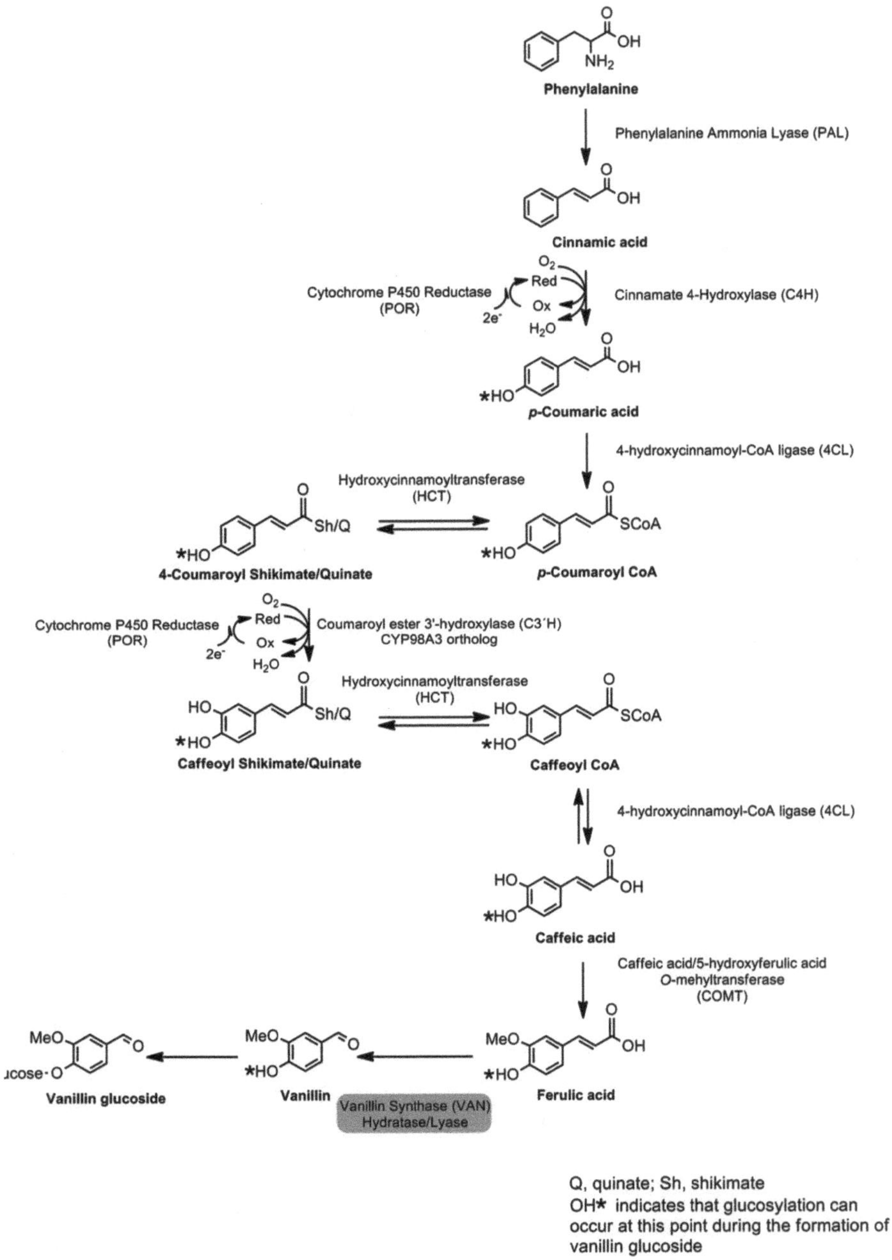

Fig. 1.5 A route to vanillin biosynthesis in pods of *V. planifolia* [8] (Figure reproduced with permission from Nature Communications)

Vanillin, obtained by bioengineered microorganisms by transforming a range of different substrates into vanillin, is entitled to the label 'natural vanillin' according to US and European legislation (EC Directive 88/388, OJ no. L 184 15/07/88) [47]. This affords significantly increased sales in the range of US\$1000 kg^{-1} for the bioengineered vanillin and enables the vanillin produced by bioengineering to compete with the chemically synthesized vanillin that currently dominates the market [48]. Several of the bioengineering approaches have been successful, and biotechnologically derived vanillin products have been available on the market for more than a decade (Fig. 1.6). Rhovanil produced by Solvay (previously known as Rhodia) was the first commercially available fermentation-derived vanillin product and was obtained by bioconversion of ferulic acid [49]. Ex-tumeric vanillin is marketed by De Monchy Aromatics and produced from curcumin [50]. Sense Capture Vanillin is obtained by bioconversion of eugenol and marketed by Mane [51]. *De novo* synthesized bio-vanillin using glucose as a precursor was commercialized in 2014 by Evolva A/S and International Flavors and Fragrances (IFF) [52].

Fermentation and bioengineering have been used to produce beers, wine, cheese, food colorants and pharmaceuticals for centuries. Vanillin has now been added to that list. Most bioengineering approaches for the synthesis of vanillin are based on the bioconversion of certain natural substances such as lignin, ferulic acid, eugenol and iso-eugenol etc., using microorganisms such as yeast, fungi and bacteria as

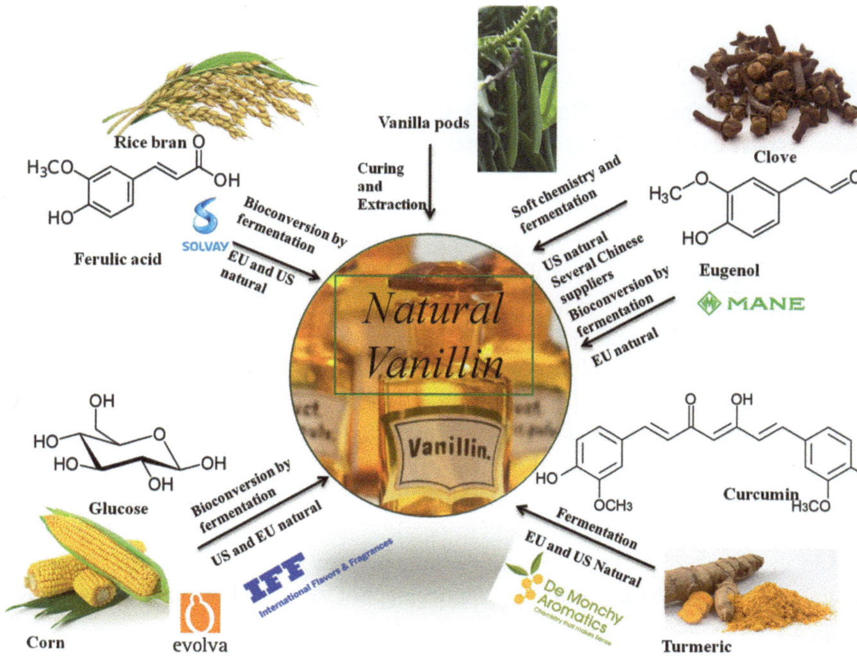

Fig. 1.6 Different commercial routes to natural vanillin [1] (Figure reproduced with permission from Molecular Plant)

production hosts by fermentation [53–55]. Microorganisms that exhibit rapid growth rates and are amenable to molecular genetics are obviously preferred. Further bioengineering has focused on increasing tolerance to high concentrations of both product and substrate. Microorganisms and fermentation ingredients, which have been given GRAS status, are preferred. GRAS is an acronym for Generally Recognized As Safe under the regulations of the US Food and Drug Administration (FDA) [56].

Microorganisms that are able to metabolize a range of different precursors into vanillin have been subjected to further bioengineering to circumvent remaining pathway bottlenecks and other drawbacks. Bioengineering includes the use of tools such as genetic engineering, enzyme optimization and cost-efficient downstream processing. However, several major yet common issues have challenged the successful use of microorganisms for efficient bioconversion of various substrates into vanillin. Bottlenecks include: (1) cytotoxicity of the flavour products obtained and of their precursors; (2) inefficient metabolic flow; and (3) costly downstream processing methods due to the physicochemical properties of the substrate and the product. The increasing knowledge of enzymes that are involved in the bioconversion of ferulic acid and other substrates into vanillin, as well as identification and characterization of the corresponding genes, offers new opportunities for more targeted bioengineering of microorganisms for vanillin production. In the following sections, the bioconversion and bioengineering of vanillin by microorganisms (bacteria, fungi and yeast) are presented and commented upon, with emphasis on the major issues encountered and the solutions obtained.

1.7 Biotechnology-Based Production of Vanillin from Eugenol, Iso-Eugenol, Ferulic Acid and Glucose

Availability of the Substrate In general, bioengineering of vanillin in microorganisms is carried out using precursors that are structurally similar to vanillin *e.g.* eugenol (2-methoxy-4-(2-propenyl)-phenol), iso-eugenol (2-methoxy-4-(1-propenyl)-phenol) or ferulic acid (4-hydroxy-3-methoxy-cinnamic acid). These compounds are also relatively cheap and easily available. Ferulic acid is one of the most abundant hydroxycinnamic acid derived products, present as a constituent of the plant cell wall and as a lignin monomer precursor [57]. Ferulic acid is widely distributed throughout the plant kingdom and was recently shown to serve an additional function as the main precursor for vanillin production in the vanilla orchid [8]. The main source of the ferulic acid for bioengineered vanillin production is agricultural waste such as sugar beet, barley and wheat bran. Ferulic acid is ester-linked to pectic side chains in beet and ether-linked to lignin in cereals [57]. Eugenol and iso-eugenol are main components of natural essential oils of clove trees [58]. The pure eugenol and iso-eugenol substrates are inexpensive and cost no more than US$ 5 kg^{-1} [59].

To reach a higher degree of sustainability, much effort has been focused on the use of agricultural waste as the main precursor for bioengineering. Several studies have attempted to remove ferulic acid from plant cell wall materials enzymatically

[60–62]. Feruloyl esterases are enzymes that are able to hydrolyze the ester bonds by which ferulic acid is attached to the cell wall polymers, and can be isolated from a wide range of fungi, yeast and bacteria [63]. Two feruloyl esterases, FaeA and FaeB, isolated from *Aspergillus niger,* are able to release ferulic acid from industrial by-products such as wheat straw, coffee pulp, apple core, maize bran, maize fiber *etc.*[64]. *A. niger* strain 1–1472 has been used to release ferulic acid from autoclaved maize bran [65].

Enzymatic hydrolysis of cell walls using a combination of commercial polysaccharide-degrading enzymes and feruloyl esterase has also been investigated [66]. Currently, these methods are not economically feasible as the commercially available polysaccharide-degrading enzymes are costly and would result in significantly increased production costs of vanillin. Ferulic acid can also be released from plant cell walls by alkaline treatment at high temperatures (85–100°C) [67, 68]. This kind of chemical release of ferulic acid would not be considered natural processing according to EU regulations, but would comply with registration as "natural" according to US legislation [47, 67].

Today, the ferulic acid used for commercial production of natural vanillin is mainly obtained as a by-product in the production of rice bran oil. The ferulic acid is liberated from the rice bran by enzymatic treatment to comply with the regulations for being classified as a natural product. The cost of naturally extracted ferulic acid is relatively high with a price around US$ 180 kg^{-1} [69].

The cost of glucose can be as low as US$ 0.30 kg^{-1}. It is the cheapest substrate used in vanillin production by bioengineering to date [53]. It is also valuable as a cheap primary energy source for the production strain. Moreover, glucose is a more attractive substrate in comparison to eugenol, ferulic acid and other phenolic compounds, because it is not toxic to the host microorganisms.

Host Microorganisms One of the key decisions in developing a vanillin bioengineering process is to choose a host strain that is highly tolerant to both the substrate and the product. Vanillin is rarely accumulated in high concentrations in living cells as it is toxic. In the vanilla plants, vanillin is glycosylated to vanillin glucoside while in many other living organisms it is expected to be oxidized to vanillic acid or reduced to vanillyl alcohol and thereby reduced in the toxicity [1, 8, 24, 53].

Studies of eugenol bioconversion and ferulic acid catabolism in *Rhodococcus sp.* I24 and *Rhodococcus sp.* PD630 have shown that *Rhodococcus sp.* I24, in contrast to *Rhodococcus sp.* PD630, can tolerate up to 2.4–3.0 mM eugenol, implying an effective eugenol catabolism naturally occurring in this strain [70].

Actinomyctetes, such as *Amycolatopsis sp.* [70] and *Streptomyces setonii* [71] are able to accumulate high concentrations of vanillin while at the same time exhibiting a high tolerance towards ferulic acid. *P. putida IE27* [72] and *Bacillus fusifomis* were reported to efficiently convert iso-eugenol into vanillin. *Bacillus fusifomis* is known to yield 32.5 g/l vanillin after 72 h incubation [73]. Similarly, the *P. putida* IE27 strain is able to produce 16.1 g/l of vanillin after 24 h incubation. The vanillin production was induced by continuously adding iso-eugenol to the cultures, which helps to prevent further oxidation of the vanillin formed into vanillic acid [72].

However, the filamentous growth of actinomycetes results in highly viscous broths, unfavourable pellet formation and a lot of fragmentation and lysis of the mycelium, thereby complicating downstream processing [74].

Compared to bacterial strains, yeast strains have not been as heavily exploited for bioengineered synthesis of vanillin. However, natural vanillin is produced via bioengineered *Saccharomyces cerevisiae* on a commercial scale, and more information on this strain is provided in the sections below. Vanillin biosynthesis in bioengineered algae and cyanobacteria is yet to be established.

Cytotoxicity The bioengineering of vanillin-producing microbial systems is challenged by the potential toxicity of precursors as well as products formed. It should be noted that compounds such as vanillin are produced in nature as part of the plants´ defense system against pathogens such as bacteria and fungi [48]. As pointed out previously, vanillin is toxic to living cells in high concentrations. Dealing with this issue is an important pre-requisite for building economically viable biotechnology-derived vanillin cell factories. In the case of *S. cerevisiae*, vanillin production beyond 0.5–1 g/l was toxic, as shown by hampered growth and low level of vanillin accumulation [53]. The natural vanillin biosynthesis pathway in the vanilla orchid *V. planifolia* has an elegant solution to cope with the toxicity issue, by glucosylation of vanillin to vanillin-β-D-glucoside (**see also Chap. 9 of this book**). The same strategy was implemented by Hansen et al. [53], in which the *A. thaliana* UDP-glucose glycosyltransferase UGT72E2 was employed to glucosylate vanillin, producing the less toxic vanillin-β-glucoside as the final product. Hansen and coworkers reported that extracellular concentration of vanillin β-D-glucoside even above 25 g/l had no effect on yeast growth [53]. Moreover, vanillin-β-glucoside has higher water solubility than vanillin and can potentially serve as a sink that can aid in directing the pathway towards vanillin synthesis.

Genes involved in metabolizing ferulic acid into vanillin have been heterologously expressed in engineered *E. coli* with high vanillin tolerance to bypass the problems related to product toxicity. This includes expression of the *Fcs* and *Ech* genes from *Amycolatopsis sp.* HR104 [75]. The vanillin-resistant mutant strain was obtained following NTG (N-methyl-N-nitro-N-nitrosoguanidine) mutagenesis and following a 48 h incubation period, as much as 1 g/l of vanillin was produced in a media containing 2 g/l of ferulic acid. To further circumvent the inhibitory effect of vanillin, XAD-2 ion-exchange resin was used to bind the vanillin formed in the medium. This increased vanillin yield to 5 g/l in 48 h when ferulic acid substrate was applied during incubation [75].

By-Products When various microorganisms metabolize eugenol, iso-eugenol and ferulic acid, vanillin is only produced as an intermediate, and is either readily reduced to vanillyl alcohol or oxidized to vanillyl acid by alcohol dehydrogenases and oxidases, respectively. General approaches used to circumvent undesired product formation are knock-outs and/or knock-downs of genes related to undesired catabolism of substrates and/or products. Several examples are listed below:

The actinomycete *Amycolatopsis sp.* ATCC 39116 is able to synthesize vanillin from ferulic acid but the vanillin formed is subjected to further undesired metabolism. Two to three times higher vanillin accumulation and a substantially reduced amount of vanillic acid was observed using the *Amycolatopsis sp.* ATCC 39116 Δvdh::Km(r) mutant when ferulic acid was provided as a substrate for biotransformation in a cultivation experiment using 2 l bioreactor scale. In the mutant strain, the *vdh* gene, which codes for the vanillin dehydrogenase activity, has been deleted [76].

Hansen et al. constructed glucose-based *de novo* vanillin biosynthesis in *S. cerevisiae*. Further metabolism of vanillin to vanillyl alcohol was circumvented by targeted deletions of alcohol dehydrogenase (ADH) encoding genes. From the tested enzymes, ADH6 was recognized as the most important enzyme catalyzing vanillin reduction in *S. cerevisiae*. The *adh6* mutants in *S. cerevisiae* grew normally under all growth conditions and showed a 50% decrease in converting vanillin to vanillyl alcohol [53].

Bacillus subtilis 3NA is a microorganism with enhanced capacity to metabolize lignin-derived compounds. The strain tolerates a high concentration of up to 20 mM of vanillin. However, *B. subtilis* 3NA further converts vanillin to vanillic acid and subsequently to guaiacol whereas ferulic acid is converted to 4-vinyl guaiacol. Gene deletion of phenolic acid decarboxylase *bsdD* resulted in an increased vanillic acid synthesis in *Bacillus subtilis* 3NA [77].

In yeast, ferulic acid is readily detoxified by the action of the decarboxylation enzymes PAD1 and FDC1, resulting in the formation of 4-vinylguaicol [78]. Mutation of *pad1* and *fdc1* is essential for improving vanillin production in yeast when ferulic acid is used as substrate [79, 80].

Inefficient Metabolic Flow One approach to bypass inefficiencies in the metabolic flux caused by inhibitory effects of substrates or accumulated intermediates is a continuous administration of the substrate to the cell culture. In the *P. putida IE27* strain, vanillin production was increased by continuous addition of iso-eugenol to the cultures, which reduced oxidation of the vanillin formed into vanillic acid. Using this method, *Bacillus fusifomis* was reported produce 32.5 g/l vanillin from isoeugenol over 72 h [72].

To reduce by-product formation, a two-step fermentation process can be carried-out using two different microbial organisms. This approach was employed by Lesage-Meessen and co-workers to optimize vanillin production by combining use of *Aspergillus niger* and *Pycnoporus cinnabarinus*. The micromycete *A. niger* metabolized ferulic acid to vanillic acid in high yield whereas the basidiomycete *P. cinnabarinus* reduced the amount of vanillic acid converted into vanillin. The vanillic acid titer from *A. niger* is reported to be 920 mg/l while vanillin titer from *P. cinnabarinus* strain was reported to be 237 mg/l [54].

The development of engineered production strains is an alternative route towards achieving efficient metabolic flow towards the desired product. Li and Frost [81] devised a route for microbial production of vanillin from glucose, in which *de novo* biosynthesis of vanillic acid in *E. coli* was combined with enzymatic *in vitro* conversion of vanillic acid to vanillin. The recombinant *E. coli* KL7 strain was engineered to dehydrate 3-dehydroshikimic acid to protocatechuic acid (3,4-dihydrobenzoic acid) by the action of 3-dehydroshikimic dehydratase (3DSD), encoded by the gene *AroZ* from the dung mold fungus *Podospora anserina*. 3-Dehydroshikimic acid is an intermediate in the shikimate pathway resulting in biosynthesis of aromatic amino acids. Protocatechuic acid was then converted to vanillic acid by a human catechol-*O*-methyltransferase (COMT). Reduction of vanillic acid to vanillin was carried out *in vitro* using a cellular extract of *Neurospora crassa*, which contained the required aromatic carboxylic acid reductase (ACAR) activity [81].

Hansen et al., reported the first example of one-cell microbial vanillin biosynthesis from glucose in the yeasts *S. cerevisiae* and *Schizosaccharomyces pombe* [53]. These strains encompass an ACAR from *Nocardia iowensis*, in combination with a phosphopantetheinyltransferase (PPtase), which is required for proper activation of the ACAR enzyme. ACAR catalyzes the ATP- and NADPH-driven reduction of protocatechuic acid to protocatechuic aldehyde and of vanillic acid into vanillin. The yeast strain utilizes the gene encoding 3DSD from *P. anserina* to mediate the formation of protocatechuic acid from 3-dehydroshikimate. From protocatechuic acid, the pathway may then proceed via vanillic acid formed by *O*-methylation catalyzed by human COMT, which is subsequently reduced to vanillin by ACAR [53]. Alternatively, protocatechuic aldehyde formed by the reduction of protocatechuic acid by the ACAR enzyme may subsequently be *O*-methylated into vanillin by COMT. To improve the metabolic flux through the *de novo* vanillin biosynthetic pathway in yeast, mutations were introduced into the production strains. These included a mutation in the AROM enzyme complex (ARO1) to increase the accumulation of 3-dehydroshikimate. This mutation resulted in an increased accumulation of protocatechuic acid, and thereby redirected the metabolic flux from aromatic amino acid production to vanillin precursor production. A more efficient and more specific ACAR enzyme, which was able to catalyse the conversion of the high concentrations of protocatechuic acid to protocatechuic aldehyde, was obtained from *Neurospora crassa*. Upon expression of the gene encoding *Nc*ACAR, the yeast efficiently catabolized the available high concentrations of protocatechuic acid. This was one of the key features for the successful generation of recombinant *S. pombe* and *S. cerevisiae* capable of *de novo* synthesizing vanillin [82].

Downstream Processing Methods Based on the Physicochemical Properties of the Substrate and the Product Commercially viable vanillin production in microorganisms is dependent on efficient low-cost downstream processing and high product recovery. Few studies have reported on the potential advantages of using various product removal techniques to prevent further metabolism of the final product, *e.g.* by binding the product to absorbent resins that can be used in the fed batch fermentation.

Topakas et al., improved vanillic acid to vanillin transformation in *P. cinnabarinus* cultures grown at bioreactor scale by absorbing the toxic vanillin produced by the hydrophobic resin Amberlite XAD-2 [83]. The use of macroporous DM11 adsorbent resins has given promising results in the fed-batch biotransformation of ferulic acid to vanillin using *Amycolatopsis sp.* strain ATCC 39116. In this study, in the presence of a surplus of DM11, continuous addition of 45 g/l ferulic acid resulted in formation of 19.2 g/l vanillin within 55 h [84]. At 20°C, vanillin concentrations above 10 g/l resulted in vanillin crystallization and this provides a convenient way of isolation.

1.8 Future Perspectives and Final Remarks

Using synthetic biology, microbial organisms have been engineered for the production of various natural food components. Vanillin is one such successful example [56]. Major constraints like substrate or product inhibition have been overcome by bioengineering. The increased knowledge of enzymes involved in bioconversion of ferulic acid and other substrates to vanillin, as well as identification and characterization of the encoding genes, offers new opportunities for improved targeted bioengineering of microorganisms for vanillin production. The recent identification of *Vp*VAN as the enzyme converting ferulic acid into vanillin in *V. planifolia* provides the option to transfer the vanilla orchid pathway for vanillin synthesis into microorganisms. Yeast and cyanobacteria are apparently the best host organisms for this purpose, as the vanilla orchid enzymes may require post-translational modifications to be fully active.

The market launch of several synthetic biology derived vanillin products has generated some media attention. Bioengineering approaches may provide a more sustainable alternative to chemical synthesis [85]. In this context it is relevant to discuss whether or not it is justified and appropriate to label biotechnologically produced flavours such as vanillin as "natural" [45]. The general public cannot be expected to understand and adapt to definitions of flavour codes that are not self-evident and obvious. The main flavour codes such as, "natural", "nature identical flavour," and "artificial" (FDA), do not offer a proper and specific description of each category when the consumer is faced to choose between commercially available products from each category in a store. This situation obviously results in unsatisfied and insecure consumers. Multinational organizations like Friends of the Earth and Greenpeace exploit the situation to establish communication platforms voicing their general resistance to all products obtained using genetic engineering, even when no genetic material is present in the commercialized product [86]. It is also clear that unconscious or conscious lack of distinction between the meanings of the words vanilla and vanillin gives rise to misinterpretations and manipulation of the available facts [87].

It is important to highlight that from a commercial point of view, as well as from the point of view of the consumer, bioengineered natural vanillin competes with chemically synthesized vanillin, which currently dominates the market [88]. And not with the vanilla extract from *V. planifolia* which contains a wide range of flavour components in addition to vanillin.

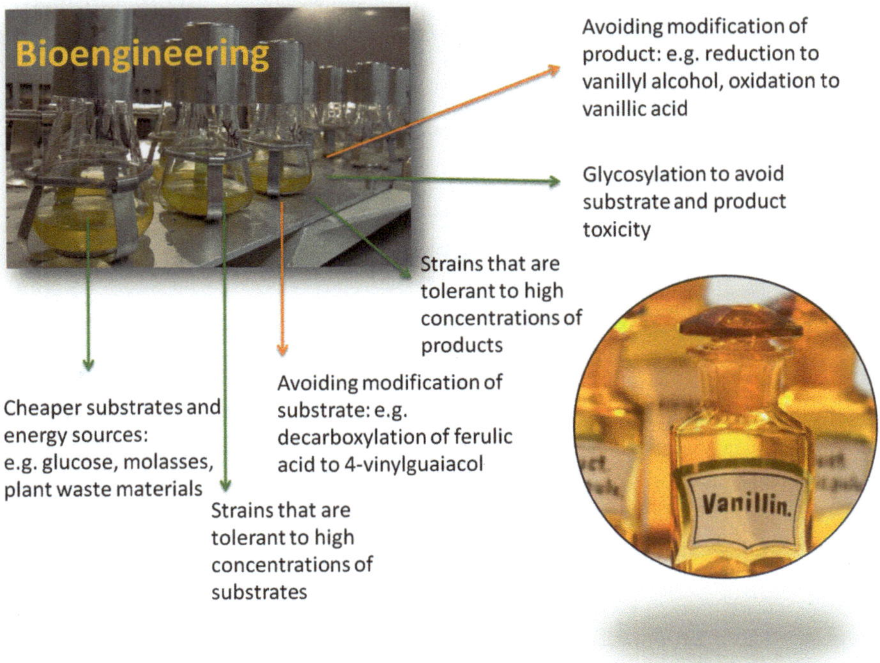

Fig. 1.7 Bioengineering approaches for vanillin production need key considerations as summarized here

In conclusion, the biotechnological production of vanillin from safe and cheap substrates by the use of food-grade production organisms and environmentally benign and economically feasible downstream processing is envisioned to result in a compatible and sustainable alternative to vanillin produced by chemical synthesis (Fig. 1.7).

References

1. Gallage NJ, Møller BL. Vanillin–bioconversion and bioengineering of the most popular plant flavor and its de novo biosynthesis in the vanilla orchid. Mol Plant. 2015;8(1):40–57.
2. Pansarin ER, et al. Pollination systems in Pogonieae (Orchidaceae: Vanilloideae): a hypothesis of evolution among reward and rewardless flowers. Flora. 2012;207(12):849–61.
3. Pansarin ER, Pansarin LM. Floral biology of two Vanilloideae (Orchidaceae) primarily adapted to pollination by euglossine bees. Plant Biol (Stuttg). 2014;16(6):1104–13.
4. Porras-Alfaro A, Bayman P. Mycorrhizal fungi of Vanilla: diversity, specificity and effects on seed germination and plant growth. Mycologia. 2007;99(4):510–25.
5. Daphna HF, Belanger F. Handbook of vanilla science and technology. Hoboken: Wiley; 2010.
6. Cameron KM, Molina MC. Photosystem II gene sequences of psbB and psbC clarify the phylogenetic position of Vanilla (Vanilloideae, Orchidaceae). Cladistics. 2006;22(3):239–48.

7. Sinha AK, Sharma UK, Sharma N. A comprehensive review on vanilla flavor: extraction, isolation and quantification of vanillin and others constituents. Int J Food Sci Nutr. 2008;59(4):299–326.
8. Gallage NJ, et al. Vanillin formation from ferulic acid in Vanilla planifolia is catalysed by a single enzyme. Nat Commun. 2014;5:1–14.
9. Odoux EaG, M. Vanilla. In: Odoux EaG M, editor. Medicinal and aromatic plants: CRC Press; 2010. p. 387.
10. Cameron KM. Utility of plastid psaB gene sequences for investigating intrafamilial relationships within Orchidaceae. Mol Phylogenet Evol. 2004;31(3):1157–80.
11. Cameron KM. On the value of nuclear and mitochondrial gene sequences for reconstructing the phylogeny of vanilloid orchids (Vanilloideae, Orchidaceae). Ann Bot. 2009;104(3):377–85.
12. Cameron KM, et al. A phylogenetic analysis of the Orchidaceae: evidence from rbcL nucleotide. Am J Bot. 1999;86(2):208–24.
13. Anuradha K, Shyamala BN, Naidu MM. Vanilla- its science of cultivation, curing, chemistry, and nutraceutical properties. Crit Rev Food Sci Nutr. 2013;53(12):1250–76.
14. Fouche JG, Jouve L. Vanilla planifolia: history, botany and culture in Reunion Island. Agronomie. 1999;19(8):689–703.
15. Rodolphe G, et al. The dynamical processes of biodiversity – Case studies of evolution and spatial distribution. In: Grillo O, editor. Biodiversity and evolution in the Vanilla Genus. Rijeka: InTech; 2011.
16. Odoux E, Brillouet JM. Anatomy, histochemistry and biochemistry of glucovanillin, oleoresin and mucilage accumulation sites in green mature vanilla pod (Vanilla planifolia; Orchidaceae): a comprehensive and critical reexamination. FRUITS. 2009;64(4):221–41.
17. Rao SR, Ravishankar GA. Vanilla flavour: production by conventional and biotechnological routes. J Sci Food Agr. 2000;80(3):289–304.
18. Jones MA, Vincente GC. Criteria for testing vanilla in relation to killing and curing methods. J Agric Res. 1949;78:425–34.
19. Sharp MD, et al. Rapid discrimination and characterization of vanilla bean extracts by attenuated total reflection infrared spectroscopy and selected Ion flow tube mass spectrometry. J Food Sci. 2012;77(3):C284–92.
20. Silva AP, et al. New insight on the genesis and fate of odor-active compounds in vanilla beans (Vanilla planifolia G. Jackson) during traditional curing. Food Res Int. 2011;44(9):2930–7.
21. De La Cruz Medina J, Jiménes GCR, García HS. Vanilla: post-harvest operations. 2009. Available from: http://www.fao.org/fileadmin/user_upload/inpho/docs/Post_Harvest_Compendium_-_Vanilla.pdf.
22. Gobley MJ. Recherches sur le principe odorant de la vanille. Pharmacie, J R Soc. 1858;34:401–5.
23. Palama TL, et al. Metabolic changes in different developmental stages of Vanilla planifolia pods. J Agric Food Chem. 2009;57(17):7651–8.
24. Boonchird C, Flegel TW. In vitro antifungal activity of eugenol and vanillin against Candida albicans and Cryptococcus neoformans. Can J Microbiol. 1982;28(11):1235–41.
25. Bomgardner MM. The problem with vanilla – After vowing to go natural, food brands face a shortage of the favored flavor. In: Chemical and Engineering news. Washington: ACS; 2017. p. 38–42.
26. Brochado AR, et al. Improved vanillin production in baker's yeast through in silico design. Microb Cell Fact. 2010;9:1–15.
27. Walton NJ, Mayer MJ, Narbad A. Molecules of interest – Vanillin. Phytochemistry. 2003;63(5):505–15.
28. Walton NJ, et al. Novel approaches to the biosynthesis of vanillin. Curr Opin Chem Biol. 2000;11(5):490–6.
29. Hocking M. Vanillin: synthetic flavoring from spent sulfite liquor (vol 74, pg 1055, 1997). J Chem Educ. 1997;74(12):1384.
30. Borregaard. Vanilla House. 2017 [cited 2017 01–08]; Available from: http://www.vanillin.com/.

31. Zenk MH. Biosynthese von vanillin in Vanilla planifolia Andr. Z Pflanzenphysiol. 1965;53:404–14.
32. Havkin-Frenkel D, Belanger F. Handbook of vanilla science and technology. Chichester: Wiley; 2010. p. 360.
33. Anwar MH. Paper Chromatography of Monohydroxyphenols in Vanilla extract. Anal Chem. 1963;35(12):1974–6.
34. Podstolski A, et al. Unusual 4-hydroxybenzaldehyde synthase activity from tissue cultures of the vanilla orchid Vanilla planifolia. Phytochemistry. 2002;61(6):611–20.
35. Negishi O, Sugiura K, Negishi Y. Biosynthesis of vanillin via ferulic acid in Vanilla planifolia. J Agric Food Chem. 2009;57(21):9956–61.
36. Rosazza JPN, et al. Review: biocatalytic transformations of ferulic acid: an abundant aromatic natural product. J Ind Microbiol Biotechnol. 1995;15(6):457–71.
37. Fritz RR, Hodgins DS, Abell CW. Phenylalanine ammonia-lyase. Induction and purification from yeast and clearance in mammals. J Biol Chem. 1976;251(15):4646–50.
38. Gabriac B, et al. Purification and immunocharacterization of a plant cytochrome P450: the cinnamic acid 4-hydroxylase. Arch Biochem Biophys. 1991;288(1):302–9.
39. Schoch G, et al. CYP98A3 from Arabidopsis thaliana is a 3 '-hydroxylase of phenolic esters, a missing link in the phenylpropanoid pathway. J Biol Chem. 2001;276(39):36566–74.
40. Lam KC, et al. Structure, function, and evolution of plant O-methyltransferases. Genome. 2007;50(11):1001–13.
41. Yang H, et al. A re-evaluation of the final step of vanillin biosynthesis in the orchid Vanilla planifolia. Phytochemistry. 2017;139:33–46.
42. Cambra I, et al. Structural basis for specificity of propeptide-enzyme interaction in barley C1A cysteine peptidases. Plos One. 2012;7(5):e37234.
43. Turk V, et al. Cysteine cathepsins: from structure, function and regulation to new frontiers. BBA Proteins Proteomics. 2012;1824(1):68–88.
44. Odoux E, Chauwin A, Brillouet JM. Purification and characterization of vanilla bean (Vanilla planifolia Andrews) beta-D-glucosidase. J Agric Food Chem. 2003;51(10):3168–73.
45. Pollan M. Why 'Natural' Doesn't Mean Anything Anymore, in The New York times magazine. New York: The New York Times Magazine; 2015.
46. Bratanova B, et al. Savouring morality. Moral satisfaction renders food of ethical origin subjectively tastier. Appetite. 2015;91:137–49.
47. FDA. CFR – Code of Federal Regulations Title 21. 2015 [cited 2016 24–08]; Available from: http://www.accessdata.fda.gov/scripts/cdrh/cfdocs/cfcfr/cfrsearch.cfm?fr=501.22.
48. Schwab W, Davidovich-Rikanati R, Lewinsohn E. Biosynthesis of plant-derived flavor compounds. Plant J. 2008;54(4):712–32.
49. Solvay. Solvay. 2017 [cited 2016 01–08-2016]; Rhovanil® vanillin]. Available from: http://www.solvay.com/en/index.html.
50. Aromatics DM vanillin natural (ex-ferulic acid) 2017 [cited 2016 01–08]; vanillin natural (ex-ferulic acid)]. Available from: http://www.demonchyaromatics.com/.
51. Mane. Sense Capture Vanillin. 2017 [cited 2016 01–08]; Available from: http://www.mane.com/.
52. Evolva. Making Vanillin in Yeast. 2017 [cited 2016 01–08]; Making Vanillin in Yeast]. Available from: http://www.evolva.com/.
53. Hansen EH, et al. De Novo biosynthesis of vanillin in fission yeast (Schizosaccharomyces pombe) and Baker's Yeast (Saccharomyces cerevisiae). Appl Environ Microbiol. 2009;75(9):2765–74.
54. LesageMeessen L, et al. A two-step bioconversion process for vanillin production from ferulic acid combining Aspergillus niger and Pycnoporus cinnabarinus. J Biotechnol. 1996;50(2–3):107–13.
55. Di Gioia D, et al. Metabolic engineering of Pseudomonas fluorescens for the production of vanillin from ferulic acid. J Biotechnol. 2011;156(4):309–16.
56. Schwab W. Genetic engineering of plants and microbial cells for flavour production. In: Berger RG, editor. Chemistry, bioprocessing and sustainability. Berlin: Springer; 2007. p. 648.

57. Hartley RD, Ford CW. Phenolic constituents of plant-cell walls and wall biodegradability. ACS Symp Ser. 1989;399:137–45.
58. Priefert H, Rabenhorst J, Steinbuchel A. Biotechnological production of vanillin. Appl Microbiol Biotechnol. 2001;56(3–4):296–314.
59. Sigma-Aldrich, product Eugenol. [cited 2016 24–08]; Available from:http://www.sigmaaldrich.com/catalog/search? term=eugenol&interface=All&N=0&mode=match%20partialmax&lang=en®ion=DK&focus=product
60. Lesage-Meessen L, et al. Fungal transformation of ferulic acid from sugar beet pulp to natural vanillin. J Sci Food Agr. 1999;79(3):487–90.
61. Bonnin E, et al. Enzymic release of cellobiose from sugar beet pulp, and its use to favour vanillin production in Pycnoporus cinnabarinus from vanillic acid. Carbohydr Polym. 2000;41(2):143–51.
62. Zheng LR, et al. Production of vanillin from waste residue of rice bran oil by Aspergillus niger and Pycnoporus cinnabarinus. Bioresour Technol. 2007;98(5):1115–9.
63. de Vries RP, et al. The faeA genes from Aspergillus niger and Aspergillus tubingensis encode ferulic acid esterases involved in degradation of complex cell wall polysaccharides. Appl Environ Microbiol. 1997;63(12):4638–44.
64. Benoit I, et al. Feruloyl esterases as a tool for the release of phenolic compounds from agro-industrial by-products. Carbohydr Res. 2006;341(11):1820–7.
65. Lesage-Meessen L, et al. A biotechnological process involving filamentous fungi to produce natural crystalline vanillin from maize bran. Appl Biochem Biotechnol. 2002;102:141–53.
66. Faulds CB, Bartolome B, Williamson G. Novel biotransformations of agro-industrial cereal waste by ferulic acid esterases. Ind Crops Prod. 1997;6(3–4):367–74.
67. Torres BR, et al. Vanillin bioproduction from alkaline hydrolyzate of corn cob by Escherichia coli JM109/pBB1. Enzyme Microb Technol. 2009;44(3):154–8.
68. Higuchi T, Ito Y, Kawamura I. P-Hydroxyphenylpropane component of grass lignin and role of tyrosine-ammonia lyase in its formation. Phytochemistry. 1967;6(6):875–81.
69. Biotech N. Ferulic acid Ex Rice Bran. 2017 [cited 2016 01–08]; Available from: http://www.novorate.com/.
70. Plaggenborg R, et al. Potential of Rhodococcus strains for biotechnological vanillin production from ferulic acid and eugenol. Appl Microbiol Biotechnol. 2006;72(4):745–55.
71. Muheim A, Lerch K. Towards a high-yield bioconversion of ferulic acid to vanillin. Appl Microbiol Biotechnol. 1999;51(4):456–61.
72. Yamada M, et al. Biotransformation of isoeugenol to vanillin by Pseudomonas putida IE27 cells. Appl Microbiol Biotechnol. 2007;73(5):1025–30.
73. Zhao LQ, et al. Biotransformation of isoeugenol to vanillin by a novel strain of Bacillus fusiformis. Biotechnol Lett. 2005;27(19):1505–9.
74. van Wezel GP, et al. Unlocking streptomyces spp. for use as sustainable industrial production platforms by morphological engineering (vol 72, pg 5283, 2006). Appl Environ Microbiol. 2006;72(10):6863.
75. Yoon SH, et al. Production of vanillin from ferulic acid using recombinant strains of Escherichia coli. Biotechnol Bioprocess Eng. 2005;10(4):378–84.
76. Fleige C, et al. Investigation of the Amycolatopsis sp. strain ATCC 39116 vanillin dehydrogenase and its impact on the biotechnical production of vanillin. Appl Environ Microbiol. 2013;79(1):81–90.
77. Graf N, Wenzel M, Altenbuchner J. Identification and characterization of the vanillin dehydrogenase YfmT in Bacillus subtilis 3NA. Appl Microbiol Biotechnol. 2016;100(8):3511–21.
78. Mukai N, et al. PAD1 and FDC1 are essential for the decarboxylation of phenylacrylic acids in Saccharomyces cerevisiae. J Biosci Bioeng. 2010;109(6):564–9.
79. Ramaen O, Sauveplane V, Pandjaitan R, Gelo-Pujic M. 2014. Improved production of vanilloids by fermentation. EP20140305939.
80. Gallage NJ, Moller BL, Hansen EH, Hansen J, 2014. Vanillin synthase. PCT/DK2013/050357.
81. Li K, Frost JW. Synthesis of vanillin from glucose. J Am Chem Soc. 1998;120(40):10545–6.

82. Joergen Hansen EHH, Sompalli HP, Sheridan JM, Heal JR, Hamilton WDO. 2013 Compositions and methods for the biosynthesis of vanillin or vanillin beta-d-glucoside. PCT7US2012/049842
83. Topakas E, et al. Bioconversion of ferulic acid into vanillic acid by the thermophilic fungus Sporotrichum thermophile. LWT - Food Sci Technol. 2003;36(6):561–5.
84. Hua DL, et al. Enhanced vanillin production from ferulic acid using adsorbent resin. Appl Microbiol Biotechnol. 2007;74(4):783–90.
85. Erika CH. Synthetic-biology firms shift focus. Nature News. 2014;505:598.
86. Rupp R. The History of Vanilla, in National Geographic. National Geographic Society; 2014.
87. Earth F.o.t. Synthetic Biology Vanillin: not natural, not sustainable, not likely to be labeled, and coming to an ice-cream cone near you [cited 2016 22–08]; Available from: http://www.foe.org/system/storage/877/eb/6/3136/synbio_vanillin_fact_sheet.pdf
88. Michael G, Nadene G. Madagascar's vanilla farmers face volatile times after poor harvest, in The Guardian. 2016. Guardian News and Media Limited: UK.

Rosmarinic Acid and Related Metabolites

2

Soheil Pezeshki and Maike Petersen

Abbreviations

ATP	Adenosine triphosphate
BAP	6-benzylaminopurine
C4H	Cinnamic acid 4-hydroxylase
CoA	Coenzyme A
DHPL	3,4-dihydroxyphenyllactic acid
DMSO	Dimethyl sulfoxide
DW	Dry weight
2,4-D	2,4-dichlorophenoxyacetic acid
3′-H	Hydroxycinnamoyl-hydroxyphenyllactate 3′-hydroxylase
3-H	Hydroxycinnamoyl-hydroxyphenyllactate 3-hydroxylase
4CL	4-coumarate CoA-ligase
HdhA	Hydroxyacid dehydrogenase
HpaBC	4-hydroxyphenylacetate 3-hydroxylase
HPPD	Hydroxyphenylpyruvate dioxygenase
HPPR	Hydroxyphenylpyruvate reductase
HQT	Hydroxycinnamoyl-CoA quinate hydroxycinnamoyltransferase
HR	Hairy roots
HST	Hydroxycinnamoyl-CoA shikimate hydroxycinnamoyltransferase
IAA	Indole-3-acetic acid
MeJA	Methyl jasmonic acid
NAA	1-naphthaleneacetic acid
NAD(P)H	Nicotinamide adenine dinucleotide (phosphate), reduced
PAL	Phenylalanine ammonia-lyase

S. Pezeshki (✉) • M. Petersen
Institut für Pharmazeutische Biologie und Biotechnologie, Philipps-Universität Marburg, Robert-Koch-Str. 4, D-35037 Marburg, Germany
e-mail: soheilpezeshki@gmail.com; petersen@staff.uni-marburg.de

W. Schwab et al. (eds.), *Biotechnology of Natural Products*,
https://doi.org/10.1007/978-3-319-67903-7_2

pHPL	4-hydroxyphenyllactic acid
pHPP	4-hydroxyphenylpyruvic acid
RA	Rosmarinic acid
RAS	Rosmarinic acid synthase, hydroxycinnamoyl-CoA:hydroxyphenyllactate hydroxycinnamoyltransferase
SA	Salicylic acid
TAL	Tyrosine ammonia-lyase
TAT	Tyrosine aminotransferase
YE	Yeast extract

2.1 Occurrence and Structures of Rosmarinic Acid and Related Metabolites

Rosmarinic acid (RA) (Table 2.1) was first described in 1958 as an ester of caffeic acid and 3,4-dihydroxyphenyllactic acid (DHPL) extracted from rosemary (*Rosmarinus officinalis*) [140], but has since then been detected in plant species across the plant kingdom from hornworts to mono- and dicotyledonous plants (for reviews see [126, 129]). Hotspots of RA presence are the sub-family Nepetoideae of the Lamiaceae and the family Boraginaceae. In other plant taxa, RA often only occurs sporadically and may not occur in all species of the same genus [126]. RA and related caffeic acid esters have been isolated from hornworts (species of the genera *Anthoceros*, *Folioceros*, *Nothothylas*, *Phaeoceros*, *Dendroceros*, *Megaceros*; [6, 155, 156, 162, 163]) as well as fern species (*Blechnum* spec.; [68, 70, 169]). Furthermore, grasses [32, 109] and species of the so-called basal orders (*Sarcandra glabra*, [191]; *Chloranthus* spec., [129]) contain RA, whereas there are – up to now – no reports from leafy mosses, liverworts and gymnosperms.

A larger number of derivatives of RA have been described, many of them occurring in *Salvia* species (Table 2.1; [18, 79, 103, 168]). These derivatives generally contain RA as core structure. Metabolites often incorrectly described as caffeic acid oligomers contain additional 4-coumaric or caffeic acid moieties or a second RA molecule. Further derivatization can occur by glycosidation (e.g. RA glucosides; [43, 95, 165]) (**see also Chap. 9 of this book**), methylation (e.g. methyl rosmarinate, methyl lithospermate [90], methylmelitric acid [105]) or the addition of ethyl and butyl or hydroxycinnamoyl moieties.

Whereas the biosynthesis of RA in Lamiaceae and Boraginaceae (e.g. *Coleus blumei*, *Salvia miltiorrhiza*, *Melissa officinalis*, *Anchusa officinalis*, *Lithospermum erythrorhizon*) is well investigated [106, 126, 128, 129], it is less well analyzed in other plant taxa. The same is true for the formation of most of the above-mentioned RA derivatives.

There are thousands of publications on the diverse biological activities of RA and derivatives such as the salvianolic acids. Recent reviews on this topic have been published by e.g. Shetty [142], Wang [168], Bulgakov et al. [18] and Amoah et al. [5], and this topic will therefore not be covered in this article.

Table 2.1 Examples for rosmarinic acid and related compounds

Structure	Common name	Reference
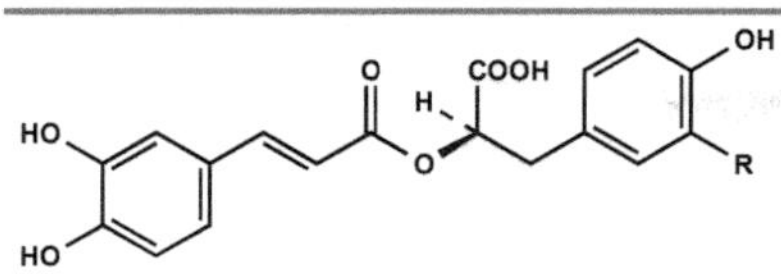	R = H isorinic acid	[139]
	R = OH rosmarinic acid	[140]
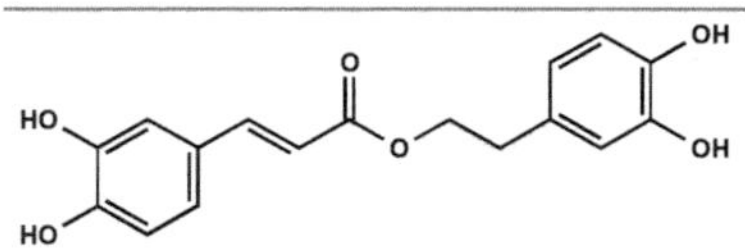	Teucrol	[45]
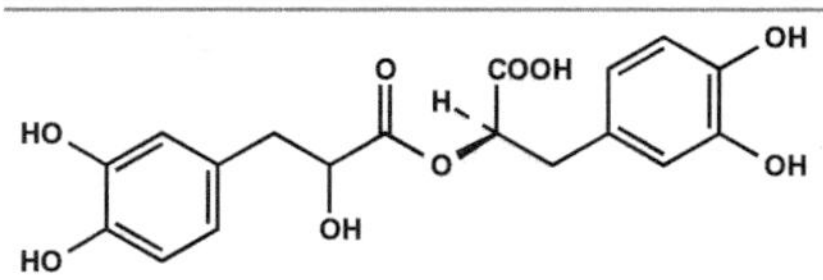	Salvianic acid C	[28]
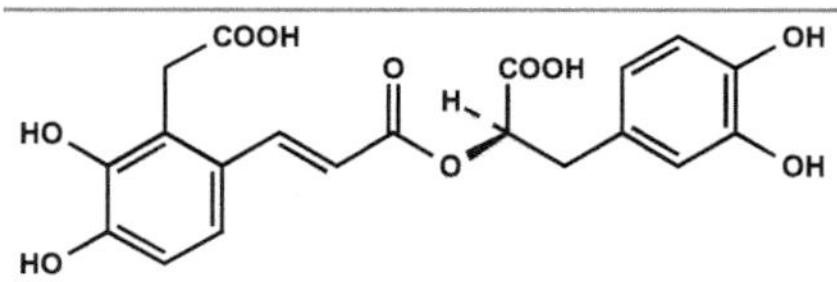	Salvianolic acid D	[4]
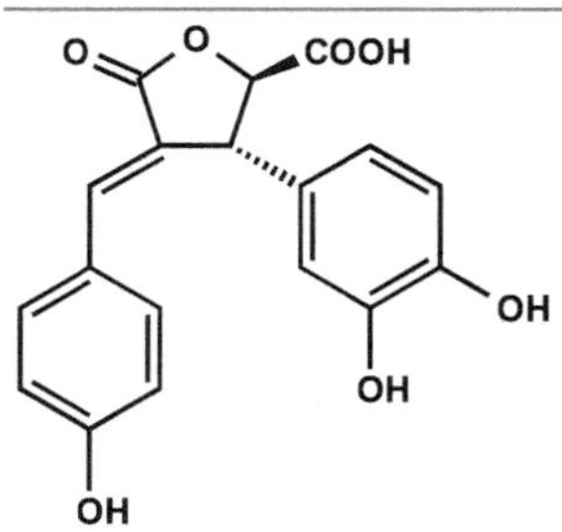	Megacerotonic acid	[155]
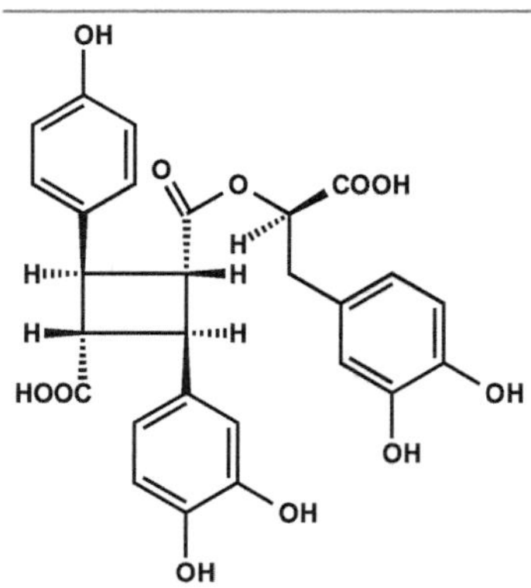	Anthocerotonic acid	[155]

(continued)

Table 2.1 (continued)

Structure	Common name	Reference
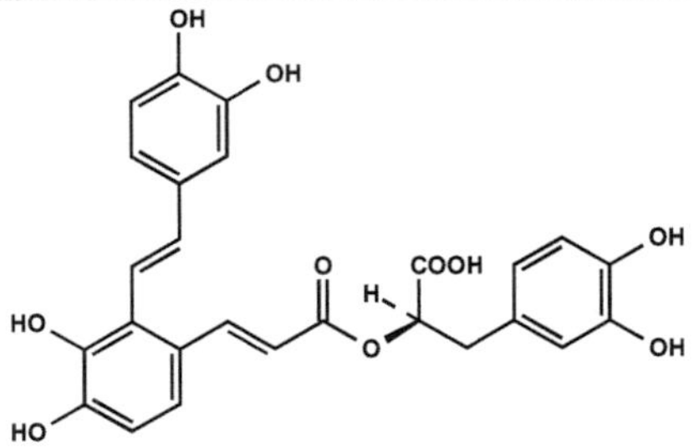	Salvianolic acid A	[96]
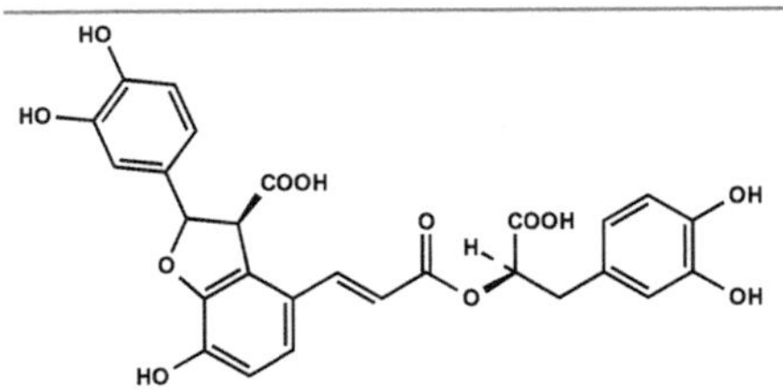	Lithospermic acid	[82, 167]
	[Monardic acid A = (7*R*,8*R*)-stereoisomer of lithospermic acid]	[114]
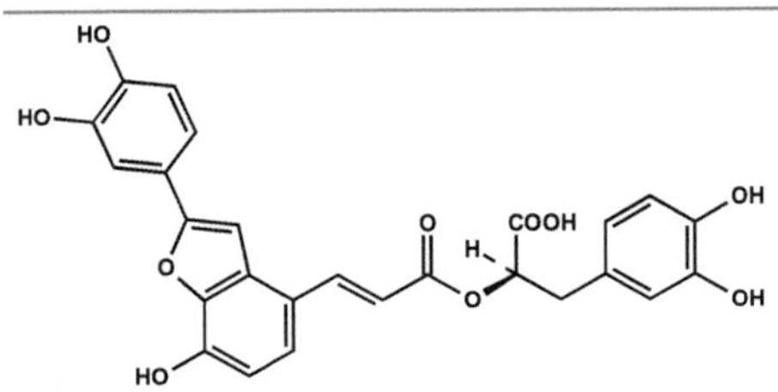	Salvianolic acid C	[3]
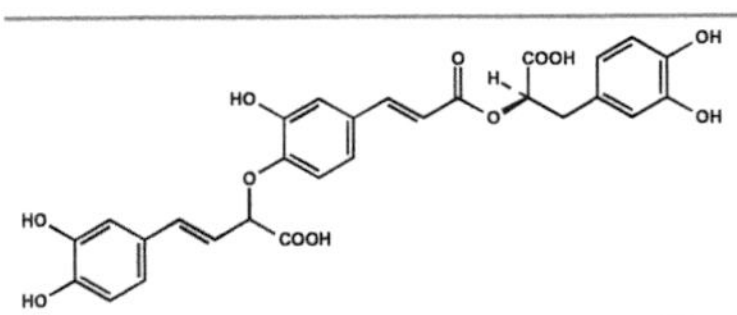	Melitric acid A	[2]
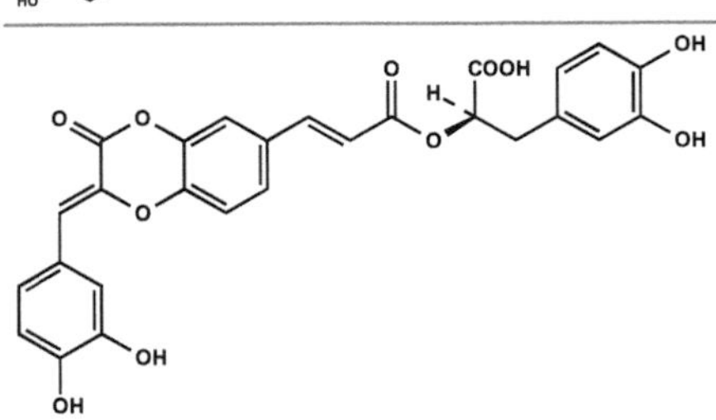	Melitric acid B	[2]
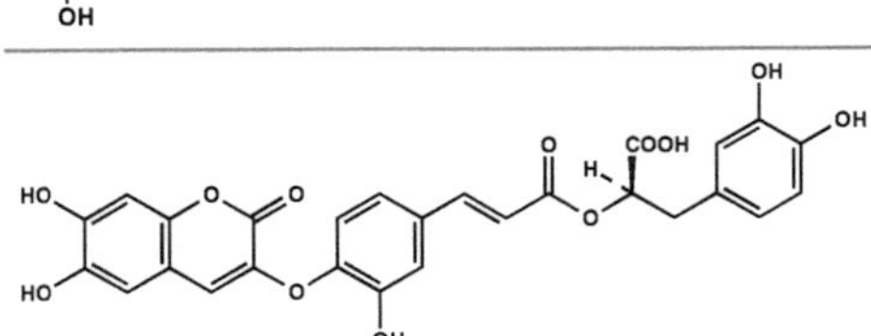	Sagecoumarin	[105]

(continued)

Table 2.1 (continued)

Structure	Common name	Reference
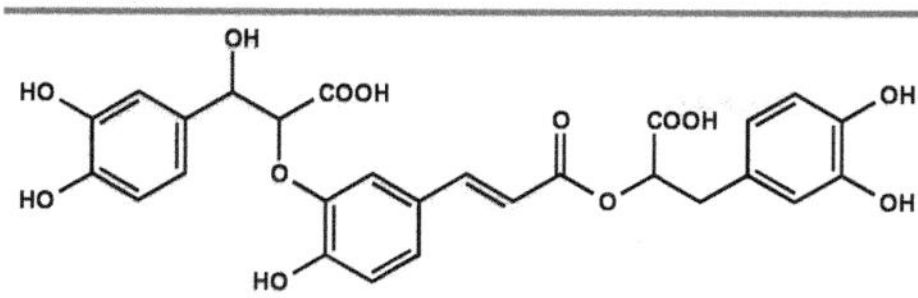	Salvianolic acid K	[102]
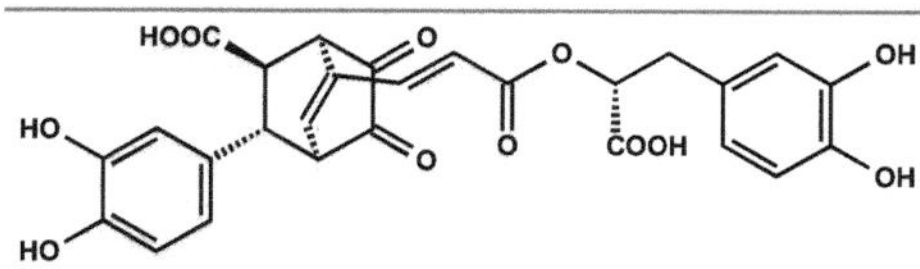	Yunnaneic acid C	[158]
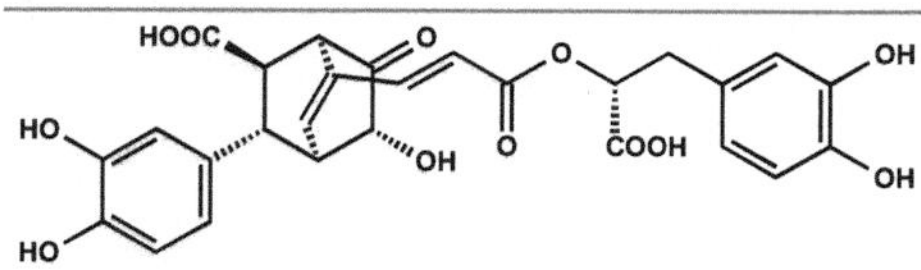	Yunnaneic acid D	[158]
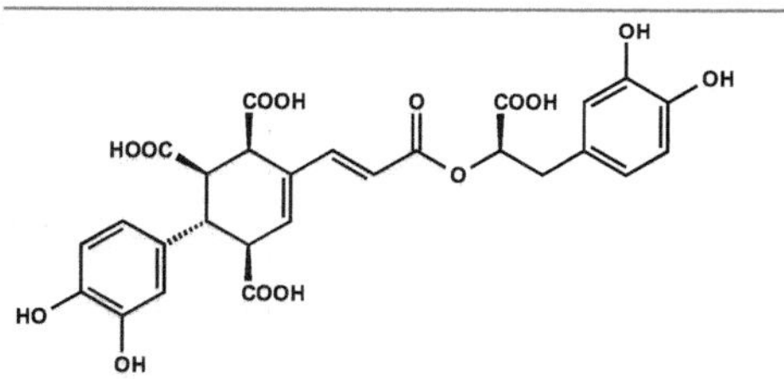	Yunnaneic acid E	[159]
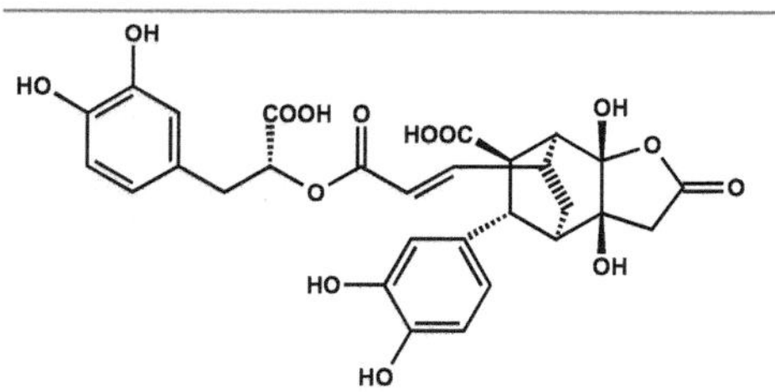	Yunnaneic acid F	[159]
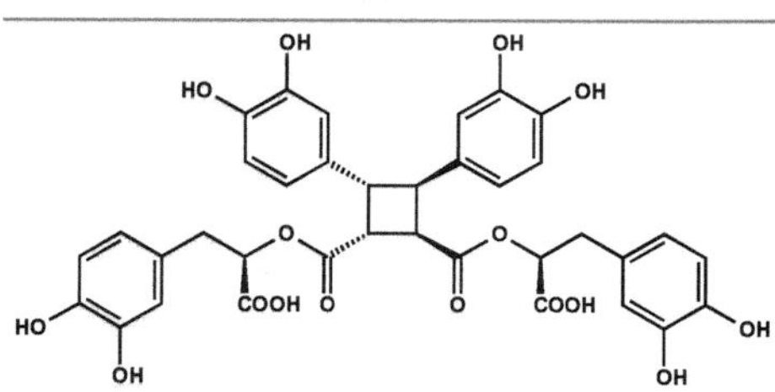	Sagerinic acid	[102]

(continued)

Table 2.1 (continued)

Structure	Common name	Reference
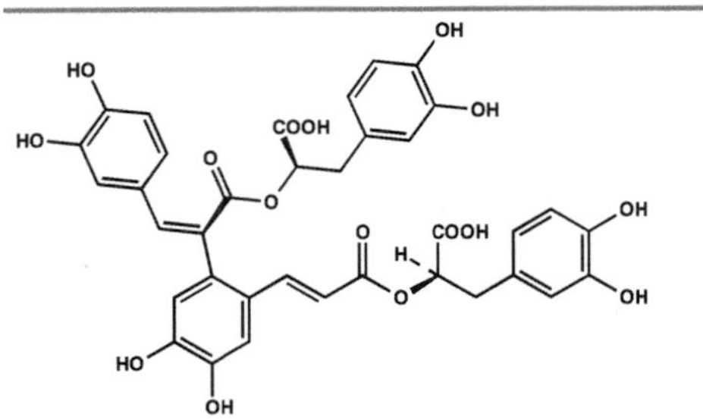	Salvianolic acid E	[4]
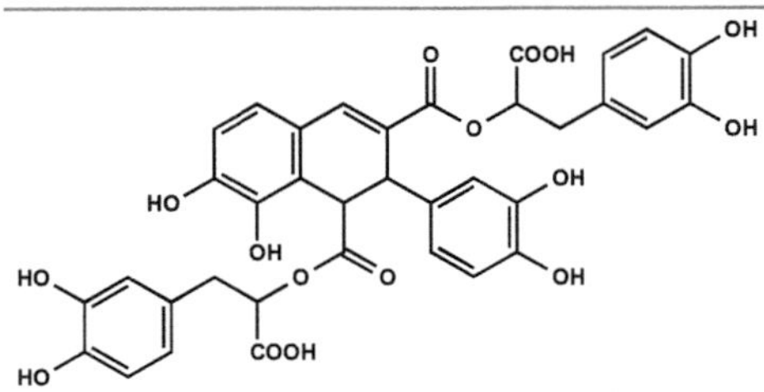	Salvianolic acid L	[104]
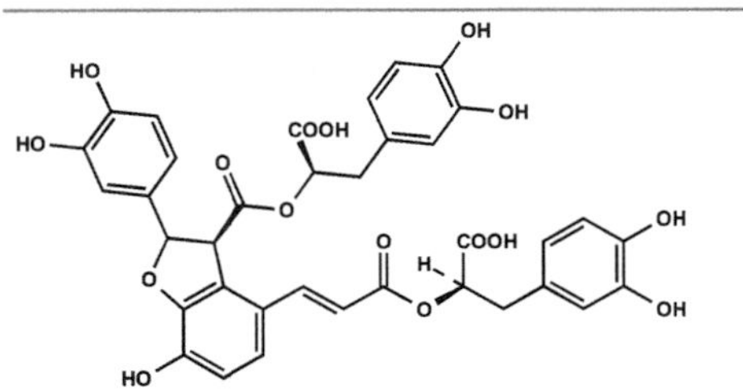	Salvianolic acid B = lithospermic acid B	[3, 157]
	[Monardic acid B = (7*R*,8*R*)-stereoisomer of lithospermic acid B]	[114]
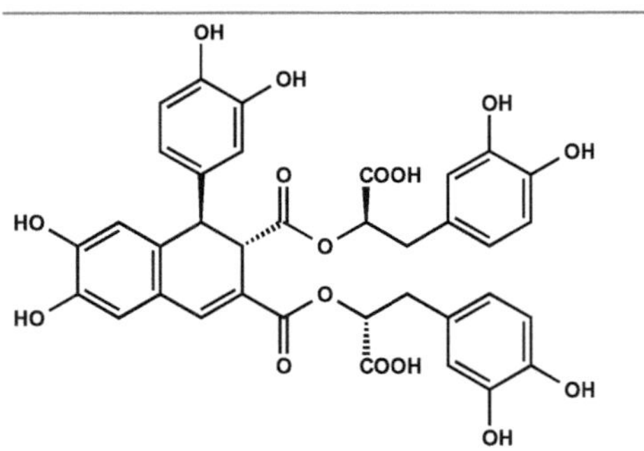	(−)-Rabdosiin	[1] [103]

(continued)

Table 2.1 (continued)

Structure	Common name	Reference
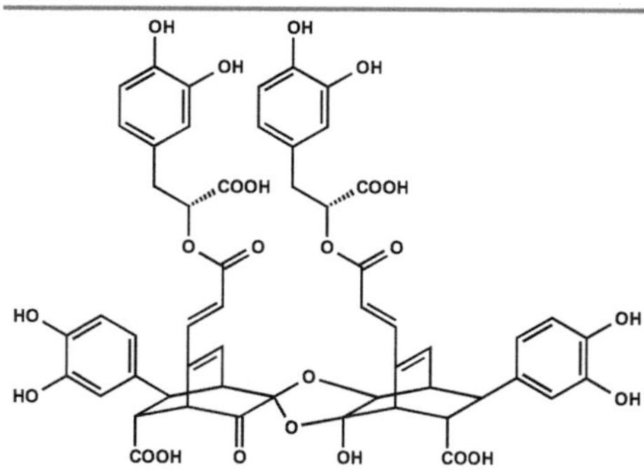	Yunnaneic acid A	[158]
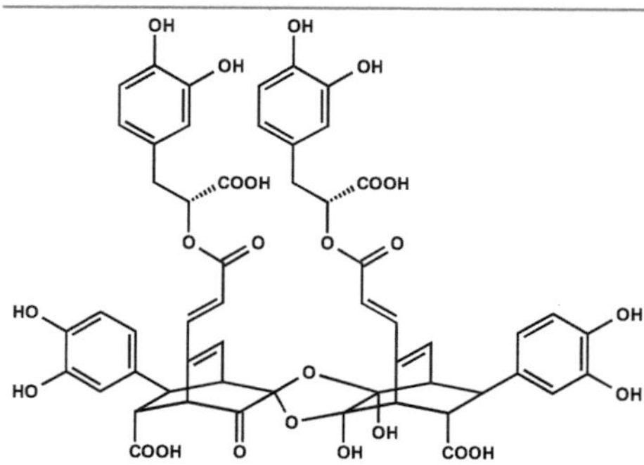	Yunnaneic acid B	[158]
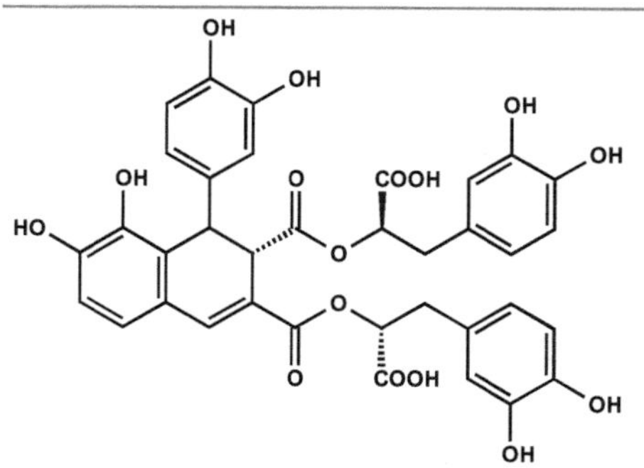	Yunnaneic acid G	[159]
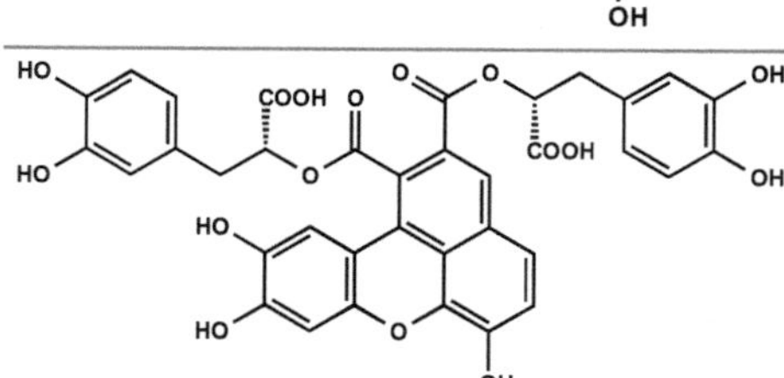	Yunnaneic acid H	[159]

(continued)

Table 2.1 (continued)

Structure	Common name	Reference
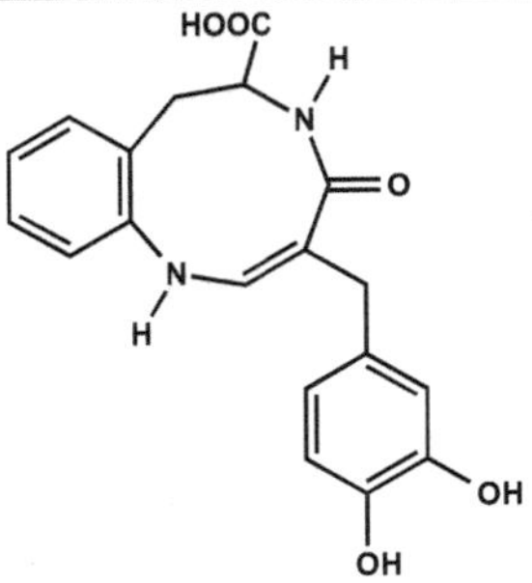	Anthocerodiazonin	[163]

2.2 Biosynthetic Pathway of Rosmarinic Acid

RA is derived from two distinct pathways: The general phenylpropanoid pathway provides the caffeic acid moiety, while DHPL is produced by a tyrosine-derived pathway [130, 131]. Both pathways are dependent on the shikimate pathway, which generates the aromatic amino acid precursors L-phenylalanine and L-tyrosine. The biosynthetic pathway (Fig. 2.1) has first been elucidated in *Coleus blumei* [130], a member of the family Lamiaceae, and *Anchusa officinalis* [36], and to a great part confirmed in *Melissa officinalis* [171, 172].

The general phenylpropanoid pathway starts with L-phenylalanine as precursor. The enzyme phenylalanine ammonia-lyase (PAL) is responsible for the transformation of the amino acid to *trans*-cinnamic acid [134]. A cytochrome P450-dependent enzyme, cinnamic acid 4-hydroxylase (C4H), introduces the first hydroxyl group to the aromatic ring in *para* position to form 4-coumaric acid [124]. Then, the ATP-dependent coenzyme A (CoA) activation of 4-coumaric acid to 4-coumaroyl-CoA is catalyzed by the enzyme 4-coumarate CoA-ligase (4CL) [81].

L-Tyrosine is the precursor in the formation of the second intermediary precursor in RA biosynthesis. Tyrosine aminotransferase (TAT) catalyzes the transamination of tyrosine and 2-oxoglutarate to 4-hydroxyphenylpyruvate (pHPP) and glutamate [36]. In a NAD(P)H-dependent step, the enzyme hydroxyphenylpyruvate reductase (HPPR) reduces pHPP to 4-hydroxyphenyllactic acid (pHPL) [69, 127].

The trans-esterification of the two precursors is catalyzed by rosmarinic acid synthase (RAS). This enzyme forms an ester of 4-coumaric acid and pHPL [127] and belongs to the BAHD acyltransferase superfamily in the subgroup hydroxycinnamoyltransferases [14]. The product 4-coumaroyl-4′-hydroxyphenyllactic acid is hydroxylated at the 3- and 3′- positions by two cytochrome P450-dependent enzyme activities, caffeoyl-4′-hydroxyphenyllactate 3′-hydroxylase and 4-coumaroyl-3′,4′-dihydroxyphenyllactate 3-hydroxylase (3'H, 3H) (Fig. 2.1) [124]. The product, RA, is then stored in the vacuole. For comprehensive reviews on biosynthesis, distribution and evolution of RA biosynthesis see e.g. Petersen and Simmonds [128] and Petersen [126].

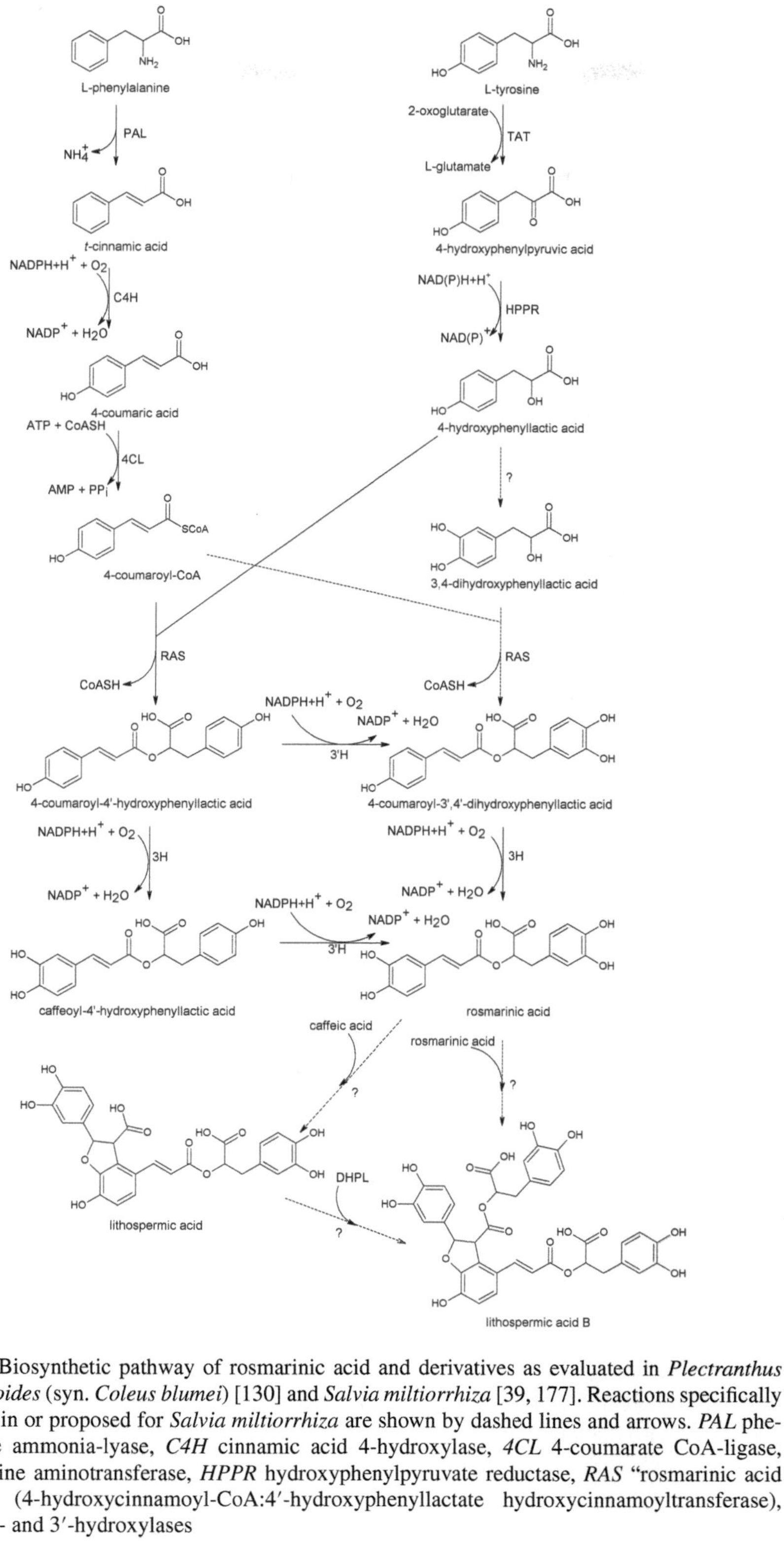

Fig. 2.1 Biosynthetic pathway of rosmarinic acid and derivatives as evaluated in *Plectranthus scutellarioides* (syn. *Coleus blumei*) [130] and *Salvia miltiorrhiza* [39, 177]. Reactions specifically described in or proposed for *Salvia miltiorrhiza* are shown by dashed lines and arrows. *PAL* phenylalanine ammonia-lyase, *C4H* cinnamic acid 4-hydroxylase, *4CL* 4-coumarate CoA-ligase, *TAT* tyrosine aminotransferase, *HPPR* hydroxyphenylpyruvate reductase, *RAS* "rosmarinic acid synthase" (4-hydroxycinnamoyl-CoA:4′-hydroxyphenyllactate hydroxycinnamoyltransferase), *3H, 3'H* 3- and 3′-hydroxylases

Di et al. [39] suggested an alternative pathway in *Salvia miltiorrhiza*. They propose that an additional hydroxylation of pHPL to DHPL occurs prior to the esterification. Accordingly, the product 4-coumaroyl-3′,4′-dihydroxyphenyllactate undergoes a single hydroxylation. While Di et al. [39] furthermore propose a direct formation of lithospermic acid B by coupling of two molecules of RA, Xiao et al. [177] suggest a sequential formation by addition of caffeic acid and DHPL in two separate reactions (Fig. 2.1).

2.3 Production of RA in Untransformed Aseptic *In Vitro* Cultures

Cell cultures of species of the families Lamiaceae and Boraginaceae have been established for the biotechnological production of RA and related compounds. However, although these efforts were successful at pilot scale, an industrial-scale production process for RA has never been established. Efforts to optimize RA production in *in vitro* cultures are summarized in the following paragraphs.

2.3.1 Species from the Family Lamiaceae

Coleus blumei **(syn.** *Solenostemon scutellarioides*, *Plectranthus scutellarioides*) The first reports on the formation of high amounts of RA in plant callus and cell suspension cultures are from 1977. Razzaque and Ellis [134] as well as Zenk et al. [187] both used *Coleus blumei*, the painted nettle, to establish suspension cultures that accumulated up to 15% of the cell dry weight (DW) as RA. The latter authors also described the influence of the sucrose concentration of the medium on the outcome of RA production as well as the incorporation of exogenously fed L-phenylalanine. The same species was used by Ulbrich et al. [164] in the first biotechnological production process, a two-phase culture system with a growth and a production phase. In the latter phase, a 5% sucrose solution was used for cultivation and a yield of 21% RA in the cell DW was achieved. Since then, *in vitro* cultures of *Coleus blumei* have been the most prominent system to elucidate the biosynthetic pathway of RA and to isolate and characterize the respective enzymes and genes (see below and review articles by [126, 128, 129]).

The influence of the carbohydrate source and concentration was investigated in more detail by Gertlowski and Petersen [54] and Petersen et al. [132]. They showed that sucrose is quickly cleaved into glucose and fructose. The optimal sucrose concentration was at 5%. Glucose as sole carbohydrate source was nearly as effective as sucrose while fructose led to a lower RA accumulation. The stimulating effect of higher sucrose concentrations on RA biosynthesis and accumulation is not due to an osmotic effect since partial replacement by mannitol could not promote RA formation. The onset of RA biosynthesis is independent of the sugar concentration and correlates with the depletion of mineral nutrients (e.g. phosphate) from the medium. Medium optimization was also done by Ju et al. [80] in order to establish a two-phase culture system with a growth phase and a production phase for callus as well as suspension cultures. Essentially the same results were obtained as described before: higher sucrose concentrations increased the formation and accumulation of RA. With 6% sucrose in the medium calli contained 33.7% RA in the DW, suspension cells accumulated RA to 10.1%.

Permeabilization was investigated in order to isolate RA from the medium instead of the cells and thus enable a continuous production process. For this purpose, Park and Martinez [118] added dimethyl sulfoxide (DMSO) to the suspension cultures. This, however, resulted in loss of cell viability. Preconditioning at a lower DMSO concentration (0.1%) ensured cell viability at higher DMSO levels (0.5–1.5%) and resulted in a prominent release of RA to the medium. With 0.5% DMSO, 2.85 g RA per 100 g cell DW was found in the medium, which was 66.4% of the total RA production.

Immobilization of *Coleus blumei* cells was performed by adding luffa cubes to a suspension culture in order to capture the cells within the sponge-like luffa material [107]. Luffa is the dry fibrous material of berry endocarp of *Luffa cylindrica*, Cucurbitaceae. After 33 days of growth, the cell-inhabited cubes were placed into a glass column and fed with medium by spraying it from the top. The cells were viable to a high percentage until 52 days but showed strongly reduced growth. RA production was higher (2% of the cell DW) than in the respective parent suspension culture (1.2%).

Approaches to further increase the production of RA were based on elicitation or transformation of *Coleus blumei* cell cultures; the latter will be described in the chapter "Hairy roots". Fungal elicitor preparations (*Pythium aphanidermatum*) or methyl jasmonate (MeJA) were added to suspension cultures of *Coleus blumei* and resulted in increased activities of some biosynthetic enzymes as well as about a three-fold RA accumulation [153]. Interestingly, an effect of the volatile MeJA could also be seen when it was applied via the gas phase.

Bauer et al. [12] investigated RA accumulation in different callus cell lines transformed by *Agrobacterium tumefaciens* and cultivated on hormone-free media. Growth and RA accumulation varied between different lines. The highest RA accumulation was 11% of the cell DW.

Whole *in vitro* grown plants of *Solenostemon scutellarioides* were investigated by Dewanjee and coworkers [37, 38]. Feeding of precursors (Phe and Tyr alone and in combination) could increase RA levels (up to 3.1-fold) as well as the activities of PAL, TAT and RAS. On the other hand, phytopathogenic fungi were applied with best results using *Alternaria alternata*. This increased RA accumulation up to 1.6-fold (18.5 mg/100 g fresh weight).

Salvia officinalis Various varieties of culinary sage (*Salvia officinalis*) were compared with respect to their RA accumulation in leaves and suspension cultures. The RA content varied between 0.89% and 7.82% of the DW, the best variety being "Dwarf". In all varieties except one, leaves contained less RA than suspension cells [174]. A similar approach was followed by Grzegorcyzk et al. [58] who compared the RA content in seed-derived and *in vitro* regenerated sage plants as well as shoot callus and cell suspension cultures after different numbers of passages. Here the highest RA content was found in suspension cells with around 1.9% of the cell DW. Hippolyte et al. [71, 72] characterized RA production in suspension cultures of sage further. The optimal sucrose concentration was 5%, which resulted in an RA content of 19% of the DW at the end of the culture period. Feeding of L-phenylalanine as precursor shortened the production period and enhanced RA production at 0.1 g/l phenylalanine in 5% sucrose medium. The highest RA accumulation of 36% of the cell DW could be observed in a low osmolarity medium (Heller-medium with 5% sucrose).

A number of studies reported on shoot cultures of sage as a source for RA. MS agar medium supplemented with different concentrations of the long chain saturated primary alcohol triacontanol showed positive effects with respect to shoot multiplication as well as diterpene and RA content. Highest RA concentrations of approximately 2% of the DW were found after addition of 20 µg/l triacontanol [59]. Shoots grown in liquid medium accumulated around 3% RA in the DW irrespective of the triacontanol concentration [55]. MeJA (50 and 100 µM) stimulated RA levels in liquid cultivated shoots even further to 4.1% of the DW on the fifth day after elicitation [56]. Shoot cultures were also cultivated in a laboratory scale sprinkle bioreactor with a 43-fold increase in biomass after 3 weeks and a RA content of 2.6% of the DW [57].

Shoot cultures on solidified MS medium were used to test the effect of sodium salicylate on diterpene (carnosol, carnosic acid) and RA production. Although the amount of diterpenes was stimulated by elicitiation, the RA levels remained largely unaffected and growth was decreased [93]. In contrast, Ejtahed et al. [46] showed a two-fold increase in RA production to 1.8% of the DW in shoot cultures after addition of 250 µM salicylic acid (SA).

Salvia miltiorrhiza *Salvia miltiorrhiza* is an important Asian medicinal plant which is very well investigated with respect to its accumulation of tanshinones and phenolic acids, among them RA, lithospermic acids and salvianolic acids (Table 2.1). Many investigations have been performed with hairy root cultures and are described below. In addition, undifferentiated cell cultures were the basis for the production of phenolic acids [42]. Morimoto et al. [112] showed that callus cultures accumulated 1.24% of the DW as RA and 0.1% as lithospermic acid B. In shoot cultures regenerated from these calli, both phenolic acids accumulated in considerably higher amounts (6.96% RA and 6.05% lithospermic acid B). The same compounds were detected in Ti-transformed suspension cells (4.59% RA and 0.81% lithospermic acid in the cell DW) [26]. In a special 6,7-V-medium, the same cell line produced phenolic acids (RA 530 mg/l and lithospermic acid B 216 mg/l) as well as tanshinones (220 mg/l), the latter being excreted to the medium to a considerable extent [21]. In an attempt to increase secondary metabolite production, yeast extract (YE) was used. This resulted in nearly doubled tanshinone production, while RA accumulation was strongly reduced [22, 23]. MeJA (10 µmol/l) also stimulated the activities of PAL and TAT as well as RA accumulation [179]. Addition of Ca^{2+} ions

(10 mM) enhanced the accumulation of RA up to 2% of the DW. This was also coupled to enhanced PAL and TAT activities [101].

Stems and leaves of *Salvia milthiorrhiza* were used by [175] to establish callus cultures and to compare their efficacy to produce RA and salvianolic acid B. Stem callus contained more phenolic acids (1.27 ± 0.38% RA and 0.87 ± 0.20% salvianolic acid B in the DW) than leaf callus (0.28 ± 0.02% RA and 0.07 ± 0.03% salvianolic acid B).

SA (22.5 mg/l) was used as elicitor to increase the production of phenolic acids. Addition of SA resulted in a doubled RA accumulation 2 days after addition (to approximately 0.03% of the DW). At the same time, H_2O_2 levels increased. External addition of H_2O_2 (10 mM) also induced RA formation and it was suggested that H_2O_2 is a mediator in elicitation processes by SA [66]. Besides increase in RA formation (to 1.1%), SA addition also induced Ca^{2+} mobilization. Extracellular addition of calcium ions (10 mM) or the calcium ionophore A23187 also enhanced RA levels [64]. SA was shown to lower the cytoplasmic pH by inhibition of the plasma membrane H^+-ATPase. The RA content was increased to about 2.25-fold of the control level [99].

Ocimum basilicum Undifferentiated *in vitro* cultures of basil (*Ocimum basilicum*) were investigated by Kintzios et al. [88]. Leaf-derived suspension cultures accumulated about 10% of the cell DW as RA. Immobilization in calcium alginate resulted in a dramatic decrease of the RA level. Immobilization in test tubes at high cell density (25×10^4 cells/ml; approximately 4 ml volume; mini-bioreactor), in contrast, resulted in highly enhanced RA production and RA concentrations of 2% of the cell DW could be achieved. RA was also determined in the medium with about 5 mg/ml in the first week of the experiment [113]. Nodal shoot explants and suspension-cultured cells of *Ocimum basilicum* were incubated in a small bioreactor by [87]. They reported increased growth and RA accumulation in the bioreactor-cultivated plant material. Highest RA levels (0.02% of the DW) were found in the organized plant material.

Addition of YE (0–5 g/l) positively influenced RA formation in basil callus cell lines from 0.67% in controls to 2.3% in the DW on medium with 5 g/l YE [63].

A red-colored cell line of basil accumulated RA and anthocyanins, both of which arise from the general phenylpropanoid pathway. Strazzer et al. [144] chose a stable anthocyanin-producing cell line that also accumulated 0.8 mg/g fresh weight RA and subjected these cells to mechanical stress (enhanced agitation) and light stress. Both treatments led to increased RA accumulation (up to 1.9 mg/g fresh weight for combined light and mechanical stress), and in parallel anthocyanin accumulation was enhanced as well. Since both biosynthetic pathways require phenylpropanoid precursors, the overall flux into the phenylpropanoid pathway must have increased. The authors also propose that both stressors might increase the formation of reactive oxygen species which can be quenched by both, RA and anthocyanins.

In vitro shoot regeneration from basil nodal explants was performed by [84]. They found highest RA levels (approximately 40 mg/g DW) in fully acclimatized plantlets. The effect of benzyladenine on RA accumulation was dependent on the basil cultivar. In the anthocyanin-producing variety, the accumulation of anthocyanin and RA were inversely correlated with the benzyladenine concentration. The same group used different culture vessel types for the micropropagation of basil shoots. A prominent difference with respect to RA accumulation (approximately 16% of the cell DW compared to 4% in other culture systems) was observed, which was inversely correlated with biomass accumulation [83].

Orthosiphon aristatus *Orthosiphon aristatus* (Java tea) was first used as suspension culture for the production of RA by Sumaryono et al. [152] and Sumaryono and Proksch [151]. These cells synthesized about 1–2 μmol RA per g fresh weight. After elicitation with YE (4–6 g/l), RA accumulation increased to 10 μmol/g fresh weight 3–4 days after elicitor addition; decarboxylated RA was found as well.

Cell cultures of *Orthosiphon aristatus* established from plants from different locations were analysed with respect to their growth characteristics. Highest RA contents ranged between 4.5% and 5.0% of the cell DW [100].

Glechoma hederacea A suspension culture of *Glechoma hederacea* accumulated up to 25.9% RA in the cell DW in CB2-medium [41] in only 7 days of culture. Besides, lower amounts of caffeic acid and chlorogenic acid were detected. This is one of the highest levels of RA accumulation described so far.

Lavandula vera Several aspects of medium optimization and elicitation have been evaluated in the course of investigations on *Lavandula vera* suspension cultures. RA was identified as the main phenolic metabolite [92]. Linsmayer and Skoog medium was used as the basic medium. Several medium components were varied and finally an optimized medium presented [77, 123]. Raising the sucrose content of the medium from 3% to 7% strongly reduced the biomass accumulation to 45% of the control but at the same time dramatically enhanced the RA yield to more than sevenfold of the control [75]. Doubling the phosphate concentration resulted in enhanced growth (131%) and enhanced RA production (206% compared to the control) [74]. Reduction of the medium's ammonium concentration to ¼ enhanced RA accumulation to 2.7 times of the control level (1.5% of the DW) but still ensured growth. Increasing the level of ammonium ions delayed the onset of RA biosynthesis and reduced the overall accumulation. Higher nitrate levels in the medium were reported to be beneficial for RA accumulation [76]. A combination of optimized medium parameters (NH_4NO_3, KNO_3, and KH_2PO_4) resulted in a 27-fold RA accumulation (17.9% of the cell DW) [123]. Feeding of the precursor phenylalanine strongly increased the amounts of caffeic acid and raised RA accumulation to 128% of the control level [119].

RA is mostly accumulated intracellularly. Adding the resin Amberlite XAD4 or a mixture of 4% polyethylene glycol and 7.5% dextran to the liquid medium as a two-phase culture system resulted in a release of RA to the extracellular phase. The total RA accumulation in presence of XAD4 was slightly increased (115% of controls), but only 6.4% of the total amount of RA was adsorbed to the resin. Cultivation with polyethylene glycol and dextran as second phase strongly reduced biomass accumulation, although the content of RA per cell remained unchanged. About 12% of the total RA amount was found in the extracellular phase [122].

Further optimization of RA production by *Lavandula vera* cell cultures was done in 3 l-bioreactors with respect to dissolved oxygen concentration, agitation speed and temperature with the result of doubling the RA production (3.5 g/l) compared to shake flask cultures [52, 120, 121] .

A selection of putatively high producing cell lines was achieved by applying a fluorinated phenylalanine derivative. As the best result, an enhanced RA accumulation from 0.5% of the cell DW to approximately 1% was observed [53].

A way of enhancing secondary metabolite production is elicitation which was also applied to *Lavandula vera* cultures. Different biotic elicitors such as bacterial homogenates and cell wall preparations did not result in increased RA accumulation [91]. An abiotic elicitor, vanadyl sulfate, was added to the culture 11 days after inoculation. The highest RA accumulation (280% of the control level) was observed with 25 mg/l vanadyl sulfate after 12 h. As an additional effect more RA was found extracellularly [49]. The addition of benzothiadiazole had only small effects, whereas elicitation with MeJA (50 μM) on day 11 enhanced RA accumulation 2.4 times [50]. Here, the best elicitation result in *Lavandula vera* suspension cultures was about 12.6% RA in the dry cell biomass (calculated with the published data).

Lavandula officinalis Common lavender cultivated as *in vitro* culture was investigated by Nitzsche et al. [115]. Suspension cultures contained about six to ten times the amount of RA as normal plants. Interestingly, here RA was also secreted to the medium, which has not been described frequently. Usually, secreted RA is quickly decomposed, e.g. by peroxidases (own unpublished observations) and thus cannot be identified as RA anymore. Application of jasmonic acid or stress by oxygen depletion changed the profile of phenolic metabolites but did not increase the RA content.

Satureja khuzistanica The Iranian species *Satureja khuzistanica* was used to establish a callus culture for RA production. On B5 medium with 5% sucrose, callus cells accumulated 7.5% RA in the DW [136]. Suspension cultures of the same species showed much higher RA contents (18% of DW) after 21 days [137]. It was shown that reducing the nitrogen content to ¼ decreased growth slightly and RA accumulation severely to 3.8% of the cell DW.

Melissa officinalis Although *Melissa officinalis*, lemon balm, is one of the most important RA-containing medicinal plants, *in vitro* cultures of *Melissa officinalis* are barely investigated. Extracts of lemon balm are used against *Herpes simplex* infection due to their content of phenolic compounds; the most important of them is RA. Besides RA, melitric acids A and B (Table 2.1) have been detected in *Melissa officinalis* [2]. Suspension cultures of lemon balm have been characterized and used as source for the isolation of cDNAs and genes for PAL, 4CL and RAS. Suspension cultures accumulated up to 6.7% of the cell DW as RA after 6 days of cultivation. The effect of increased sucrose concentrations was not as prominent as observed for suspension cultures of *Coleus blumei* [171, 172]. Hot water extracts of whole *in vitro* cultured lemon balm plants were analysed by Barros et al. [9]. They showed a wide variety of phenolic acids in the plant material, of which sagerinic acid (Table 2.1) was dominant followed by lithospermic acid and RA (which commonly is named as the dominant phenolic acid). Attempts to increase the RA content in *Melissa officinalis* shoot cultures by treatment with 200 ppb ozone for 3 h resulted in a transient increase of the RA content (30 mg/g fresh weight) at 2 h after starting the ozone treatment [161] .

Ocimum sanctum Holy basil (*Ocimum sanctum*, syn. *O. tenuiflorum*) is cultivated for medicinal and religious purposes because of its essential oil composed of several phenolic compounds (e.g. eugenol, isoeugenol, estragol). It also contains other phenolic antioxidants, mainly RA (0.012–0.025% of the DW). Callus cultures derived from different plant organs showed RA concentrations of 0.14–0.27% of the DW [65].

Rabdosia rubescens The effect of the sucrose concentration and the ratio of NO_3^- to NH_4^+ on specialized metabolism and plant regeneration were tested by Dong et al. [40]. The best result with respect to RA was achieved with 5% sucrose and a NO_3^-/NH_4^+ ratio of 2:1.

Agastache rugosa The effect of MeJA on RA accumulation was investigated in suspension cultures of *Agastache rugosa* (Korean or Indian mint). 50 μM MeJA proved to be optimal for the stimulation of RA accumulation from 7.8 to 36.6% of the cell DW. Also other phenolic acids were present in higher levels. The expression levels of PAL, C4H and 4CL correlated well with the increase in the RA level [86].

2.3.2 Species of the Family Boraginaceae

Anchusa officinalis Anchusa officinalis was one of the first species taken into culture for the production of RA and the investigation of its biosynthesis. Suspension cultures accumulated up to 6% of the cell DW as RA and the accumulation phase correlated with the linear growth phase. Early biosynthetic investigations established that 20–30% of exogenously applied, radioactively labelled phenylalanine or tyrosine was incorporated into RA [33]. Microspectrophotometric investigations suggested that RA is accumulated in the vacuoles [20]. Ellis [44] also studied the accumulation of RA in clonal cell lines derived from single cells with known productivity. This showed that high-producing mother cells did not result in high-producing clonal cell lines. After several subcultures each cell line established a quite stable level of RA production, which was not related to the RA production level of the mother cell. The optimization of the culture medium with respect to macronutrients (sucrose, alternative sugars, nitrate, phosphate and Ca^{2+}) and phytohormones (2,4-D, NAA, IAA and 2-chloro-4-fluorophenoxyacetic acid as auxins; BAP, kinetin and zeatin as cytokinins) was undertaken by De Eknamkul and Ellis [34, 35]. Surprisingly, a combination of all optimized levels of single macronutrients did not result in increased growth and RA production. Variations of phytohormone contents were performed in standard B5-medium. The highest RA levels of 12% of the DW were achieved in medium with 0.25 mg/l NAA as auxin, while medium with 2,4-D showed a decrease in RA accumulation. In contrast to previous results, the onset of RA synthesis was shifted to the exponential growth phase.

Su and Humphrey [145] established a high density culture of *Anchusa officinalis* with perfusion and tested several growth media. Using this technique the RA yield was doubled in comparison to control cultures. This, however, was only based on higher cell densities (38 g DW/l compared to approximately 14 g/l) while the RA content in the cells (approximately 3.3% of the DW) decreased slightly. The principle of perfusion culture was transferred to a membrane-aerated bioreactor. Here, the cell density was at 26 g/l and the calculated cellular RA content approximately 4.6% of the DW [146]. Optimization of the perfusion strategy in shake flasks led to higher productivity with respect to RA. The best result was obtained by growing the culture as batch culture in B5 medium with 3% sucrose and 0.25 mg/l NAA for 10 days, followed by perfusing the culture with B5 medium containing 6% sucrose and the same NAA concentration at a constant perfusion rate of 0.1/day. The obtained cell density was 35 g/l and 11.3% RA were found in the cell DW [149]. This procedure has been transferred to a stirred-tank bioreactor with similar productivity. However, the suspension cells proved to be very sensitive to agitation, aeration conditions and the dissolved oxygen concentration [148]. The inoculum size strongly influenced the productivity with best results at 4 g DW/l [147]. The results with a perfusion culture of *Anchusa officinalis* have been summarized by Su et al. [150].

Lithospermum erythrorhizon Suspension cultures of *Lithospermum erythrorhizon* were mainly investigated with respect to their accumulation of the red pigment shikonin. However, unpigmented cell cultures also accumulate RA (0.55% of the DW) and lithospermic acid [48]. Interestingly, the accumulation of phenolic acids and shikonin cannot occur under the same culture conditions but require different culture media. Elicitors such as YE and MeJA were added to increase the RA amount up to 0.22% of the cell fresh weight [110, 111]. Elicited cell cultures were mainly used to investigate the biosynthetic pathway for RA in this species. Besides RA, Yamamoto et al. [181, 182] identified RA-related compounds in *Lithospermum erythrorhizon* such as rhabdosiin, lithospermic acid and lithospermic acid B as well as lithospermic acid B glucoside (Table 2.1). Among these, lithospermic acid B was the predominant compound. Addition of MeJA or YE strongly increased the formation of RA by factors of 10- and 4-fold of the control cells, respectively. At the same time, the activities of PAL, 3H and 3'H were increased while RAS activity remained at a rather low level [117].

2.3.3 Non-vascular Plant Species

Anthoceros agrestis **(Anthocerotaceae)** The occurrence of RA in non-vascular plants like the hornworts was first described by Takeda et al. [155, 156]. Hornworts are among the earliest land plants to evolve. Nevertheless, hornworts contain RA and related compounds like anthocerotonic acid, megacerotonic acid and anthocerodiazonin (Table 2.1) as well as other phenolic compounds [162, 163]. Cell cultures of *Anthoceros agrestis* have been established by Binding and Mordhorst [15] and further investigated with respect to RA accumulation and biosynthesis by Petersen and coworkers (e.g. [125, 165]). Not all enzymes found in Lamiaceae for RA biosynthesis have to date been found in *Anthoceros* as well and thus the biosynthetic pathway is still under investigation. Suspension cultures of *Anthoceros agrestis* can accumulate quite high levels of RA. Pezeshki has measured up to 9% RA in the cell DW in a hormone-free B5-derived medium with 1% sucrose after 2 weeks of cultivation [133], whereas higher sugar content (2%) resulted in a lower RA accumulation. In the latter medium, however, an accumulation of RA 3′-*O*-β-D-glucoside at the beginning of the culture period was observed [165] (**see also Chap. 9 of this book**). With respect to the intracellular RA concentrations *Anthoceros agrestis* suspension cultures are in no way inferior to cell cultures of many higher plant species. It must, however, be mentioned that the cell mass increase of these cultures is lower.

2.4 Production of Rosmarinic Acid in Hairy Roots

Hairy roots have become a common type of axenic plant *in vitro* culture due to their easy maintenance and rapid biomass increase. Usually hairy roots are established by infecting plant material with *Rhizobium rhizogenes* (formerly *Agrobacterium rhizogenes*) strains, which transfer genes of their Ri plasmid to the plant cells. These are stably integrated into the plant genome and direct the plant cells to produce roots. The developing roots often carry high numbers of root hairs that give the roots a "hairy" appearance [108]. In recent years, efforts have been made to optimize the production of plant metabolites in hairy root cultures of plants that contain the very same metabolites or to insert new pathways for small molecules or proteins of interest into model plants [51, 62, 141, 160].

Hairy root cultures of members of both, the Boraginaceae and Lamiaceae, have been used for the production of RA and other caffeic acid derivatives. As of April 2016, 36 scientific articles had been published on this topic.

2.4.1 Hairy Roots of Lamiaceae Species

The production of RA in hairy root cultures of plants in the Lamiaceae is well documented (Tables 2.2 and 2.3). Hairy roots have several advantages with respect to undifferentiated cell cultures or plants. The hairy root material contains mostly the same metabolites as the source plant but is more stable than undifferentiated plant

Table 2.2 Hairy root cultures of Lamiaceae species established for the production of RA and related compounds. For experiments with *Salvia miltiorrhiza* see Table 2.3

Plant species	Compound	Experiment	Reference
Agastache foeniculum	RA	Establishment of HR, 4-fold higher production of RA (0.02% DW) than in non-transformed roots	[116]
Coleus blumei	RA	Establishment of HR and normal roots, comparison of biomass and RA content. HR had 2.8-fold higher RA content (5% DW). Effects of MeJA and YE on HR	[11]
Coleus blumei	RA, caffeic acid, chlorogenic acid	Transformation of HR with the *Arabidopsis thaliana* PAL gene under the control of the constitutive CaMV 35S promotor decreases the formation of RA and chlorogenic acid, but enhances caffeic acid levels	[10]
Coleus blumei	RA and other phenolic acids	Endogenously synthesized elicitor β-cryptogein causes excretion of RA from the cells to the medium	[166]
Coleus blumei	RA	Establishment of HR, RNAi-mediated suppression of HPPR or RAS reduced RA by 92%, overexpression led to RA levels of 176% of the control HR lines (1.73% DW)	[73]
Coleus blumei	RA	Comparison of different tissues revealed high stability of production of RA in HR	[13]
Coleus forskohlii	RA and other natural products	Analysis of various media for HR cultures with respect to biomass and RA accumulation. Comparison of the elicitors YE, SA and MeJA. Increase of RA content up to 3.4-fold higher with MeJA than control	[98]
Dracocephalum kotschyi	RA and flavonoids	Establishment of HR, up to 15-times higher production of RA in HR than in non-transformed roots (max. 0.15% DW)	[47]
Dracocephalum moldavica	RA	Establishment of HR, analysis of different media, tenfold higher RA content than in untransformed roots (7.8% DW)	[173]
Hyssopus officinalis	RA and other phenolic acids	Comparison of different media with respect to RA amount. Highest value was 6% DW, 60% higher than in callus, cell suspension culture and 1 year old field plant roots. Detection of nine other phenolic acids in HR	[89]

(continued)

Table 2.2 (continued)

Plant species	Compound	Experiment	Reference
Nepeta cataria	RA	Elicitation of HR cultures with auxins and polyamines led to increase in biomass and RA accumulation (1.92% DW)	[185]
Ocimum basilicum	RA and related phenolic acids	Comparison of various clones of HR cultures with respect to RA amount. Highest amount was 14.1% DW	[154]
Ocimum basilicum	RA	Increased production of RA in HR and elicited HR compared with untreated or untransformed roots. Exudation of RA into medium upon treatment with *Pythium ultimum*	[8]
Ocimum basilicum	RA and other antioxidants	Production of RA as major antioxidant, dependent on cultivar (up to 7.6% DW)	[143]
Salvia officinalis	RA	Comparison of two lines of HR cultures transformed with different strains of *Agrobacterium* and with untransformed HR, up to 2.3-fold increase in RA accumulation (approx. 4.5% DW)	[60]
Salvia officinalis	RA	Comparison of shoot and HR cultures with respect to accumulation of antioxidants and biomass (shoot 2.6%, HR 3.5% DW)	[57]
Salvia wagneriana	RA	Establishment of culture, no elicitation of RA with JA	[135]

DW dry weight, *HR* hairy roots, *JA* jasmonic acid, *MeJA* methyl jasmonic acid, *SA* salicylic acid, *YE* yeast extract

cells. Moreover, the yield of RA and other caffeic acid derivatives can be increased by eliciting with e.g. MeJA or SA.

A problem remains during the downstream processing of the phenolic acids: the extraction of RA from cells and organs is a tedious process. For biotechnological use, exudation of the phenolic metabolites into the medium would be an important step for a simpler and cheaper production. Two publications deal with this problem. In 2002, Bais and coworkers treated hairy roots of RA-producing *Ocimum basilicum* with *Pythium ultimum*. Upon this fungal *in situ* challenge, the hairy roots produced droplets on the roots tips with concentrated RA solutions. This behavior was absent with other fungi or in untreated roots. It has been hypothesized that this strategy might be useful for the plant root to prevent infections with soil pathogens, as RA showed effective antimicrobial activity [8].

The oomycete *Phytophthora cryptogea* produces ß-cryptogein. This proteinaceous elicitor causes activation of phenylpropanoid metabolism via stimulation of calcium-dependent pathways. By transforming *Rhizobium rhizogenes* with the coding sequence for β-cryptogein, Vuković and her colleagues obtained modified

Coleus blumei hairy root clones with the ability to produce the elicitor endogenously. These cultures were able to secrete RA and caffeic acid into the culture medium [166].

A modulation of product amounts can not only be achieved by changing the medium or eliciting the root culture, but also by manipulating the expression pattern of genes for enzymes of RA biosynthesis. Hücherig and Petersen [73] used techniques of RNAi suppression and overexpression with a constitutive promotor to modulate gene expression for HPPR and RAS in hairy roots of *Coleus blumei*. They showed that the insertion of interfering hairpin RNA of both genes led to decreased expression values of HPPR and RAS and accordingly to reduced RA accumulation. One HPPR-RNAi-line accumulated only about 8% of the RA amount found in control lines (1.73% of DW in controls). In contrast, an overexpression of these genes led to a 1.8-fold increase in RA accumulation compared to control lines.

By far the most publications on RA production in hairy roots are dealing with the plant *Salvia miltiorrhiza* (Table 2.3), the red or Chinese sage, named for its red ochre-colored roots. It is an important plant in traditional Chinese medicine, also known as Danshen, Dan Shen or Tan Shen. Two substance groups dominate the constituents of the plant extracts, namely phenolic acids (RA, lithospermic acids, and salvianolic acids) and diterpenes (tanshinones). Danshen is employed for the treatment of various diseases associated with malfunctioning blood flow, cardiovascular and cerebrovascular diseases, such as coronary heart disease, hypertension, angina pectoris, ischemic strokes and hyperlipidemia. It is used in various phytopharmaceutical forms, for oral application or injection, as solids, liquids or aerosols, as single preparation or in combination with other drugs. Clinical and pharmacological studies of bioactive metabolites isolated from Danshen have focused on Danshensu, which is DHPL, salvianolic acid B and tanshinone IIA [30, 190].

Xiao et al. [177] investigated the production of lithospermic acid B in hairy roots. It has been hypothesized that lithospermic acid B is directly derived from RA. After elicitation of hairy root cultures of *S. miltiorrhiza* with silver ions (Ag^+), they investigated accumulation of RA, lithospermic acid B and intermediates of the RA biosynthetic pathway as well as gene expression of enzymes involved in this pathway and found an inverse proportionality of RA and lithospermic acid accumulation after elicitation. This finding, combined with metabolic profiling and gene activity measurements, led to the conclusion that RA is the precursor of lithospermic acid B.

Other publications presented a genetic engineering approach to stimulate the accumulation of phenolic acids in *S. miltiorrhiza*. Xiao et al. [178] used an overexpression/suppression approach to manipulate the expression patterns of genes of the RA biosynthetic pathway. The upregulation of the single genes for *c4h*, *tat* and *hppr* as well as suppression of the *4-hydroxyphenylpyruvate dioxygenase* gene (*hppd*) led to an increase of RA, lithospermic acid B or both. The gene product HPPD participates in the tyrosine catabolic pathway by catalyzing the conversion of 4-hydroxphenylpyruvate to homogentisate. A co-overexpression of *tat* and *hppr* resulted in the highest accumulation of both RA and lithospermic acid, 4.3 and 3.2-fold higher than in the wild type, respectively.

Table 2.3 Hairy root cultures of *Salvia miltiorrhiza* (Lamiaceae) established for the production of RA and related compounds

Compounds	Experiment	Reference
Lithospermic acid B, RA and related compounds	Establishment of HR, comparison of different media with respect to accumulation of lithospermic acid B (between 0.73 and 1.61% DW) and RA (0.48% DW) and increase of biomass	[25]
Phenolic acids	Methyl viologen inhibited biomass production and decreased content of phenolic acids in HR	[24]
Phenolic acids, RA and lithospermic acid B, other natural products	Increase of phenolic acids and other natural products and biomass upon elicitation with YE (up to 2.89% lithospermic acid B and 2.98% RA in DW)	[27]
RA and related compounds	Comparison of two elicitors, YE and silver ions. Increase of RA accumulation and gene expression for enzymes of RA biosynthesis for both elicitors, effects with YE higher (up to 8% DW)	[183]
RA and lithospermic acid B	Elicitation of HR with MeJA increased RA and lithospermic acid B approx. 2–8-fold higher than untreated control (RA up to 6.02% DW, lithospermic acid B up to 19.3% DW). Gene expression was elevated for RA biosynthesis genes	[176]
RA, salvianolic acid B, DHPL (danshensu)	Dependence of phenolic acid content on concentration of MeJA elicitor in medium and growth stage of HR. Accumulation in DW, RA 14.35%, salvianolic acid B 1.59%, DHPL 0.51%	[29]
RA and lithospermic acid B	Elicitation of HR cultures with Ag^+ led to approx. 3-fold increase of lithospermic acid B (to 18.8% DW), while RA content decreased. Analysis of gene expression and intermediates suggest RA as precursor for lithospermic acid B	[177]
RA and lithospermic acid B	Overexpression of genes of RA biosynthesis and suppression of genes for by-products led to 3.2–4.3-fold increase of phenolic acids in HR compared to untransformed wildtypes	[178]
RA, salvianolic acids	Effects of MeJA and YE on accumulation of RA and salvianolic acids. MeJA elevated the accumulation of salvianolic acid B up to 7.11% and RA up to 3.38% DW, YE increased RA content up to 5.71% DW but suppressed salvianolic acids	[189]
Salvianolic acids and RA	Effects of sugar and other nutrients of the medium on accumulation of salvianolic acids in whole plants, seedlings and HR	[170]
Phenolic acids, RA and lithospermic acid B	Effects of various concentrations of abscisic acid and fluridone on growth and accumulation of phenolic acids in HR	[31]

(continued)

Table 2.3 (continued)

Compounds	Experiment	Reference
RA, lithospermic acid B and other natural products	Overexpression of allene oxide cyclase promoted biosynthesis of natural products in HR, RA increased 2.1-fold compared to wildtype (up to 0.28% DW), lithospermic acid B accumulated 1.8-fold more than wildtype and 2.3-fold more than blank vector control (up to 1.90% DW)	[61]
RA, salvianolic acid B and caffeic acid	Gene expression study of RA biosynthetic genes after elicitation with MeJA, LC-MS-analysis of phenolic acids. Both expression and accumulation were elevated several hours after induction	[97]
RA, lithospermic acid B	[Ring-^{13}C]-labeled phenylalanine and UPLC/Q-TOF measurement to analyze the biosynthetic pathway of phenolic acids	[39]
RA, salvianolic acid B and tanshinones	Endophytic bacteria decrease the production of phenolic acids and biomass and increase the production of tanshinones	[184]
RA and salvianolic acid B	Treatment of HR with MeJA and fungal extracts, expression and activity analysis of phenylpropanoid and tyrosine-derived pathway (RA max 4.5% DW)	[188]
RA, salvianolic acid B and tanshinones	Silver ions as elicitor for secondary metabolites, analysis of gene expression (up to 1.5% DW)	[180]
RA and salvianolic acid B	Study on transcription factors for RA biosynthesis	[67]

DW dry weight, *HR* hairy roots, *JA* jasmonic acid, *MeJA* methyl jasmonic acid, *SA* salicylic acid, *YE* yeast extract

Using the overexpression of allene oxide cyclase, Gu et al. [61] were also able to enhance the accumulation of secondary metabolites, namely tanshinone IIA, RA and lithospermic acid B in hairy roots of *Salvia miltiorrhiza*. Allene oxide cyclase catalyzes a reaction in the pathway toward jasmonates, which are a group of phytohormones that are induced in response to various stresses [16]. Jasmonates are known to trigger plant defence mechanisms, especially the production of secondary metabolites. MeJA, for instance, is an important elicitor. Overexpression of the *Salvia miltiorrhiza* allene oxide cyclase gene in hairy root cultures led also to an increase in RA biosynthetic genes encoding PAL, HPPR and 4CL.

2.4.2 Hairy Roots of Boraginaceae Species

Two species of the Boraginaceae family (*Lithospermum erythrorhizon* and *Eritrichium sericeum*) have been used to establish hairy roots for the production of RA and related compounds (Table 2.4). *Lithospermum erythrorhizon*, the purple gromwell, accumulates RA, lithospermic acid and rabdosiin, a condensation product of two molecules of RA (Table 2.1). Another interesting natural substance from this species is shikonin, a prenylated naphthoquinone. The content of phenolic

Table 2.4 Hairy root cultures of Boraginaceae species established for the production of RA and related compounds

Plant species	Compounds	Experiment	Reference
Lithospermum erythrorhizon	Lithospermic acid B, RA, rabdosiin and other natural products	Analysis of HR in M-9 medium for caffeic acid derivatives and other natural products	[181]
Eritrichium sericeum and *Lithospermum erythrorhizon*	Rabdosiin and RA	Presence of *rolC* in HR inhibits production of phenolic acids compared to control cultures. MeJA-triggered (1 μM) *E. sericeum* HR can accumulate up to 3.41% and 6.92% of the DW as rabdosiin and RA, respectively	[19]

DW dry weight, *HR* hairy roots, *MeJA* methyl jasmonic acid

compounds in hairy roots was considerably lower than in suspension cultures and RA was hardly detectable [181]. Thus, this culture system is inferior for biotechnological uses.

Bulgakov et al. [19] described interesting effects of the agrobacterial *rolC* gene. This gene is located on the Ri plasmid, which is transferred during infection of the plant with *Rhizobium rhizogenes*. RolC causes inhibition of phenolic acid production (namely RA and rabdosiin) in *Lithospermum erythrorhizon* and *Eritrichium sericeum* callus and hairy root cultures, leading to depletion of both substances to a level two- to three-fold lower than in untransformed plant material. Yet, the effects are reversible with cantharidin, an inhibitor of serine/threonine phosphatases, which has led to the hypothesis, that *rolC* affects shikimate metabolism via a set of regulatory phosphatases, which in return can be affected by cantharidin. This finding was unexpected because several publications had demonstrated that transgenic hairy roots bearing the *rolC* gene can produce more secondary metabolites without further treatment than untransformed cultures.

2.5 Production of Rosmarinic Acid and Related Caffeic Acid Esters in Microorganisms

In recent years, several efforts to introduce a biosynthetic pathway for RA and related phenolic metabolites into *Escherichia coli* have been reported. The first step was taken by Kim et al. [85] who inserted coding sequences for 4CL and hydroxycinnamoyl-CoA shikimate hydroxycinnamoyltransferase (HST) or hydroxycinnamoyl-CoA quinate hydroxycinnamoyltransferase (HQT) into *E. coli* and fed different hydroxycinnamic acids to the bacteria. These were capable to produce hydroxycinnamoylshikimate or hydroxycinnamoylquinate, metabolites closely related to RA. They circumvented the necessity to introduce enzymes necessary to hydroxylate the benzene ring, which, in plants, requires cytochrome P450s.

Therefore, their approach can be viewed as a biotransformation rather than a *de novo* synthesis of hydroxycinnamic acid esters. To increase the amount of acceptor substrates, the authors mutated different enzymes of the shikimate pathway, leading either to the production of quinate or shikimate esters [85].

The next step was taken by Bloch and Schmidt-Dannert in 2014 (Fig. 2.2). They took advantage of the fact that RAS, the key enzyme for RA production and responsible for esterification of 4-coumaroyl-CoA and pHPL, can also use caffeoyl-CoA and DHPL as substrates, since the enzyme has a broad substrate promiscuity regarding the hydroxylation in *meta* position [94, 138]. In plants, these hydroxyl groups are added after the RAS reaction by cytochrome P450 reactions. The engineered pathway starts for both, the acceptor and the donor, with pHPP from the bacterial shikimate pathway. The acceptor molecule DHPL is produced by addition and overexpression of two enzymes, a dehydrogenase (HdhA; hydroxyacid dehydrogenase from *Lactobacillus delbrueckii* ssp. *bulgaricus*) and a hydroxylase complex (HpaBC; 4-hydroxyphenylacetate 3-hydroxylase from *E. coli*), using $FADH_2$ (and $NAD(P)H + H^+$) as cofactors. The donor is synthesized by using three enzymes. In the bacterial pathway to aromatic amino acids, pHPP is transaminated to tyrosine. An inserted tyrosine ammonia-lyase (TAL from *Rhodobacter sphaeroides*) deaminates tyrosine to 4-coumaric acid, which is hydroxylated with the HpaBC complex described above to build caffeic acid. After CoA activation with an inserted 4CL (At4CL2 from *Arabidopsis thaliana*), an introduced RAS (CbRAS from *Coleus blumei*) produces RA. Alongside RA, isorinic acid (ester of caffeic acid and pHPL) was observed. The introduction of RAS from other plants species, namely *Lavandula angustifolia* or *Melissa officinalis*, resulted in higher production of RA and isorinic acid (1.8 ± 0.3 μM RA, 5.3 ± 0.7 μM isorinic acid with MoRAS; approximately 2.5 mg phenolic acids/l). Both metabolites were released into the medium and the amount of product was increased when appropriate precursors were fed to the medium. The authors stated, however, that an industrial use of this modified *E. coli* needs either feeding of expensive precursors like pHPL or DHPL, which would elevate production costs into unprofitable ranges or to use bacterial strains with upregulated shikimic acid and tyrosine biosynthetic pathways, so that the precursors would be produced autotrophically [17].

A similar approach was followed by Jiang et al. [78] using a tyrosine-overproducing *E. coli* strain as a platform [7]. Furthermore, coding sequences for an *Arabidopsis thaliana* 4CL, a mutated D-lactate dehydrogenase (LDH^{Y52A}) from *Lactobacillus pentosus* [186], the HpaBC complex from *E. coli* BW25113 and a synthetic CbRAS sequence (optimized for expression in *E. coli*) were used. The final transformed *E. coli* strain was able to produce approximately 133 mg RA per litre of culture besides approximately 55 mg/l caffeoyl-phenyllactate.

Recently, Zhuang et al. [192] achieved the formation of 18 RA analogues by feeding *E. coli* BLRA1 transformed with a 4CL from *Arabidopsis thaliana* and RAS from *Coleus blumei* with different donor substrates (4-coumaric acid, caffeic acid, ferulic acid, 3,4-dihydroxyphenylpropanoic acid and 4-hydroxyphenylpropanoic acid) and various acceptors (pHPL, DHPL, phenyllactic acid, mandelic acid and tyrosol).

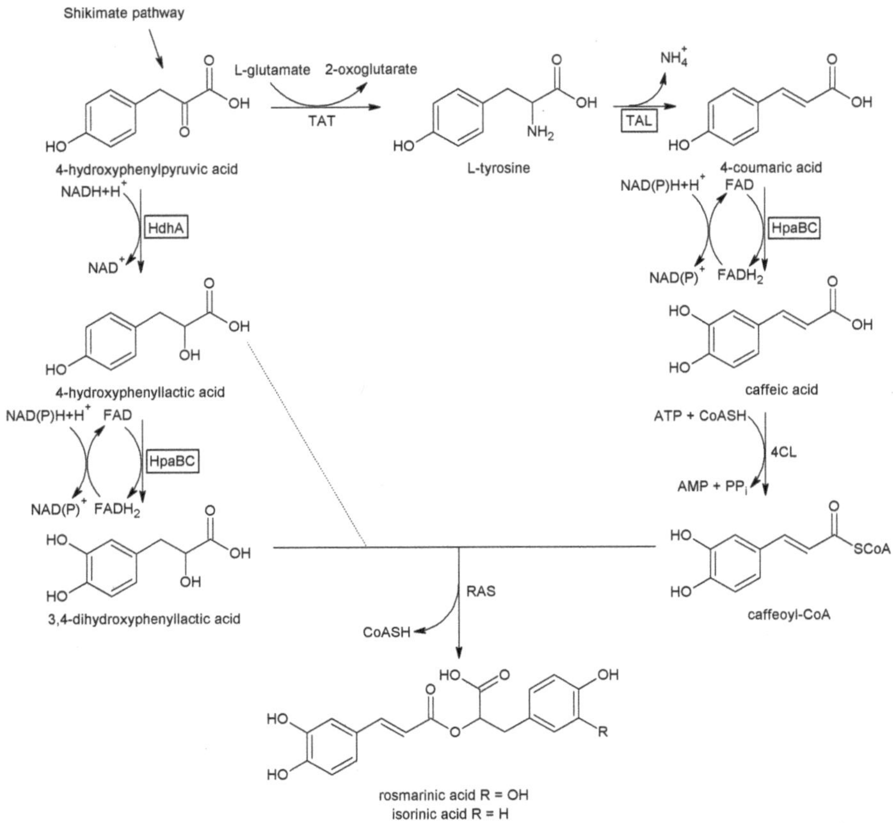

Fig. 2.2 Formation of isorinic acid and rosmarinic acid in *Escherichia coli* as established by Bloch and Schmidt-Dannert [17]. *HdhA* hydroxyacid dehydrogenase from *Lactobacillus delbrueckii* ssp. *bulgaricus*, *HpaBC* 4-hydroxyphenylacetate 3-hydroxylase from *E. coli*, *TAT* tyrosine aminotransferase (endogenous), *TAL* tyrosine ammonia-lyase from *Rhodobacter sphaeroides*, *4CL* 4-coumarate CoA-ligase (At4CL2) from *Arabidopsis thaliana*, *RAS* rosmarinic acid synthase from *Coleus blumei*. Microbial enzymes are marked by boxes

2.6 *In Vitro* Formation of Non-natural Hydroxycinnamic Acid Esters and Amides by "Rosmarinic Acid Synthase"

RAS is the essential ester-forming enzyme in the biosynthetic pathway towards RA [130]. *In vivo*, this enzyme couples a hydroxycinnamoyl unit activated as CoA thioester (4-coumaroyl-CoA, caffeoyl-CoA) to the aliphatic OH-group of a phenylpyruvate derivate. RAS proteins from lavender as well as *Coleus blumei* heterologously expressed in *Escherichia coli* displayed unexpected substrate promiscuity. The recombinant proteins were shown to form esters as well as amides and accepted a considerable variety of compounds leading to products that had not yet been described, e.g. hydroxycinnamoyl-D-phenylalanine, hydroxycinnamoyl-D-tyrosine, hydroxycinnamoyl-D-DOPA, hydroxycinnamoyl-phenethylamine, hydroxycinnamoyl-tyramine, and hydroxycinnamoyl-tryptamine [94, 138].

2.7 Conclusion and Outlook

RA and related metabolites are among those specialized metabolites in plants that are produced at the highest levels. Often, the contents in undifferentiated cells, such as callus and suspension cells, are considerably higher (sometimes exceeding 30% of the DW) than in the source plants. Undifferentiated cells, however, often lose their production capacity with increasing numbers of subcultivations. This disadvantage is less pronounced in differentiated organs. Here HR cultures are the most often established production systems. Up to now, however, the RA production levels in HR are lower (<20% of the DW) than in undifferentiated cells. Many production systems have been established, mostly at laboratory scale. With the exception of early attempts in the 1980s [164], these have not been developed further to semi-industrial or industrial scale. This may be due to the lack of commercial demand for these phenolic acids, since, despite the many biological effects of RA and related phenolic acids, medicinal applications have not been developed, perhaps with exception of *Salvia miltiorrhiza* and its extracted ingredients as traditional Chinese medicines.

Very recent approaches have shown that RA and similar metabolites can also be produced in genetically modified *E. coli*. Here, a combination of bacterial and plant genes have been used and the necessity of membrane-bound cytochrome P450 enzymes circumvented. The amount of RA produced in prokaryotic systems (133 mg/l; [78]) is, however until now, not competitive with plant cell cultures (e.g. 6.4 g/l in *Salvia officinalis* [72]).

References

1. Agata I, Hatano T, Nishibe S, Okuda T. Rabdosiin, a new rosmarinic acid dimer with a lignan skeleton from *Rabdosia japonica*. Chem Pharm Bull. 1988;36:3223–5.
2. Agata I, Kusakabe H, Hatano T, Nishibe S, Okuda T. Melitric acids A and B, new trimeric caffeic acid derivatives from *Melissa officinalis*. Chem Pharm Bull. 1993;41:1608–11.
3. Ai C, Li L. Stereostructure of salvianolic acid B and isolation of salvianolic acid C from *Salvia miltiorrhiza*. J Nat Prod. 1988;51:145–9.
4. Ai C, Li L. Salvianolic acids D and E: two new depsides from *Salvia miltiorrhiza*. Planta Med. 1992;58:197–9.
5. Amoah SKS, Sandjo LP, Kratz JM, Biavatti MW. Rosmarinic acid – pharmaceutical and clinical aspects. Planta Med. 2016;82:388–406.
6. Asakawa Y. Chemical constituents of the bryophytes – chemical constituents of Anthocerotae (hornworts). In: Herz W, Kirby GW, Moore RE, Steglich W, Tamm C, editors. Progress in the chemistry of organic natural products. Wien/New York: Springer; 1995. p. 460–4.
7. Bai YF, Bi HP, Zhuang YB, Liu C, Cai T, Liu XN, Zhang XL, Liu T, Ma YH. Production of salidroside in metabolically engineered *Escherichia coli*. Sci Rep. 2014;4:6640.
8. Bais HP, Walker TS, Schweizer HP, Vivanco JM. Root specific elicitation and antimicrobial activity of rosmarinic acid in hairy root cultures of *Ocimum basilicum*. Plant Physiol Biochem. 2002;40:983–95.
9. Barros L, Duenas M, Dias MI, Sousa MJ, Santos-Buelga C, Ferreira ICFR. Phenolic profiles of cultivated, in vitro cultured and commercial samples of *Melissa officinalis* L. infusions. Food Chem. 2013;136:1–8.

10. Bauer N, Fulgosi H, Jelaska S. Overexpression of phenylalanine ammonia-lyase in transgenic roots of *Coleus blumei* alters growth and rosmarinic acid synthesis. Food Technol Biotechnol. 2011;49:24–31.
11. Bauer N, Kiseljak D, Jelaska S. The effect of yeast extract and methyl jasmonate on rosmarinic acid accumulation in *Coleus blumei* hairy roots. Biol Plant. 2009;53:650–6.
12. Bauer N, Leljak-Levanic D, Jelaska S. Rosmarinic acid synthesis in transformed callus culture of *Coleus blumei* benth. Z Naturforsch. 2004;59c:554–60.
13. Bauer N, Vukovic R, Likic S, Jelaska S. Potential of different *Coleus blumei* tissues for rosmarinic acid production. Food Technol Biotechnol. 2015;53:3–10.
14. Berger A, Meinhard J, Petersen M. Rosmarinic acid synthase is a new member of the superfamily of BAHD acyltransferases. Planta. 2006;224:1503–10.
15. Binding H, Mordhorst G. Gametophyte regeneration and apospory from archegoniate protoplasts under conditions devised for higher plants. Bot Acta. 1991;104:330–5.
16. Bishop G, Sakakibara H, Seo M, Yamaguchi S. Biosynthesis of hormones. In: Buchanan BB, Gruissem W, Jones RL, editors. Biochemistry & molecular biology of plants. Chichester: Wiley; 2015. p. 769–833.
17. Bloch SE, Schmidt-Dannert C. Construction of a chimeric biosynthetic pathway for the de novo biosynthesis of rosmarinic acid in *Escherichia coli*. Chembiochem. 2014;15:2393–401.
18. Bulgakov VP, Inyushkina YV, Fedoreyev SA. Rosmarinic acid and its derivatives: biotechnology and applications. Crit Rev Biotechnol. 2012;32:203–17.
19. Bulgakov VP, Veselova MV, Tchernoded GK, Kiselev KV, Fedoreyev SA, Zhuravlev YN. Inhibitory effect of the *Agrobacterium rhizogenes* rolC gene on rabdosiin and rosmarinic acid production in *Eritrichium sericeum* and *Lithospermum erythrorhizon* transformed cell cultures. Planta. 2005;221:471–8.
20. Chaprin N, Ellis BE. Microspectrophotometric evaluation of rosmarinic acid accumulation in single cultured plant cells. Can J Bot. 1984;62:2278–82.
21. Chen H, Chen F. Kinetics of cell growth and secondary metabolism of a high-tanshinone-producing line of the Ti transformed *Salvia miltiorrhiza* cells in suspension culture. Biotechnol Lett. 1999;21:701–5.
22. Chen H, Chen F. Effects of yeast elicitor on the growth and secondary metabolism of a high-tanshinone-producing line of the Ti transformed *Salvia miltiorrhiza* cells in suspension culture. Process Biochem. 2000a;35:837–40.
23. Chen H, Chen F. Effect of yeast elicitor on the secondary metabolism of Ti-transformed *Salvia miltiorrhiza* cell suspension cultures. Plant Cell Rep. 2000b;19(7):710.
24. Chen H, Chen F. Induction of phytoalexin formation in crown gall and hairy root culture of *Salvia miltiorrhiza* by methyl viologen. Biotechnol Lett. 2000c;22:715–20.
25. Chen H, Chen F, Zhang YL, Song JY. Production of lithospermic acid B and rosmarinic acid in hairy root cultures of *Salvia miltiorrhiza*. J Ind Microbiol Biotechnol. 1999b;22:133–8.
26. Chen H, Chen F, Zhang YL, Song JY. Production of rosmarinic acid and lithospermic acid B in Ti-transformed *Salvia miltiorrhiza* cell suspension cultures. Process Biochem. 1999a;34:777–84.
27. Chen H, Chen F, Chiu FCK, Lo CMY. The effect of yeast elicitor on the growth and secondary metabolism of hairy root cultures of *Salvia miltiorrhiza*. Enzyme Microb Technol. 2001;28:100–5.
28. Chen Z, Gu W, Huang H, Yang X, Sun C, Chen W, Dong Y, Ma H. Studies on the water-soluble phenolic acid constituents of *Salvia miltiorrhiza*. Yaoxue Tongbao. 1981;16:24–5.
29. Chen Z, Zhang J, Zhao S, Wang Z, Hu Z. Effects of methyl jasmonate on accumulation of phenolic acids in *Salvia miltiorrhiza* hairy root. Zhongguo Yaoxue Zazhi. 2010;45:970–4.
30. Cheng TO. Cardiovascular effects of Danshen. Int J Cardiol. 2007;121:9–22.
31. Cui B, Liang Z, Liu Y, Liu F, Zhu J. Effects of ABA and its biosynthetic inhibitor fluridone on accumulation of phenolic acids and activity of PAL and TAT in hairy root of *Salvia miltiorrhiza*. Zhongguo Zhongyao Zazhi. 2012;37:754–9.
32. da Silva Pinto L, de Souza FHD, Nascimento IR, Lopes LMX. Phenylpropanoids from *Paspalum atratum* (Poaceae). Biochem Syst Ecol. 2015;63:68–71.

33. De Eknamkul W, Ellis BE. Rosmarinic acid production and growth characteristics of *Anchusa officinalis* cell suspension cultures. Planta Med. 1984;50:346–50.
34. De Eknamkul W, Ellis BE. Effects of macronutrients on growth and rosmarinic acid formation in cell suspension cultures of *Anchusa officinalis*. Plant Cell Rep. 1985a;4:46–9.
35. De Eknamkul W, Ellis BE. Effects of auxins and cytokinins on growth and rosmarinic acid formation in cell suspension cultures of *Anchusa officinalis*. Plant Cell Rep. 1985b;4:50–3.
36. De Eknamkul W, Ellis BE. Tyrosine aminotransferase: the entrypoint enzyme of the tyrosine-derived pathway in rosmarinic acid biosynthesis. Phytochemistry. 1987;26:1941–6.
37. Dewanjee S, Gangopadhyay M, Das U, Sahu R, Khanra R. Enhanced rosmarinic acid biosynthesis in *Solenostemon scutellarioides* culture: a precursor-feeding strategy. Nat Prod Res. 2014b;28:1691–8.
38. Dewanjee S, Gangopadhyay M, Das U, Sahu R, Samanta A, Banerjee P. Signal transducer and oxidative stress mediated modulation of phenylpropanoid pathway to enhance rosmarinic acid biosynthesis in fungi elicited whole plant culture of *Solenostemon scutellarioides*. Enzyme Microb Technol. 2014a;66:1–9.
39. Di P, Zhang L, Chen J, Tan H, Xiao Y, Dong X, Zhou X, Chen W. ^{13}C tracer reveals phenolic acids biosynthesis in hairy root cultures of *Salvia miltiorrhiza*. ACS Chem Biol. 2013;8:1537–48.
40. Dong C, Su X, Wang W. Effects of sucrose concentration, nitrogen on the growth and main secondary metabolites accumulated of regeneration plant of *Rabdosia rubescens* (Hemsl.) Hara. Xibei Zhiwu Xuebao. 2009;29:494–8.
41. Döring AS, Petersen M. Production of caffeic, chlorogenic and rosmarinic acids in plants and suspension cultures of *Glechoma hederacea*. Phytochem Lett. 2014;10:cxi–cxvii.
42. Dreger M, Krajewska-Patan A, Gorska-Paukszta M, Pieszak M, Buchwald W, Mikolajczak P. Production of the secondary metabolites in *Salvia miltiorrhiza* in vitro. Herba Polonica. 2010;56:78–90.
43. Duan Y, Dai Y, Gao H, Ye W, Yao X. Chemical constituents of *Sarcandra glabra*. Zhongcaoyao. 2010;41:29–32.
44. Ellis BE. Characterization of clonal cultures of *Anchusa officinalis* derived from single cells of known productivity. J Plant Physiol. 1985;119:149–58.
45. El-Mousallamy AMD, Hawas UW, Hussein SAM. Teucrol, a decarboxyrosmarinic acid and its 4′-O-triglycoside, teucroside from *Teucrium pilosum*. Phytochemistry. 2000;55:927–31.
46. Ejtahed RS, Radjabian T, Hoseini Tafreshi SA. Expression analysis of phenylalanine ammonia lyase gene and rosmarinic acid production in *Salvia officinalis* and *Salvia virgata* shoots under salicylic acid elicitation. Appl Biochem Biotechnol. 2015;176:1846–58.
47. Fattahi M, Nazeri V, Torras-Claveria L, Sefidkon F, Cusido RM, Zamani Z, Palazon J. A new biotechnological source of rosmarinic acid and surface flavonoids: hairy root cultures of *Dracocephalum kotschyi* Boiss. Ind Crop Prod. 2013;50:256–63.
48. Fukui H, Yazaki K, Tabata M. Two phenolic acids from *Lithospermum erythrorhizon* cell suspension cultures. Phytochemistry. 1984;23:2398–9.
49. Georgiev M, Kuzeva S, Pavlov A, Kovacheva E, Ilieva M. Enhanced rosmarinic acid production by *Lavandula vera* MM cell suspension culture through elicitation with vanadyl sulfate. Z Naturforsch. 2006b;61C:241–4.
50. Georgiev MI, Kuzeva SL, Pavlov AI, Kovacheva EG, Ilieva MP. Elicitation of rosmarinic acid by *Lavandula vera* MM cell suspension culture with abiotic elicitors. World J Microbiol Biotechnol. 2007a;23:301–4.
51. Georgiev MI, Pavlov AI, Bley T. Hairy root type plant in vitro systems as sources of bioactive substances. Appl Microbiol Biotechnol. 2007b;74:1175–85.
52. Georgiev M, Pavlov A, Ilieva M. Rosmarinic acid production by *Lavandula vera* MM cell suspension: the effect of temperature. Biotechnol Lett. 2004;26:855–6.
53. Georgiev M, Pavlov A, Ilieva M. Selection of high rosmarinic acid producing *Lavandula vera* MM cell lines. Process Biochem. 2006a;41:2068–71.
54. Gertlowski C, Petersen M. Influence of the carbon source on growth and rosmarinic acid production in suspension cultures of *Coleus blumei*. Plant Cell Tiss Org. 1993;34:183–90.

55. Grzegorczyk I, Wysokinska H. Liquid shoot culture of *Salvia officinalis* L. for micropropagation and production of antioxidant compounds; effect of triacontanol. Acta Soc Bot Pol. 2008;77:99–104.
56. Grzegorczyk I, Wysokinska H. The effect of methyl jasmonate on production of antioxidant compounds in shoot cultures of *Salvia officinalis* L. Herba Pol. 2009;55:238–43.
57. Grzegorczyk I, Wysokinska H. Antioxidant compounds in *Salvia officinalis* L. shoot and hairy root cultures in the nutrient sprinkle bioreactor. Acta Soc Bot Poloniae. 2010;79:7–10.
58. Grzegorcyzk I, Bilichowski I, Mikiciuk-Olasik E, Wysokinska H. In vitro cultures of *Salvia officinalis* L. as a source of antioxidant compounds. Acta Soc Bot Poloniae. 2005;74:17–21.
59. Grzegorcyzk I, Bilichowski I, Mikiciuk-Olasik E, Wysokinska H. The effect of triacontanol on shoot multiplication and production of antioxidant compounds in shoot cultures of *Salvia officinalis* L. Acta Soc Bot Poloniae. 2006;75:11–5.
60. Grzegorczyk I, Krolicka A, Wysokinska H. Establishment of *Salvia officinalis* L. hairy root cultures for the production of rosmarinic acid. Z Naturforsch. 2006;61c:351–6.
61. Gu XC, Chen JF, Xiao Y, Di P, Xuan HJ, Zhou X, Zhang L, Chen WS. Overexpression of allene oxide cyclase promoted tanshinone/phenolic acid production in *Salvia miltiorrhiza*. Plant Cell Rep. 2012;31:2247–59.
62. Guillon S, Trémouillaux-Guiller J, Pati PK, Rideau M, Gantet P. Hairy root research: recent scenario and exciting prospects. Curr Opin Plant Biol. 2006;9:341–6.
63. Guirgis AA, Abd El-Kawi MA, Abbas HN, Araffa AMS, Maksoud AI. High rosmarinic acid content in induced mutants and in *in vitro* elicited sweet basil (*Ocimum basilicum* L.) callus. Asian J Plant Sci. 2007;6:1058–64.
64. Guo H, Zhu N, Deyholos MK, Liu J, Zhang X, Dong J. Calcium mobilization in salicylic acid-induced *Salvia miltiorrhiza* cell cultures and its effect on the accumulation of rosmarinic acid. Appl Biochem Biotechnol. 2015;175:2689–702.
65. Hakkim FL, Shankar CG, Girija S. Chemical composition and antioxidant property of holy basil (*Ocimum sanctum* L.) leaves, stems, and inflorescence and their *in vitro* callus cultures. J Agric Food Chem. 2007;55:9109–17.
66. Hao W, Guo H, Zhang J, Hu G, Yao Y, Dong J. Hydrogen peroxide is involved in salicylic acid-elicited rosmarinic acid production in *Salvia miltiorrhiza* cell cultures. Sci World J. 2014, Article ID 843764, 7 pages.
67. Hao G, Jiang X, Feng L, Tao R, Li Y, Huang L. Cloning, molecular characterization and functional analysis of a putative R2R3-MYB transcription factor of the phenolic acid biosynthetic pathway in *S. miltiorrhiza* Bge. f. *alba*. Plant Cell Tiss Org Cult. 2016;124:151–68.
68. Harborne JB. Caffeic acid ester distribution in higher plants. Z Naturforsch. 1966;21b:604–5.
69. Häusler E, Petersen M, Alfermann AW. Hydroxyphenylpyruvate reductase from cell suspension cultures of *Coleus blumei* Benth. Z Naturforsch. 1991;46c:371–6.
70. Häusler E, Petersen M, Alfermann AW. Rosmarinsäure in *Blechnum*-Spezies. In: Haschke HP, Schnarrenberger C, editors. Botanikertagung, vol. 1992. Berlin: Akademie-Verlag; 1992. p. 507.
71. Hippolyte I, Marin B, Baccou JC, Jonard R. Influence du milieu d'entretien et de la concentration en saccharose sur la production d'acide rosmarinique par des suspensions cellulaires de Sauge (*Salvia officinalis* L.). CR Acad Sci Paris. 1991;313:365–71.
72. Hippolyte I, Marin B, Baccou JC, Jonard R. Growth and rosmarinic acid production in cell suspension cultures of *Salvia officinalis* L. Plant Cell Rep. 1992;11:109–12.
73. Hücherig S, Petersen M. RNAi suppression and overexpression studies of hydroxyphenylpyruvate reductase (HPPR) and rosmarinic acid synthase (RAS) genes related to rosmarinic acid biosynthesis in hairy root cultures of *Coleus blumei*. Plant Cell Tiss Org Cult. 2013;113:375–85.
74. Ilieva M, Pavlov A. Rosmarinic acid by *Lavandula vera* MM cell suspension: phosphorus effect. Biotechnol Lett. 1996;18:913–6.
75. Ilieva M, Pavlov A. Rosmarinic acid production by *Lavandula vera* MM cell-suspension culture. Appl Microbiol Biotechnol. 1997;47:683–8.

76. Ilieva M, Pavlov A. Rosmarinic acid production by *Lavandula vera* MM cell suspension culture: nitrogen effect. World J Microbiol Biotechnol. 1999;15:711–4.
77. Ilieva-Stoilova MP, Pavlov AI, Kovatcheva-Apostolova EG. Further research into *Lavandula* species: cell cultures of *L. vera* and rosmarinic acid production. Med Aromat Plants Ind Profiles. 2002;29:214–26.
78. Jiang J, Bi H, Zhuang Y, Liu S, Liu T, Ma Y. Engineered synthesis of rosmarinic acid in *Escherichia coli* resulting production of a new intermediate, caffeoyl-phenyllactate. Biotechnol Lett. 2016;38:81–8.
79. Jiang RW, Lau KM, Hon PM, Mak TCW, Woo KS, Fung KP. Chemistry and biological activities of caffeic acid derivatives from *Salvia miltiorrhiza*. Curr Med Chem. 2005;12:237–46.
80. Ju Q, Liu G, Liu X, Han X, Liu H. Influence of growth regulators and sucrose concentrations on growth and rosmarinic acid production in calli and suspension cultures of *Coleus blumei*. Nat Prod Res. 2009;23:127–37.
81. Karwatzki B, Petersen M, Alfermann AW. Transient activity of enzymes involved in the biosynthesis of rosmarinic acid in cell cultures of *Coleus blumei*. Planta Med. 1989;55:663–4.
82. Kelley CJ, Mahajan JR, Brooks LC, Neubert LA, Breneman WR, Carmack M. Polyphenolic acids of *Lithospermum ruderale* Dougl. ex Lehm. (Boraginaceae). 1. Isolation and structure determination of lithospermic acid. J Org Chem. 1975;40:1804–15.
83. Kiferle C, Lucchesini M, Maggini R, Pardossi A, Mensuali-Sodi A. *In vitro* culture of sweet basil: gas exchanges, growth, and rosmarinic acid production. Biol Plantarum. 2014;58:601–10.
84. Kiferle C, Lucchesini M, Mensuali-Sodi A, Maggini R, Raffaelli A, Pardossi A. Rosmarinic acid content in basil plants grown in vitro and in hydroponics. Cent Eur J Biol. 2011;6:946–57.
85. Kim BG, Jung WD, Mok H, Ahn JH. Production of hydroxycinnamoyl-shikimates and chlorogenic acid in *Escherichia coli*: production of hydroxycinnamic acid conjugates. Microb Cell Factories. 2013b;12:15.
86. Kim YB, Kim JK, Uddin M, Xu H, Park WT, Tuan PA, Li X, Chung E, Lee JH, Park SU. Metabolomics analysis and biosynthesis of rosmarinic acid in *Agastache rugosa* Kuntze treated with methyl jasmonate. PLoS One. 2013a;8:e64199.
87. Kintzios S, Kollias H, Straitouris E, Makri O. Scale-up micropropagation of sweet basil (*Ocimum basilicum* L.) in an airlift bioreactor and accumulation of rosmarinic acid. Biotechnol Lett. 2004;26:521–3.
88. Kintzios S, Makri O, Panagiotopoulos E, Scapeti M. *In vitro* rosmarinic acid accumulation in sweet basil (*Ocimum basilicum* L.). Biotechnol Lett. 2003;25:405–8.
89. Kochan E, Wysokinska H, Chmiel A, Grabias B. Rosmarinic acid and other phenolic acids in hairy roots of *Hyssopus officinalis*. Z Naturforsch. 1999;54c:11–6.
90. Kohda H, Takeda O, Tanaka S, Yamasaki K, Yamashita A, Kurokawa T, Ishibashi S. Isolation of inhibitors of adenylate cyclase from Dan-shen, the root of *Salvia miltiorrhiza*. Chem Pharm Bull. 1989;37:1287–90.
91. Kojuharova L, Pavlov A, Ilieva M, Mincheva M, Kovatcheva E. Enhancement of phenolic compounds synthesis from plant cell culture of *Lavandula vera* MM. Proceedings of bioprocess engineering '95 1995:24–27.
92. Kovatcheva E, Pavlov A, Koleva I, Ilieva M, Mihneva M. Rosmarinic acid from *Lavandula vera* MM cell culture. Phytochemistry. 1996;43:1243–4.
93. Kracun-Kolarevic M, Dmitrovic S, Filipovic B, Peric M, Misic D, Simonovic A, Todorovic S. Influence of sodium salicylate on rosmarinic acid, carnosol and carnosic acid accumulation by *Salvia officinalis* L. shoots grown in vitro. Biotechnol Lett. 2015;37:1693–701.
94. Landmann C, Hücherig S, Fink B, Hoffmann T, Dittlein D, Coiner HA, Schwab W. Substrate promiscuity of a rosmarinic acid synthase from lavender (*Lavandula angustifolia* L.). Planta. 2011;234:305–20.
95. Lee JH, Park KH, Lee MH, Kim HT, Seo WD, Kim JY, Baek IY, Jang DS, Ha TJ. Identification, characterisation, and quantification of phenolic compounds in the antioxidant activity-containing fraction from the seeds of Korean perilla (*Perilla frutescens*) cultivars. Food Chem. 2013;136:843–52.

96. Li L, Tan R, Chen W. Salvianolic acid A, a new depside from roots of *Salvia miltiorrhiza*. Planta Med. 1984;50:227–8.
97. Li W, Gao W, Zhao J, Cui G, Shao A, Huang L. Research on mechanism of secondary metabolites of phenolic acids in *Salvia miltiorrhiza* hairy root induced by jasmonate. Zhongguo Zhongyao Zazhi. 2012;37:13–6.
98. Li W, Koike K, Asada Y, Yoshikawa T, Nikaido T. Rosmarinic acid production by *Coleus forskohlii* hairy root cultures. Plant Cell Tiss Org Cult. 2005;80:151–5.
99. Li X, Guo H, Qi Y, Liu H, Zhang X, Ma P, Liang Z, Dong J. Salicylic acid-induced cytosolic acidification increases the accumulation of phenolic acids in *Salvia miltiorrhiza* cells. Plant Cell Tiss Org Cult. 2016;126:333–41.
100. Liang LF, Keng CL, Lim BP. Selection of cell lines for the production of rosmarinic acid from cell suspension cultures of *Orthosiphon stamineus* Benth. In Vitro Cell Dev-Pl. 2006;42:538–42.
101. Liu L, Dong J, Zhang J, Dang X, Xing B, Yang X. Effects of calcium on synthesis of rosmarinic acid and related enzymes in suspension cultures of *Salvia miltiorrhiza*. Shengwu Gongcheng Xuebao. 2012;28:1359–69.
102. Lu Y, Foo LY. Rosmarinic acid derivatives from *Salvia officinalis*. Phytochemistry. 1999;51:91–4.
103. Lu Y, Foo LY. Polyphenolics of *Salvia* – a review. Phytochemistry. 2002;59:117–40.
104. Lu Y, Foo LY. Salvianolic acid L, a potent phenolic antioxidant from *Salvia officinalis*. Tetrahedron Lett. 2001;42:8223–5.
105. Lu Y, Foo LY, Wong H. Sagecoumarin, a novel caffeic acid trimer from *Salvia officinalis*. Phytochemistry. 1999;52:1149–52.
106. Ma XH, Ma Y, Tang JF, He YL, Liu YC, Ma XJ, Shen Y, Cui GH, Lin HX, Rong QX, Guo J, Huang LQ. The biosynthetic pathways of tanshinones and phenolic acids in *Salvia miltiorrhiza*. Molecules. 2015;20:16235.
107. Martinez BC, Park CH. Immobilization of *Coleus blumei* cells in a column reactor using a spray feeding system. Biotechnol Tech. 1994;8:301–6.
108. Mehrotra S, Srivastava V, Rahman LU, Kukreja AK. Hairy root biotechnology – indicative timeline to understand missing links and future outlook. Protoplasma. 2015;252:1189–201.
109. Míka V, Kubáň V, Klejdus B, Odstrčilová V, Nerušil P. Phenolic compounds as chemical markers of low taxonomic levels in the family Poaceae. Plant Soil Environ. 2005;51:506–12.
110. Mizukami H, Ogawa T, Ohashi H, Ellis BE. Induction of rosmarinic acid biosynthesis in *Lithospermum erythrorhizon* cell suspension cultures by yeast extract. Plant Cell Rep. 1992;11:480–3.
111. Mizukami H, Tabira Y, Ellis BE. Methyl jasmonate-induced rosmarinic acid biosynthesis in *Lithospermum erythrorhizon* cell suspension cultures. Plant Cell Rep. 1993;12:706–9.
112. Morimoto S, Goto Y, Shoyama Y. Production of lithospermic acid B and rosmarinic acid in callus tissue and regenerated plantlets of *Salvia miltiorrhiza*. J Nat Prod. 1994;57:817–23.
113. Moschopoulou G, Kintzios S. Achievement of thousand-fold accumulation of rosmarinic acid in immobilized cells of sweet basil (*Ocimum basilicum* L.) by ten-fold increase of the volume of the immobilization matrix. J Biol Res. 2011;15:59–65.
114. Murata T, Oyama K, Fujiyama M, Oobayashi B, Umehara K, Miyase T, Yoshizaki F. Diastereomers of lithospermic acid and lithospermic acid B from *Monarda fistulosa* and *Lithospermum erythrorhizon*. Fitoterapia. 2013;91:51–9.
115. Nitzsche A, Tokalov SV, Gutzeit HO, Ludwig-Müller J. Chemical and biological characterization of cinnamic acid derivatives from cell cultures of lavender (*Lavandula officinalis*) induced by stress and jasmonic acid. J Agric Food Chem. 2004;52:2915–23.
116. Nourozi E, Hosseini B, Hassani A. A reliable and efficient protocol for induction of hairy roots in *Agastache foeniculum*. Biologia. 2014;69:870–9.
117. Ogata A, Tsuruga A, Matsuno M, Mizukami H. Elicitor-induced rosmarinic acid biosynthesis in *Lithospermum erythrorhizon* cell suspension cultures: activities of rosmarinic acid synthase and the final two cytochrome P450-catalyzed hydroxylations. Plant Biotechnol. 2004;21:393–6.

118. Park CH, Martinez BC. Enhanced release of rosmarinic acid from *Coleus blumei* permeabilized by dimethyl sulfoxide (DMSO) while preserving cell viability and growth. Biotechnol Bioeng. 1992;40:459–64.
119. Pavlov A, Ilieva M. The influence of phenylalanine on accumulation of rosmarinic and caffeic acids by *Lavandula vera* MM cell culture. World J Microbiol Biotechnol. 1999;15:397–9.
120. Pavlov A, Georgiev MI, Ilieva MP. Production of rosmarinic acid by *Lavandula vera* MM cell suspension in bioreactor: effect of dissolved oxygen concentration and agitation. World J Microbiol Biotechnol. 2005a;21:389–92.
121. Pavlov AI, Georgiev MI, Panchev IN, Ilieva MP. Optimization of rosmarinic acid production by *Lavandula vera* MM plant cell suspension in a laboratory bioreactor. Biotechnol Prog. 2005b;21:394–6.
122. Pavlov A, Ilieva M, Mincheva M. Release of rosmarinic acid by *Lavandula vera* MM cell suspension in two-phase culture systems. World J Microbiol Biotechnol. 2001;17:417–21.
123. Pavlov AI, Ilieva MP, Panchev IN. Nutrient medium optimization for rosmarinic acid production by *Lavandula vera* MM cell suspension. Biotechnol Prog. 2000;16:668–70.
124. Petersen M. Cytochrome P450-dependent hydroxylation in the biosynthesis of rosmarinic acid in *Coleus*. Phytochemistry. 1997;45:1165–72.
125. Petersen M. Cinnamic acid 4-hydroxyase from cell cultures of the hornwort *Anthoceros agrestis*. Planta. 2003;217:96–101.
126. Petersen M. Rosmarinic acid: new aspects. Phytochem Rev. 2013;12:207–27.
127. Petersen M, Alfermann AW. Two new enzymes of rosmarinic acid biosynthesis from cell cultures of *Coleus blumei*: hydroxyphenylpyruvate reductase and rosmarinic acid synthase. Z Naturforsch. 1988;43c:501–4.
128. Petersen M, Simmonds MSJ. Rosmarinic acid. Phytochemistry. 2003;62:121–5.
129. Petersen M, Abdullah Y, Benner J, Eberle D, Gehlen K, Hücherig S, Janiak V, Kim KH, Sander M, Weitzel C, Wolters S. Evolution of rosmarinic acid biosynthesis. Phytochemistry. 2009;70:1663–79.
130. Petersen M, Häusler E, Karwatzki B, Meinhard J. Proposed biosynthetic pathway for rosmarinic acid in cell cultures of *Coleus blumei* Benth. Planta. 1993;189:10–4.
131. Petersen M, Häusler E, Meinhard J, Karwatzki B, Gertlowski C. The biosynthesis of rosmarinic acid in suspension cultures of *Coleus blumei*. Plant Cell Tiss Org Cult. 1994;38:171–9.
132. Petersen M, Szabo E, Meinhard J, Karwatzki B, Gertlowski C, Kempin B, Fuss E. Biosynthesis and accumulation of rosmarinic acid in suspension cultures of *Coleus blumei*. Plant Cell Tiss Org Cult. 1995;43:89–92.
133. Pezeshki S. Biosynthese von Kaffeesäuremetaboliten im Ackerhornmoos *Anthoceros agrestis* und im Kleinen Blasenmützenmoos *Physcomitrella patens*. PhD thesis, Philipps-Universität Marburg. 2016.
134. Razzaque A, Ellis BE. Rosmarinic acid production in *Coleus* cell cultures. Planta. 1977;137:287–91.
135. Ruffoni B, Bertoli A, Pistelli L, Pistelli L. Micropropagation of *Salvia wagneriana* Polak and hairy root cultures with rosmarinic acid production. Nat Prod Res. 2016;4:1–7.
136. Sahraroo A, Babalar M, Mirjalili MH, Moghaddam MRF, Ebrahimi SN. *In-vitro* callus induction and rosmarinic acid quantification in callus culture of *Satureja khuzistanica* Jamzad (Lamiaceae). Iranian J Pharm Res. 2014;13:1447–56.
137. Sahraroo A, Mirjalili MH, Corchete P, Babalar M, Moghaddam MRF. Establishment and characterization of a *Satureja khuzistanica* Jamzad (Lamiaceae) cell suspension culture: a new in vitro source of rosmarinic acid. Cytotechnology. 2015;68:1415–24.
138. Sander M, Petersen M. Distinct substrate specificities and unusual substrate flexibilities of two hydroxycinnamoyltransferases, rosmarinic acid synthase and hydroxycinnamoyl-CoA:shikimate hydroxycinnamoyltransferase, from *Coleus blumei* Benth. Planta. 2011;233:1157–71.
139. Satake T, Kamiya K, Saiki Y, Hama T, Fujimoto Y, Kitanaka S, Kimura Y, Uzawa J, Endang H, Umar M. Studies on the constituents of fruits of *Helicteres isora* L. Chem Pharm Bull. 1999;47:1444–7.

140. Scarpati ML, Oriente G. Isolamento e costituzione dell'acido rosmarinico (dal *rosmarinus off.*). Ric Sci. 1958;28:2329–33.
141. Shanks JV, Morgan J. Plant 'hairy root' culture. Curr Opin Biotechnol. 1999;10:151–5.
142. Shetty K. Rosmarinic acid biosynthesis and mechanism of action. Funct Foods Biotechnol. 2007;165:187–207.
143. Srivastava S, Conlan XA, Adholeya A, Cahill DM. Elite hairy roots of *Ocimum basilicum* as a new source of rosmarinic acid and antioxidants. Plant Cell Tiss Org Cult. 2016;126:19–32.
144. Strazzer P, Guzzo F, Levi M. Correlated accumulation of anthocyanins and rosmarinic acid in mechanically stressed red cell suspensions of basil (*Ocimum basilicum*). J Plant Physiol. 2011;168:288–93.
145. Su WW, Humphrey AE. Production of rosmarinic acid in high density perfusion cultures of *Anchusa officinalis* using a high sugar medium. Biotechnol Lett. 1990;12:793–8.
146. Su WW, Humphrey AE. Production of rosmarinic acid from perfusion culture of *Anchusa officinalis* in a membrane-aerated bioreactor. Biotechnol Lett. 1991;13:889–92.
147. Su WW, Lei F. Rosmarinic acid production in perfused *Anchusa officinalis* culture: effect of inoculum size. Biotechnol Lett. 1993;15:1035–8.
148. Su WW, Lei F, Kao NP. High density cultivation of *Anchusa officinalis* in a stirred-tank bioreactor with in situ filtration. Appl Microbiol Biotechnol. 1995;44:293–9.
149. Su WW, Lei F, Su LY. Perfusion strategy for rosmarinic acid production by *Anchusa officinalis*. Biotechnol Bioeng. 1993;42:884–90.
150. Su WW, Asali EC, Humphrey AE. *Anchusa officinalis*: production of rosmarinic acid in perfusion cell cultures. In: Bajaj YPS, editor. Biotechnology in agriculture and forestry, medicinal and aromatic plants VI. Berlin/Heidelberg: Springer; 1994. p. 1–20.
151. Sumaryono W, Proksch P. *Orthosiphon aristatus* (Java tea): *in vitro* culture and elicitation of rosmarinic acid in cell suspensions. In: Biotechnology in agriculture and forestry, medicinal and aromatic plants V. Berlin\Heidelberg: Springer; 1993. p. 287–99.
152. Sumaryono W, Proksch P, Hartmann T, Nimtz M, Wray V. Induction of rosmarinic acid accumulation in cell suspension cultures of *Orthosiphon aristatus* after treatment with yeast extract. Phytochemistry. 1991;30:3267–71.
153. Szabo E, Thelen A, Petersen M. Fungal elicitor preparations and methyl jasmonate enhance rosmarinic acid accumulation in suspension cultures of *Coleus blumei*. Plant Cell Rep. 1999;18:485–9.
154. Tada H, Murakami Y, Omoto T, Shimomura K, Ishimaru K. Rosmarinic acid and related phenolics in hairy root cultures of *Ocimum basilicum*. Phytochemistry. 1996;42:431–4.
155. Takeda R, Hasegawa J, Shinozaki M. The first isolation of lignans, megacerotonic acid and anthocerotonic acid, from non-vascular plants, Anthocerotae (hornworts). Tetrahedron Lett. 1990a;31:4159–62.
156. Takeda R, Hasegawa J, Shinozaki K. Phenolic compounds from Anthocerotae. In: Zinsmeister HD, Mues R, editors. Bryophytes: their chemistry and chemical taxonomy. Proceedings of the phytochemical Society of Europe, vol. 29. Oxford: Oxford Science Publications; 1990b. p. 201–7.
157. Tanaka T, Morimoto S, Nonaka G, Nishioka I, Yokozawa T, Chung HY, Oura H. Magnesium and ammonium-potassium lithospermates B, the active principles having a uremia-preventive effect from *Salvia miltiorrhiza*. Chem Pharm Bull. 1989;37:340–4.
158. Tanaka T, Nishimura A, Kouno I, Nonaka G, Young TJ. Isolation and characterization of yunnaneic acids A-D, four novel caffeic acid metabolites from *Salvia yunnanensis*. J Nat Prod. 1996;59(9):843.
159. Tanaka T, Nishimura A, Kouno I, Nonaka G, Yang CR. Four new caffeic acid metabolites, yunnaneic acids E-H, from *Salvia yunnanensis*. Chem Pharm Bull. 1997;45:1596–600.
160. Tian L. Using hairy roots for production of valuable plant secondary metabolites. In: Krull R, Bley T, editors. Filaments in bioprocesses. Advances in biochemical engineering/biotechnology, vol. 149. Cham: Springer International Publishing; 2015. p. 275–324.

161. Tonelli M, Pellegrini E, D'Angiolillo F, Petersen M, Nali C, Pistelli L, Lorenzini G. Ozone-elicited secondary metabolites in shoot cultures of *Melissa officinalis* L. Plant Cell Tiss Org Cult. 2015;120:617–29.
162. Trennheuser F. Phytochemische Untersuchung und *in vitro* Kultur ausgewählter Vertreter der Anthocerotopsida. PhD thesis, University of Saarbrücken. 1992.
163. Trennheuser F, Burkhard G, Becker H. Anthocerodiazonin, an alkaloid from *Anthoceros agrestis*. Phytochemistry. 1994;37:899–903.
164. Ulbrich B, Wiesner W, Arens H. Large-scale production of rosmarinic acid from plant cell cultures of *Coleus blumei* Benth. In: Neumann KH, Barz W, Reinhard E, editors. Primary and secondary metabolism of plant cell cultures. Berlin\Heidelberg: Springer; 1985. p. 293–303.
165. Vogelsang K, Schneider B, Petersen M. Production of rosmarinic acid and a new rosmarinic acid 3′-*O*-β-D-glucoside in suspension cultures of the hornwort *Anthoceros agrestis* Paton. Planta. 2006;223:369–73.
166. Vukovic R, Bauer N, Curkovic-Perica M. Genetic elicitation by inducible expression of β-cryptogein stimulates secretion of phenolics from *Coleus blumei* hairy roots. Plant Sci. 2013;199–200:18–28.
167. Wagner H, Wittmann D. Zur chemischen Struktur der Lithospermsäure aus *Lithospermum officinale* L. Tetrahedron Lett. 1975;(8):547–50.
168. Wang BQ. *Salvia miltiorrhiza*: chemical and pharmacological review of a medicinal plant. J Med Plants Res. 2010;4:2813–20.
169. Wang CZ, Davin LB, Lewis NG. Stereoselective phenolic coupling in *Blechnum spicant*: formation of 8-2′ linked (−)-cis-blechnic, (−)-trans-blechnic and (−)-brainic acids. Chem Commun. 2001:113–4.
170. Wang CL, Liang ZS, Li DR, Liu Y, Liu FH. Effects of growth-regulating substances and soluble sugar contents on accumulation of salvianolic acids in *Salvia miltiorrhiza* Bunge. Zhiwu Shengli Xuebao. 2012;48:181–8.
171. Weitzel C, Petersen M. Enzymes of phenylpropanoid metabolism in the important medicinal plant *Melissa officinalis* L. Planta. 2010;232:731–42.
172. Weitzel C, Petersen M. Cloning and characterisation of rosmarinic acid synthase from *Melissa officinalis* L. Phytochemistry. 2011;72:572–8.
173. Weremczuk-Jezyna I, Grzegorczyk-Karolak I, Frydrych B, Krolicka A, Wysokinska H. Hairy roots of *Dracocephalum moldavica*: rosmarinic acid content and antioxidant potential. Acta Physiol Plant. 2013;35:2095–103.
174. Whitaker RJ, Hashimoto T, Evans DA. Production of the secondary metabolite, rosmarinic acid, by plant cell suspension cultures. Ann N Y Acad Sci. 1984;435:364–6.
175. Wu CF, Karioti A, Rohr D, Bilia AR, Efferth T. Production of rosmarinic acid and salvianolic acid B from callus culture of *Salvia miltiorrhiza* with cytotoxicity towards acute lymphoblastic leukemia cells. Food Chem. 2016;201:292–7.
176. Xiao Y, Gao S, Di P, Chen J, Chen W, Zhang L. Methyl jasmonate dramatically enhances the accumulation of phenolic acids in *Salvia miltiorrhiza* hairy root cultures. Physiol Plant. 2009;137:1–9.
177. Xiao Y, Gao SH, Di P, Chen JF, Chen WS, Zhang L. Lithospermic acid B is more responsive to silver ions (Ag^+) than rosmarinic acid in *Salvia miltiorrhiza* hairy root cultures. Biosci Rep. 2010;30:33–40.
178. Xiao Y, Zhang L, Gao S, Saechao S, Di P, Chen J, Chen W. The c4h, tat, hppr and hppd genes prompted engineering of rosmarinic acid biosynthetic pathway in *Salvia miltiorrhiza* hairy root cultures. PLoS One. 2011;6:e29713.
179. Xing BY, Dang XL, Zhang JY, Wang B, Chen ZY, Dong JE. Effects of methyl jasmonate on the biosynthesis of rosmarinic acid and related enzymes in *Salvia miltiorrhiza* suspension cultures. Zhiwu Shengli Xuebao. 2013;49:1326–32.
180. Xing B, Yang D, Guo W, Liang Z, Yan X, Zhu Y, Liu Y. Ag^+ as a more effective elicitor for production of tanshinones than phenolic acids in *Salvia miltiorrhiza* hairy roots. Molecules. 2015;20:309–24.

181. Yamamoto H, Inoue K, Yazaki K. Caffeic acid oligomers in *Lithospermum erythrorhizon* cell suspension cultures. Phytochemistry. 2000a;53:651–7.
182. Yamamoto H, Yazaki K, Inoue K. Simultaneous analysis of shikimate-derived secondary metabolites in *Lithospermum erythrorhizon* cell suspension cultures by high-performance liquid chromatography. J Chromatogr B. 2000b;738:3–15.
183. Yan Q, Shi M, Ng J, Wu JY. Elicitor-induced rosmarinic acid accumulation and secondary metabolism enzyme activities in *Salvia miltiorrhiza* hairy roots. Plant Sci. 2006;170:853–8.
184. Yan Y, Zhang S, Zhang J, Ma P, Duan J, Liang Z. Effect and mechanism of endophytic bacteria on growth and secondary metabolite synthesis in *Salvia miltiorrhiza* hairy roots. Acta Physiol Plant. 2014;36:1095–105.
185. Yang YK, Lee SY, Park WT, Park NI, Park SU. Exogenous auxins and polyamines enhance growth and rosmarinic acid production in hairy root cultures of *Nepeta cataria* L. plant. OMICS. 2010;3:190–3.
186. Yao YF, Wang CS, Qiao J, Zhao GR. Metabolic engineering of *Escherichia coli* for production of salvianic acid A via an artificial biosynthetic pathway. Metab Eng. 2013;19:79–87.
187. Zenk MH, El-Shagi H, Ulbrich B. Production of rosmarinic acid by cell-suspension cultures of *Coleus blumei*. Naturwiss. 1977;64:585–6.
188. Zhang S, Yan Y, Wang B, Liang Z, Liu Y, Liu F, Qi Z. Selective responses of enzymes in the two parallel pathways of rosmarinic acid biosynthetic pathway to elicitors in *Salvia miltiorrhiza* hairy root cultures. J Biosci Bioeng. 2014;117:645–51.
189. Zhao S, Zhang J, Yang L, Wang Z, Hu Z. Determination and biosynthesis of multiple salvianolic acids in hairy roots of *Salvia miltiorrhiza*. Yaoxue Xuebao. 2011;46:1352–6.
190. Zhou L, Zuo Z, Chow MSS. Danshen: an overview of its chemistry, pharmacology, pharmacokinetics, and clinical use. J Clin Pharmacol. 2005;45:1345–59.
191. Zhu L, Li Y, Yang J, Zuo L, Zhang D. Studies on chemical constituents of *Sarcandra glabra*. Zhongguo Zhongyao Zazhi. 2008;33:155–7.
192. Zhuang Y, Jiang J, Bi H, Yin H, Liu S, Liu T. Synthesis of rosmarinic acid analogues in *Escherichia coli*. Biotechnol Lett. 2016;38:619–27.

Bioproduction of Resveratrol

3

Jian Wang, Yaping Yang, and Yajun Yan

3.1 Introduction

Resveratrol (3, 5, 4′-trihydroxy-*trans*-stilbene) is an important member of the stilbene class of plant-derived polyphenolic metabolites [1]. It naturally occurs in many different plants, including grapes, various berries, peanuts, and their derived food products, such as wine and juice [2, 3]. In plants, as a defense compound, resveratrol has been shown to exert biological effects against pathogenic infection and injury [4]. Especially, resveratrol has been regarded as an important health-promoting nutraceutical in red wine [5, 6]. Numerous studies have reported that resveratrol exhibits protective effects against some cancers and possesses a beneficial effect on the cardiovascular system [7]. Pre-clinical tests have demonstrated that resveratrol is a promising pharmaceutical candidate compound with considerable pharmacological potential to be used for the treatment of neurodegenerative diseases [8, 9]. Resveratrol has also been demonstrated to exert a positive effect on the lifespan of different organisms [10]. Additionally, resveratrol exhibits a high level of antioxidant activity, which exceeds the activity of vitamin E [3]. Thus, resveratrol is currently being developed and sold as over-the-counter ingredient of nutritional supplements, pharmaceuticals, nutraceuticals and cosmetics [11]. These potential applications of resveratrol have triggered new research interests toward engineering the biosynthesis of this value-added compound.

J. Wang • Y. Yang
College of Engineering, University of Georgia, Athens, GA 30602, USA

Y. Yan (✉)
BioChemical Engineering Program, College of Engineering, University of Georgia, Athens, GA 30602, USA
e-mail: yajunyan@uga.edu

W. Schwab et al. (eds.), *Biotechnology of Natural Products*,
https://doi.org/10.1007/978-3-319-67903-7_3

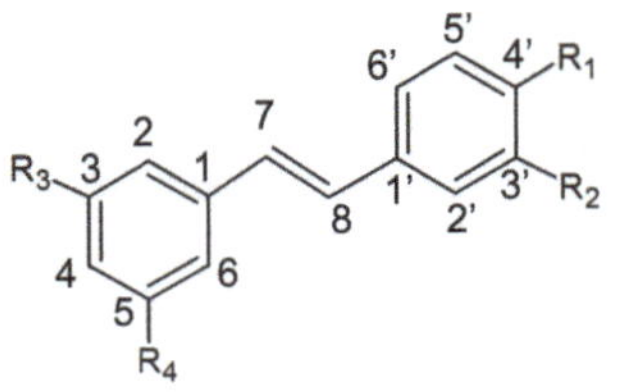

Fig. 3.1 Resveratrol and its stilbene analogs

3.2 Biosynthesis of Resveratrol and Its Analogs

The potential therapeutic value of resveratrol has stimulated the exploration of resveratrol-related natural products. Natural resveratrol derivatives include deoxy derivatives such as pinosylvin, hydroxylated derivatives such as piceatannol, methylated derivatives such as pinostilbene (3,4′-dihydroxy-5-methoxy-*trans*-stilbene), pterostilbene (3,5-dimethoxy-4′-hydroxy-*trans*-stilbene), 3,4′,5-trimethoxystilbene, and desoxyrhapotigenin (3,5-dihydroxy-4′-methoxy-*trans*-stilbene), and glucosylated derivatives such as piceid (resveratrol-3-*O*-glucoside) and resveratroloside (resveratrol-4′-*O*-glucoside) (Fig. 3.1) [12, 13] (**see also Chap. 9 of this book**). Interestingly, the naturally occurring hydroxylated, methylated and glucosylated resveratrol derivatives exhibit higher biological properties and oral bioavailability than the parent compound resveratrol [14–16]. Among them, *trans*-resveratrol trimethylether and pinostilbene appeared to be the most potent compounds, which were reported to be up to 100-fold more cytotoxic than resveratrol in cancer cell lines [12, 17]. Structure-activity relationship studies with resveratrol analogs revealed that the phenolic hydroxyl or methyl groups are the major structural determinants of the molecule's activity [18, 19]. Thus, the chemical scaffold of resveratrol has been utilized as the starting backbone to synthesize new resveratrol analogs for improving the chemopreventive activities of resveratrol [20, 21]. Because of its advantageous biological activities over resveratrol, pterostilbene has been approved by the FDA to be generally recognized as safe (GRAS) status as a food ingredient, which provides new commercial opportunities in natural food and beverage processing [22].

The biosynthesis of resveratrol compounds is initiated from phenylpropanoid acids including cinnamic acid, *p*-coumaric acid, caffeic acid and ferulic acid that are derived from aromatic amino acids (Fig. 3.2) [23]. L-phenylalanine and L-tyrosine

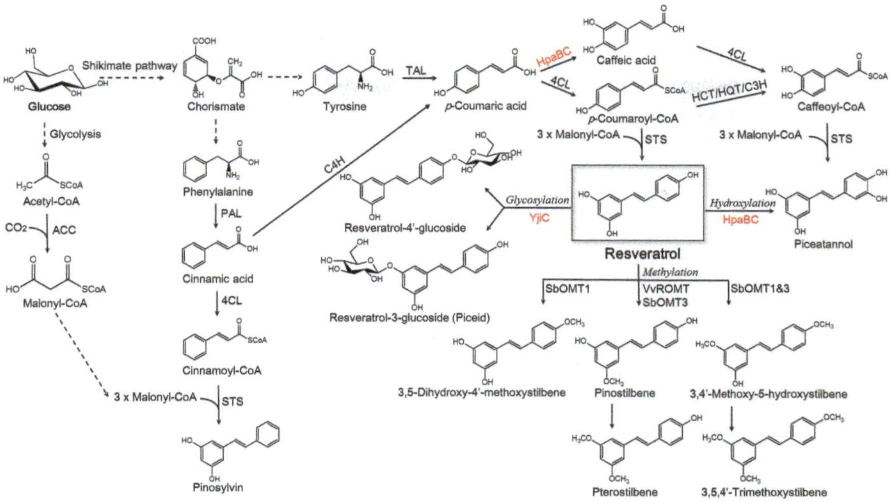

Fig. 3.2 Biosynthetic pathway and metabolism of resveratrol and its stilbene derivatives. The enzymes in black are plant-derived enzymes that involve in natural biosynthetic pathways, and enzymes in red denote bacterial enzymes that are utilized for respective reactions. *ACC* acetyl-CoA carboxylase, *PAL* phenylalanine ammonia-lyase, *TAL* tyrosine ammonia-lyase, *C4H* cinnamate-4-hydroxylase, *C3H* coumaroylquinate (coumaroylshikimate) 3′-monooxygenase, *HCT* hydroxycinnamoyl-CoA: shikimate hydroxycinnamoyl transferase, *HQT* hydroxycinnamoyl-CoA quinate hydroxycinnamoyl transferase, *4CL* 4-coumarate: coenzyme A ligase, *STS* stilbene synthase, *YjiC* UDP-glycosyltransferase from *Bacillus* species, *SbOMT1 and SbOMT3* *O*-methyltransferase 1 and 3 from *Sorghum bicolor*, *VvROMT* resveratrol *O*-methyltransferase from *V. vinifera*

are converted to the phenylpropanoid acids cinnamic acid and *p*-coumaric acid, respectively, via phenylalanine ammonia lyase (PAL) and tyrosine ammonia lyase (TAL), which are then converted to cinnamoyl-CoA and *p*-coumaroyl-CoA by 4-coumaroyl-CoA ligase (4CL). Then malonyl-CoA is condensed with cinnamoyl-CoA or *p*-coumaroyl-CoA to produce pinosylvin and resveratrol, respectively, via stilbene synthases (STSs) [24]. Resveratrol can serve as a scaffold to generate several derivatives through the action of tailoring enzymes like hydroxylases, *O*-methyltransferases and glucosyltransferases, whereby diverse resveratrol analogs are generated (Fig. 3.2) [25] (**see also Chap. 9 of this book**).

3.3 Bioproduction of Resveratrol in Microorganisms

As a high-value phytochemical found in many plants, resveratrol production has been engineered in various transgenic plants like *Polygonum cuspidatum*, *Arabidopsis thaliana*, *Nicotiana tabacum*, *Arachis hypogea*, *Vitis amurensis* and especially *V. vinifera* [26–29]. Industrially, resveratrol is mainly extracted from plants, which suffers from several limitations like low yields, impurities, and long turn-around time involving the growth period required by the plants. Moreover, resveratrol compounds are mostly produced in response to stress situations such as fungal infection or injury, or induced by elicitors [27, 30]. The chemical synthesis is also a widely reported method for resveratrol production with high yield, whereas the complexity of synthesis process and the generation of unwanted byproducts greatly limit the application of this technique.

With the advance of metabolic engineering, microorganisms like yeast and *Escherichia coli* can be equipped with resveratrol biosynthetic genes, which represents an alternative strategy and has achieved remarkable progress in resveratrol bioproduction [11, 31]. Specifically, besides efficient biotransformation of fed precursors to resveratrol, *de novo* production of resveratrol from renewable carbon sources has also been realized in microorganisms [32–35]. Thus, microbial synthesis has attracted much attention due to its high purity and productivity. The well-known manufacturer, the Swiss company (Evolva), possesses a high purity (98%) and high capacity of 40 to 50 tons per year for resveratrol production [11]. However, cost-effective production of resveratrol is still required for the increasing market demand. Thus, in this chapter, we will introduce the recent progress on the bioproduction of resveratrol in microbial hosts.

3.3.1 Yeast

Saccharomyces cerevisiae, the most commonly used yeast, is a well-studied eukaryotic model microorganism that is a Generally Regarded As Safe (GRAS) microbe for industrial applications like baking, brewing and winemaking. It is an ideal microbial host for production of valuable phytochemicals because of its ease of expressing plant-derived enzymes (like cytochrome P450 enzymes) and its tractability of genetic manipulations [36]. Many plant-derived chemicals like flavonoids, stilbenoids, benzylisoquinoline alkaloids and terpenoids have been produced via metabolic engineering in *S. cerevisiae* [36, 37]. Thus, yeasts like *S. cerevisiae* have been generally recognized as industrially robust hosts for plant-derived natural compounds.

Resveratrol pathway has been engineered in yeast by introducing two enzymes, 4-coumarate: coenzyme A ligase (4CL) from plant or bacterial origin and stilbene synthase (STS) from various plant species (Table 3.1). When 4CL2 from *N. tabacum* and STS from *V. vinifera* were integrated into the genome of *S. cerevisiae* and co-expressed, 6 mg/L resveratrol was produced via feeding of *p*-coumaric acid [38]. Expression of 4CL1 from *A. thaliana* and STS from *A. hypogaea* in *S. cerevisiae* W303-1A achieved resveratrol production with a titer of 3.1 mg/L and 14.4% (mol/mol) conversion yield from 15.3 mg/L *p*-coumaric acid [40]. This established the commercial viability for resveratrol production in a food-grade yeast, although the titer was still far from large-scale industrialization. Several studies have been performed to increase resveratrol production via either engineering a fusing enzyme 4CL::STS or creating synthetic scaffolds of 4CL and STS via small peptide ligands,

Table 3.1 Biosynthesis of resveratrol in microorganisms

Microbial hosts	Introduced genes	Fed precursors	Titers (mg/L)	References
Yeast				
S. cerevisiae CEN.PK	4CL2 (*Nicotiana tabacum*) STS (*Vitis vinifera*)	*p*-Coumaric acid	6	[38]
Industrial yeast	4CL1 (*Arabidopsis thaliana*) STS (*Vitis vinifera*)	*p*-Coumaric acid	391	[39]
S. cerevisiae W303-1A	4CL1 (*Arabidopsis thaliana*) STS (*Arachis hypogaea*)	*p*-Coumaric acid	3.1	[40]
S. cerevisiae WAT11	TAL (*Rhodobacter sphaeroides*) 4CL::STS, 4CL1 (*Arabidopsis thaliana*)-STS (*Vitis vinifera*) fusion enzyme	Tyrosine	1.9	[41]
S. cerevisiae WAT11	4CL1 (*Arabidopsis thaliana*) STS (*Vitis vinifera*)	*p*-Coumaric acid	14.4	[42]
S. cerevisiae CEN. PK102-5B	TAL (*Herpetosiphon aurantiacus*) 4CL1 (*Arabidopsis thaliana*) STS (*Vitis vinifera*)	Glucose	415.65	[32]
S. cerevisiae CEN. PK102-5B	TAL (*Herpetosiphon aurantiacus*) 4CL1 (*Arabidopsis thaliana*) STS (*Vitis vinifera*)	Ethanol	531.41	[32]
E. coli				
E. coli BL21 (DE3)	4CL2 (*Nicotiana tabacum*) STS (*Vitis vinifera*)	*p*-Coumaric acid	16	[38]
E. coli BW27784	4CL1 (*Arabidopsis thaliana*) STS (*Arachis hypogaea*)	*p*-Coumaric acid	105	[43]
E. coli BLR (DE3)	4CL (*Lithospermum erythrohizon*) STS (*Arachis hypogaea*) ACC (*Corynebacterium glutamicum*)	*p*-Coumaric acid	171	[44]
E. coli BL21 (DE3)	4CL::STS, 4CL(*Arabidopsis thaliana*)-STS (*Arachis hypogaea*) fusion enzyme	*p*-Coumaric acid	80.5	[45]
E. coli BW25113	4CL2 (*Petroselinum crispum*) STS (*Vitis vinifera*)	*p*-Coumaric acid	268.20	[46]
E. coli BW27784	4CL1 (*Arabidopsis thaliana*) STS (*Vitis vinifera*)	*p*-Coumaric acid	2340	[47]
E. coli BW27784 (DE3)	4CL1 (*Arabidopsis thaliana*) STS (*Vitis vinifera*)	*p*-Coumaric acid	1600	[48]

(continued)

Table 3.1 (continued)

Microbial hosts	Introduced genes	Fed precursors	Titers (mg/L)	References
E. coli BL21 (DE3)	PAL (*Rhodotorula rubra*) 4CL (*Lithospermum erythrorhizon*) STS (*Arachis hypogaea*)	Tyrosine	37	[49]
E. coli BL21 (DE3)	TAL (*Rhodotorula glutinis*) 4CL (*Petroselinum crispum*) STS (*Vitis vinifera*)	Tyrosine	35	[50]
E. coli BL21(DE3)	TAL (*Saccharothrix espanaensis*) 4CL (*Arabidopsis thaliana*) STS (*Arachis hypogaea*)	Tyrosine	114.4	[23]
E. coli C41(DE3)	TAL (*Saccharothrix espanaensis*) 4CL (*Streptomyces coelicolor*) STS (*Arachis hypogaea*)	Tyrosine	1.4	[51]
E. coli C41 (DE3)	TAL (*Saccharothrix espanaensis*) 4CL (*Streptomyces coelicolor*) STS (*Arachis hypogaea*)	Glucose	5.2	[33]
E. coli BW25113 (DE3)	TAL (*Rhodotorula glutinis*) 4CL (*Petroselinum crispum*) STS (*Vitis vinifera*)	Glucose	4.6	[35]
Other bacteria				
Corynebacterium glutamicum DelAro[3]	4CL (*Petroselinum crispum*) STS (*Arachis hypogaea*)	*p*-Coumaric acid	158	[34]
Corynebacterium glutamicum DelAro[4]	TAL (*Flavobacterium johnsoniae*) 4CL (*Petroselinum crispum*) STS (*Arachis hypogaea*)	Glucose	59	[34]

which led to resveratrol production enhancement with a titer of 5.3 mg/L and 14.4 mg/L, respectively [42, 52]. Specifically, when a codon-optimized tyrosine ammonia lyase (TAL) from *Rhodobacter sphaeroides* was introduced into an *S. cerevisiae* strain expressing the 4CL::STS fusion construct, 1.9 mg/L resveratrol could be produced from 12 mg/L tyrosine [41]. Based on that, introducing the *araE* gene encoding a arabinose transporter from *E. coli* could facilitate transporting resveratrol and further produce up to 3.1 mg/L resveratrol [41]. Fermentation of an industrial *S. cerevisiae* expressing 4CL1 from *A. thaliana* and STS from *V. vinifera*

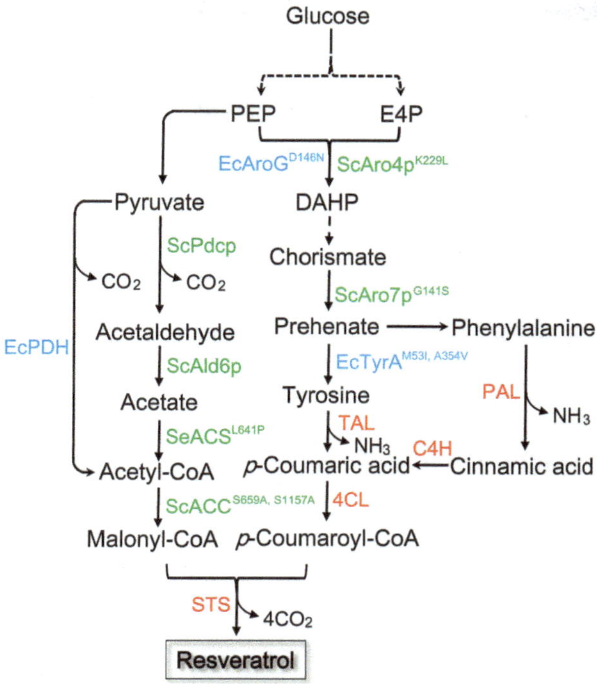

Fig. 3.3 Metabolic engineering of resveratrol production in yeast and *E. coli*. Critical endogenous enzymes from *E. coli (blue)*, *S. cerevisiae* or *Salmonella enterica (green)* and heterologous enzymes for resveratrol biosynthesis *(red)* are included. *EcAroG*^{D164N} feedback inhibition resistant 3-deoxy-D-arabino-heptulosonate-7-phosphate (DAHP) synthase, *EcTyrA*$^{M53I, A354V}$ feedback inhibition resistant chorismate mutase/prephenate dehydrogenase, *EcPDH* pyruvate dehydrogenase, *ScAro4p*^{K229L} feedback inhibition resistant DAHP synthase, *ScAro7p*^{G141S} feedback inhibition resistant chorismate mutase, *ScPdcP* pyruvate decarboxylase, *ScAld6p* aldehyde dehydrogenase, *SeACS*^{L641P} acetylation-resistant acetyl-CoA synthetase variant from *S. enterica*, *ScACC*$^{S659A, S1157A}$ inactivation-resistant acetyl-CoA carboxylase, *PAL* phenylalanine ammonia lyase, *TAL* tyrosine ammonia lyase, *C4H* cinnamate-4-hydroxylase, *4CL* 4-coumaroyl-CoA ligase, *STS* stilbene synthase

in rich medium yielded a significantly higher resveratrol titer of 391 mg/L with optimal supply of the precursor, *p*-coumaric acid (2.46 g/L) [39].

Recently, *de novo* production of resveratrol from glucose or ethanol without costly supplementation of precursors like *p*-coumaric acid or tyrosine has been achieved in *S. cerevisiae* (Fig. 3.3). First, introduction of the resveratrol biosynthetic pathway consisting of TAL from *Herpetosiphon aurantiacus*, 4CL1 from *A. thaliana* and STS from *V. vinifera* enabled production of 2.73 mg/L resveratrol from glucose. Subsequently, the formation of the pivotal precursor *p*-coumaric acid was increased by enhancing the endogenous shikimate pathway (over-expression of feedback-insensitive *ARO*4, *ARO*7 mutant ARO4^{K229L} and ARO7^{G141S}) and malonyl-CoA (by over-expression of a post-translational de-regulated acetyl-CoA carboxylase mutant (ACC$^{S659A, S1157A}$) (Fig. 3.3). Finally, multiple copies of the resveratrol pathway genes were integrated into the genome of the engineered *S. cerevisiae*. Fed-batch fermentation of the final strain resulted in a resveratrol titer of 415.65 and 531.41 mg/L with glucose or ethanol as carbon source, respectively [32].

3.3.2 *E. coli*

E.coli is one of the most widely used microbial organisms for metabolic engineering owing to favorable characteristics like fast growth, amenability of genetic manipulation, and compatibility of diverse synthetic biology tools. *E. coli* is sometimes advantageous over yeast in producing certain plant-derived natural products including most flavonoids, stilbenoids, benzylisoquinoline alkaloids and terpenoids [37, 53]. Metabolic engineering in *E. coli* for resveratrol production is feasible because the availability of both precursors, tyrosine and malonyl-CoA, can be significantly enhanced via genetic modification of the host cells. However, most work for resveratrol production in *E. coli* is still based on the biotransformation of fed precursors like *p*-coumaric acid and tyrosine (Table 3.1).

In contrast to yeast, which is sensitive to high concentration of *p*-coumaric acid, *E. coli* is much more tolerant to *p*-coumaric acid (>3 g/L) [40, 54]. When 1 mM *p*-coumaric acid was fed to *E. coli* cultures expressing 4CL1 from *A. thaliana* and STS from *A. hypogaea*, 105 mg/L of the stilbene resveratrol was produced with a conversion yield of 46% (mol/mol) [43]. Fusion of 4CL1 from *A. thaliana* and STS from *A. hypogaea* led to a final resveratrol titer (when 1 mM substrate *p*-coumaric acid was fed to *E. coli*) of 80.5 mg/L with a conversion yield of 35.3% (mol/mol) [45]. Generally, when *p*-coumaric acid was supplemented, increasing the availability of malonyl-CoA further enhanced the production of resveratrol. Over-expression of ACC from *Corynebacterium glutamicum* in *E. coli* carrying 4CL from *Lithospermum erythrohizon* and STS (grown with supplemental *p*-coumaric acid) from *A. hypogaea* increased the intracellular pool of malonyl-CoA and thus the production of resveratrol to a titer of 171 mg/L and a conversion yield of 74.9% (mol/mol) [44]. This implied that an improvement of resveratrol production can be achieved by redirecting more malonyl-CoA into the resveratrol biosynthetic pathway. Some other strategies have also been exerted to enhance resveratrol production to a great extent via increasing malonyl-CoA in *E. coli*. Expression of different combinations of 4CL and STS in two different *E. coli* strain backgrounds revealed that expression of 4CL from *A. thaliana* and STS from *V. vinifera* in *E. coli* BW27784 produced the highest titer of resveratrol (1.3 g/L). Based on the best combination, addition of cerulenin, a specific inhibitor of the fatty acid biosynthesis pathway, further increased the production of resveratrol to 2.3 g/L [47]. To redirect carbon flux into malonyl-CoA, phosphoglycerate kinase (PGK), glyceraldehyde-3-phosphate dehydrogenase (GapA) and genes coding for the components of the pyruvate dehydrogenase complex (PDH) were over-expressed in *E. coli* BW27784 (DE3) *ΔfumC* mutant. The final recombinant strain carrying 4CL from *A. thaliana* and STS from *V. vinifera* resulted in resveratrol production titers of 1.6 g/L in shake flask experiments without adding expensive inhibitors of fatty acid metabolism like cerulenin [48]. Most recently, anti-sense RNA (asRNA)-based repression of *fab* genes has increased the concentration of intracellular malonyl-CoA whereas knock-out of most *fab* genes are lethal to host cells [46]. When 4CL2 from *Petroselinum crispum* and STS from *V. vinifera* were co-expressed with anti-sense RNA targeting *fabD* (*asfabD*) in *E. coli* BW25113, 268.2 mg/L resveratrol was produced, which is a 1.70-fold increase compared with the control strain without *asfabD* [46].

Expression of three enzymes, TAL, 4CL and STS, enabled *E. coli* to produce resveratrol from tyrosine. Expression of PAL from *Rhodotorula rubra*, 4CL from *Lithospermum erythrorhizon* and STS from *A. hypogaea*, or TAL from *R. glutinis*, 4CL from *P. crispum* and STS from *V. vinifera* in *E. coli* BL21 (DE3) produced 37 and 35 mg/L resveratrol, respectively [49, 50]. Over-expression of TAL from *Saccharothrix espanaensis*, 4CL from *A. thaliana* and STS from *A. hypogaea* in *E. coli* achieved the highest production of resveratrol, with a titer of 114.4 mg/L from tyrosine [23]. As tyrosine is a native amino acid that can be overproduced in *E. coli*, *de novo* production of resveratrol from a simple carbon source is feasible. As a proof-of-concept demonstration, expression of the artificial pathway harboring TAL from *S. espanaensis*, 4CL from *Streptomyces coelicolor* and STS from *A. hypogaea* in *E. coli* produced 1.4 mg/L resveratrol [51]. When codon-optimized TAL and STS were used, the resveratrol titer was further increased to 5.2 mg/L [33]. Recently, a site-specific integration strategy was utilized to chromosomally insert resveratrol biosynthetic pathway containing genes TAL from *R. glutinis*, 4CL from *P. crispum* and STS from *V. vinifera* into *E. coli* BW25113 (DE3) at the loci of *tyrR* and *trpED*. The resulting final strain was capable of producing 4.6 mg/L resveratrol from glucose [35]. We can therefore propose that resveratrol production in *E. coli* can be further improved if aromatic pathway flux and the malonyl-CoA pool are both enhanced.

3.3.3 Other Bacteria

Besides the widely used microbial hosts yeast and *E. coli*, other industrially available bacteria like *Corynebacterium glutamicum* have also been harnessed as a chassis for resveratrol production (Table 3.1). In order to construct a suitable *C. glutamicum* platform strain for resveratrol production, four gene clusters (*phd*, *cat*, *ben* and *pca*) comprising 20 genes involved in the catabolism of aromatic compounds were deleted to obtain *C. glutamicum* Del^{Aro3}. Expression of the plant-derived and codon-optimized genes 4CL (from *Petroselinum crispum*) and STS (from *A. hypogaea*) with additional supplementation of the antibiotic cerulenin (to downregulate lipid biosynthesis from malonyl-CoA), led to resveratrol production with a titer of 158 mg/L from supplemented *p*-coumaric acid. Additional engineering via deletion of *qsuB* (encoding a putative dehydroshikimate dehydratase) and over-expression of *aroH* from *E. coli* and TAL from *Flavobacterium johnsoniae* enabled resveratrol production directly from glucose with a titer of 59 mg/L, which is up-to-date the highest production titer of resveratrol from a simple carbon source in microorganisms [34]. Caffeic acid, added in the same way, resulted in the production of the expected dihydroxylated stilbene, piceatannol. The quantity of the stilbene piceatannol produced from caffeic acid was also relatively high, around 13 mg/L [43].

3.4 Bioproduction of Resveratrol Analogs and Derivatives

Resveratrol is one of the few stilbenes that has been widely investigated and metabolically engineered in microbes. However, other stilbenes like pinosylvin, piceatannol, as well as methylated and glucosylated resveratrols have also been produced in microbial hosts.

The resveratrol analog pinosylvin is derived from cinnamic acid, while piceatannol is derived from caffeic acid or Dopa (Fig. 3.1). Microbial based production of pinosylvin or piceatannol has been developed and mainly focused on the biotransformation from phenylpropenoic acids or aromatic amino acids. *E. coli* expressing PAL from *R. rubra*, 4CL from *L. erythrorhizon* and STS from *A. hypogaea* produced 20 mg/L pinosylvin when phenylalanine was supplemented [49]. Co-expression of PAL, 4CL and STS in *E. coli* BL 21(DE3) achieved *de novo* biosynthesis of pinosylvin from glucose with a titer of 13.3 mg/L [23]. Total biosynthesis of pinosylvin was enabled, via *in vivo* evolution of the STS from *Pinus strobus* for improved activity and addition of the fatty acid production inhibitor cerulenin for increasing intracellular malonyl-CoA, to allow the production of 70 mg/L pinosylvin from glucose in *E. coli* [55]. Most recently, 47.49 mg/L pinosylvin was produced from glycerol via optimization of the expression of the pinosylvin pathway and clustered regularly interspaced short palindromic repeats interference (CRISPRi)-mediated repression of the *fabD* gene [56]. Piceatannol is a hydroxylated resveratrol, which can be produced by hydroxylation from resveratrol. STS shows catalytic promiscuity towards caffeic acid, though the activity to caffeic acid is relatively low compared with native *p*-coumaric acid [43]. To date, several monooxygenases including a bacterial P450 monooxygenase CYP102A1 from *Bacillus megaterium*, a tyrosinase MelC2 from melanin-forming *Streptomyces avermitilis* MA4680 and a two-component flavin-dependent monooxygenase HpaBC from *Pseudomonas aeruginosa* have been reported to efficiently hydroxylate resveratrol to piceatannol [57–59]. *E. coli* BL21(DE3) expressing a C3H from *S. espanaensis* NRRL 15764 produced 65.4 mg/L piceatannol from 100 mg/L resveratrol [23]. When combined with the resveratrol biosynthetic pathway via expression of four genes (TAL, 4CL, C3H and STS), 21.5 mg/L piceatannol was produced from glucose [23].

The methylated derivatives of resveratrol, including desoxyrhapotigenin (3,5-dihydroxy-4′-methoxystilbene), pinostilbene (3,4′-dihydroxy-5-methoxystilbene), pterostilbene (3,5-dimethoxy-4′-hydroxystilbene), 3,4′, -dimethoxy-5-hydroxystilbene and 3,4′,5-trimethoxystilbene, have been produced via expression of a resveratrol *O*-methyltransferase (ROMT) gene [22, 33, 60]. At least five resveratrol *O*-methyltransferases have been expressed and characterized, namely SbOMT1 and

SbOMT3 from *Sorghum bicolor* [61, 62], VvROMT from *V. vinifera* [63], VrROMT from *V. riparia* [60], and OsPMT from *Oryza sativa* [49, 60–63]. Expression of the codon-optimized SbROMT3 in *E. coli* led to the production of pinostilbene (mono-methylated resveratrol derivative) (34 mg/L) from 1 mM resveratrol, with a very small amount of pterostilbene (di-methylated resveratrol derivative) (0.16 mg/L) [60]. However, co-expression of 4CL::STS and VvROMT in *E. coli* led to the production of pterostilbene (50 mg/L) as the major product and pinostilbene as the minor product when fed with *p*-coumaric acid [22]. OsPMT, a pinosylvin methyltransferase from *O. sativa*, turned out to have a rather relaxed substrate specificity towards pinosylvin and resveratrol [49]. Expression of PAL from *R. rubra*, 4CL from *L. erythrorhizon* and OsPMT in *E. coli* led to the production of pinostilbene (18 mg/L) and pterostilbene (5.8 mg/L) when fed with tyrosine, and pinosylvin monomethyl ether (27 mg/L) and pinosylvin dimethyl ether (27 mg/L) when fed with phenylalanine [49]. Recently, incorporation of two resveratrol *O*-methyltransferase genes SbOMT1 and SbOMT3 in *E. coli* established the artificial biosynthetic pathway for total biosynthesis of methylated resveratrol analogues, including pinostilbene, pterostilbene, 3,5-dihydroxy-4′-methoxystilbene, 3,4′-dimethoxy-5-hydroxystilbene, and 3,5,4′-trimethoxystilbene from glucose [33].

Resveratrol glucoside derivatives, like piceid (resveratrol-3-*O*-glucoside) and resveratroloside (resveratrol-4′-*O*-glucoside), are endowed with beneficial advantages like improved bioavailability and solubility [13]. *E. coli* expressing a glucosyltransferase (GT) from *Phytolacca americana* (*Pa*GT3) produced both of the resveratrol glucoside derivatives with a ratio of resveratroloside: piceid of 10:3 (mol/mol) [64]. Recently, the expression of a codon-optimized TAL from *S. espanaensis*, CCL from *S. coelicolor*, codon-optimized STS from *A. hypogaea* and an additional UDP-glycosyltransferase YjiC from *Bacillus* species in *E. coli* enabled the *de novo* production of piceid and resveratroloside from a simple sugar medium [65] (**see also Chap. 9 of this book**).

3.5 Strategies in Metabolic Engineering of Resveratrol Production

Microbial-based metabolic engineering for resveratrol production has achieved great progress in recent years, thus validating the concept of microbial factories for plant-derived valuable phytochemicals from renewable carbon sources. However, there is still a need for improvements to make microbial-based resveratrol production industrially and commercially feasible. Precursor availability and low stilbene synthase activity in the heterologous hosts are the main bottlenecks during total biosynthesis of resveratrol. Generally, increasing the precursor supply (namely aromatic amino acids and malonyl-CoA) via genetic manipulation of host strains and improving the activity of key enzymes via protein engineering are the main strategies for elevating the productivity of microbial factories.

3.5.1 Pathway Engineering to Increase Precursor Supply

Significant efforts have been devoted to optimizing the production of aromatic amino acids or their derived phenylpropenoic acids in microbes especially yeast and *E. coli* [66–72]. Aromatic amino acids are produced from shikimate pathway in bacteria, and especially *E. coli* can be easily engineered to overproduce aromatic amino acids. Metabolic engineering of the shikimate pathway in *E. coli* has firstly focused on enhancing carbon flux toward chorismate, a branch point to Phe, Tyr and Trp (Fig. 3.3). Increasing E4P supply and PEP availability are the two main approaches to enhance chorismate production. Over-expression of transketolases, especially (TktA, encoded by *tktA*), is an efficient approach for increasing the supply of E4P, while over-expression of native PEP synthetase *ppsA* redirects pyruvate to PEP for aromatic amino acid biosynthesis [69, 73]. Repression or disruption of the global regulator gene *csrA* results in increased levels of PEP and thus increased production of aromatic amino acids [73, 74]. Additionally, over-expression of feedback-inhibition-resistant aromatic pathway enzymes like $AroG^{D164N}$ (DAHP synthase) and $TyrA^{M53I, A354V}$ (chorismate mutase/prephenate dehydrogenase), and/or deletion of aromatic amino acid specific repressor gene *tyrR* or *trpR* further enhance the production of aromatic compounds [69]. Heterologous expression of PAL or TAL leads to enhanced production of phenylpropenoic acids like cinnamic acid or *p*-coumaric acid and incorporation of an appropriate hydroxylase results in the production of caffeic acid [54, 68]. Similar approaches have been successful in yeast. A *p*-coumaric acid overproducing yeast has been established via overexpressing TAL from *F. johnsoniaeu*, DAHP synthase $ARO4^{K229L}$, chorismate mutase $ARO7^{G141S}$, and *E. coli* shikimate kinase II (*aroL*), while at the same time deleting phenylpyruvate decarboxylase ARO10 and pyruvate decarboxylase PDC5. This final strain produced the highest titer of *p*-coumaric acid of 1.93 g/L from glucose, which could serve as a platform yeast host for producing *p*-coumaric acid-derived natural products, including resveratrol.

The other substrate for resveratrol biosynthesis, malonyl-CoA, is also involved in fatty acid biosynthesis in bacteria. The cellular concentration of malonyl-CoA is maintained at a very low level in *E. coli* [75]. Increasing the flux to malonyl-CoA via carboxylation of acetyl-CoA and inhibiting the consumption of malonyl-CoA via repression of fatty acid biosynthesis are the main two strategies to enhance malonyl-CoA availability (Fig. 3.4). Over-expression of acetyl-CoA carboxylase (ACC) alone resulted in a threefold increase in cellular malonyl-CoA concentration [76]. Over-expression of ACC, along with deletion of competing pathways leading to the byproducts acetate (*pta* and *ackA*) and ethanol (*adhE*) as well as over-expression of an acetate assimilation enzyme (*acs*), led to a 15-fold elevated cellular malonyl-CoA level in *E. coli* [76]. As for blocking the malonyl-CoA consumption pathway, direct knockouts of *fab* genes are lethal to cells [46]. Thus, inhibition of fatty acid biosynthesis is preferred and has been achieved by three major approaches. First, the addition of cerulenin, a covalent inhibitor of FabB and FabF, can greatly facilitate the accumulation of malonyl-CoA and thus enhance the production of resveratrol [34, 47, 77, 78]. Second, repression of the *fab* operon, especially the

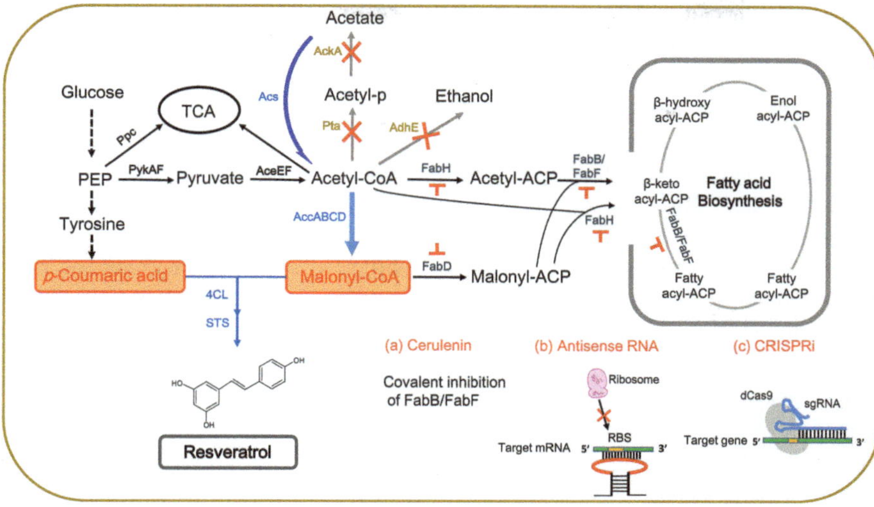

Fig. 3.4 Strategies for increasing the malonyl-CoA pool. (**a**) Cerulenin-mediated inhibition of the *fab* operon. (**b**) Antisense RNA-mediated inhibition of the *fab* operon. (**c**) Systematic inhibition of competing pathways via CRISPRi-mediated repression

fabD genes via antisense RNA, can enhance the accumulation of malonyl-CoA and its derived natural compounds like naringenin and resveratrol [46, 79]. Recently, the CRISPRi tool has been implemented in *E. coli* to systematically repress multiple genes and direct carbon flux to malonyl-CoA, thereby achieving a 7.4-fold increase in naringenin production with a final titer of 421.6 mg/L [80]. The CRISPRi-based downregulation of the *fabD* gene along with an introduced pinosylvin biosynthetic pathway in *E. coli* led to a 1.9-fold increase of pinosylvin production with a final titer of 47.5 mg/L from 0.5 mM cinnamic acid [56].

3.5.2 Protein Engineering

In addition to pathway engineering, protein engineering of STS, which catalyzes the first committed step in stilbene biosynthesis, provides new possibilities to increase the production titer for the stilbene backbone. Protein engineering and mutagenesis of 4CL and STS have indeed been applied to improve resveratrol production capabilities in microbes. It had been hypothesized that colocalization of the two enzyme active sites should improve the catalytic efficiency of the enzymatic reactions. The unnatural fusion of 4CL from *A. thaliana* and STS from *V. vinifera* (4CL::STS) is an example to prove the validity of the hypothesis. Introduction of the engineered 4CL::STS in *S. cerevisiae* resulted in a 15-fold increase of resveratrol levels (with a final titer of 5.25 mg/L) compared to yeast expressing the individual enzymes [52, 81]. Yeast carrying the codon-optimized TAL from *R. sphaeroides* in combination with fused 4CL::STS greatly increased resveratrol biosynthesis, reaching a titer of

1.06 mg/L without adding L-tyrosine and 1.90 mg/L with tyrosine added [41]. A similar fusion enzyme 4CL::STS (4CL from *A. thaliana* and STS from *A. hypogaea*) was also created and enabled *E. coli* to produce 80.5 mg/L resveratrol from 1 mM *p*-coumaric acid [45]. An alternative and compromised "fusion" strategy has been established using synthetic scaffolds to spatially recruit pathway enzymes in a designable manner [82]. 4CL1 with SH3 domain (Src homology 3 domain from the adaptor protein CRK) and STS with PDZ domain (PSD95/DlgA/Zo-1 domain from the adaptor protein syntrophin) were recruited and optimized in yeast cells, which led to a fivefold improvement of resveratrol production (6.7 mg/L) over the non-scaffolded control, and a 2.7-fold increase over the fusion enzyme strategy [42].

Enzyme evolution via random mutagenesis and phenotype screening is another, but sometimes labor-intensive, option to generate enzyme variants with improved performance. To adapt the STS of *P. strobus* for optimal expression in *E. coli*, error-prone PCR was performed to randomly mutate the codon-optimized STS and screen STS library with increased fluorescence of pinosylvin. This led to the discovery of two amino acid substitutions (T248A and Q361R) that can increase the pinosylvin production titer to 70 mg/L from glucose in the presence of cerulenin [55]. A more rational approach for protein engineering is via structure-based modeling and site-directed mutagenesis of enzymes. However, due to the paucity of protein structural information, this has met with limited success. Intriguingly, the mutation of STS from *V. vinifera*, which shows high structural similarity with STS from *A. hypogaea* (for which a crystal structure has been resolved), expanded the biosynthetic scope of STS for polyketides production [83]. Feeding non-natural substrates to wild type VvSTS or derived variants with an altered substrate binding and/or cyclization pocket, produced 7 and 9 non-natural polyketides, respectively. This implies the possibility of producing various kinds of natural and non-natural polyketides with pharmaceutical potential by mutagenesis of stilbene synthases.

3.6 Conclusion and Perspective

Plant derived phytochemicals are a valuable arsenal for pharmaceutical and nutrition additives. Resveratrol and its derivatives have been extensively investigated because of purported health-promoting effects. To achieve a cost-effective production of resveratrol, metabolic engineering, especially the microbial-based bioproduction, is an appealing approach and has achieved great progress in recent years. The amenability of genetic manipulation and tolerance of heterologous pathways make microbes capable of producing resveratrol by way of harnessing and redirecting their native metabolic networks. Systematic metabolic engineering, including genetic manipulation, pathway optimization and protein engineering, enabled microbial hosts to produce resveratrol and various derivatives. Via further optimization of the fermentation processes, metabolic engineering makes it possible to maximize the production of resveratrol in microbes to an industrial scale. Especially, with the recent development of novel synthetic biology strategies and generation of new stilbene synthase variants, an improved production of natural resveratrol compounds or more bioactive non-natural resveratrol analogs can be achieved via microbial platforms. The remarkable progress of resveratrol bioproduction in the last decades highlights the opportunity of microbial-based production of phytochemicals to fulfill the great demands for value-added nutraceuticals to fortify human against diseases.

References

1. Sun P, Liang JL, Kang LZ, Huang XY, Huang JJ, Ye ZW, et al. Increased resveratrol production in wines using engineered wine strains *Saccharomyces cerevisiae* EC1118 and relaxed antibiotic or auxotrophic selection. Biotechnol Prog. 2015;31(3):650–5.
2. Ferreira S, Domingues F. The antimicrobial action of resveratrol against *Listeria monocytogenes* in food-based models and its antibiofilm properties. J Sci Food Agric. 2016;96(13):4531–5.
3. Kiselev KV, Shumakova OA, Maniakhin A. Effect of plant stilbene precursors on the biosynthesis of resveratrol in *Vitis amurensis* Rupr. Cell cultures. Prikl Biokhim Mikrobiol. 2013;49(1):61–6.
4. Jeandet P, Hebrard C, Deville MA, Cordelier S, Dorey S, Aziz A, et al. Deciphering the role of phytoalexins in plant-microorganism interactions and human health. Molecules. 2014;19(11):18033–56.
5. Shin SY, Jung SM, Kim MD, Han NS, Seo JH. Production of resveratrol from tyrosine in metabolically engineered *Saccharomyces cerevisiae*. Enzyme Microb Technol. 2012;51(4):211–6.
6. Latruffe N, Rifler JP. Bioactive polyphenols from grapes and wine emphasized with resveratrol. Curr Pharm Des. 2013;19(34):6053–63.
7. Xia N, Daiber A, Forstermann U, Li H. Antioxidant effects of resveratrol in the cardiovascular system. Br J Pharmacol. 2016;174(12):1633–46.
8. Tellone E, Galtieri A, Russo A, Giardina B, Ficarra S. Resveratrol: a focus on several neurodegenerative diseases. Oxidative Med Cell Longev. 2015;doi:10.1155/2015/392169.
9. Rege SD, Geetha T, Griffin GD, Broderick TL, Babu JR. Neuroprotective effects of resveratrol in Alzheimer disease pathology. Front Aging Neurosci. 2014;6:218.
10. Bhullar KS, Hubbard BP. Lifespan and healthspan extension by resveratrol. BBA-Mol Basis Dis. 2015;1852(6):1209–18.
11. Mei Y-Z, Liu R-X, Wang D-P, Wang X, Dai C-C. Biocatalysis and biotransformation of resveratrol in microorganisms. Biotechnol Lett. 2015;37(1):9–18.
12. Wang TT, Schoene NW, Kim YS, Mizuno CS, Rimando AM. Differential effects of resveratrol and its naturally occurring methylether analogs on cell cycle and apoptosis in human androgen-responsive LNCaP cancer cells. Mol Nutr Food Res. 2010;54(3):335–44.
13. Su D, Cheng Y, Liu M, Liu D, Cui H, Zhang B, et al. Comparision of piceid and resveratrol in antioxidation and antiproliferation activities *in vitro*. PLoS One. 2013;8(1):e54505.
14. Martinez-Marquez A, Morante-Carriel JA, Ramirez-Estrada K, Cusido RM, Palazon J, Bru-Martinez R. Production of highly bioactive resveratrol analogues pterostilbene and piceatannol in metabolically engineered grapevine cell cultures. Plant Biotechnol J. 2016;14(9):1813–25.
15. Muzzio M, Huang Z, SC H, Johnson WD, McCormick DL, Kapetanovic IM. Determination of resveratrol and its sulfate and glucuronide metabolites in plasma by LC-MS/MS and their pharmacokinetics in dogs. J Pharm Biomed Anal. 2012;59:201–8.
16. Gambini J, Ingles M, Olaso G, Lopez-Grueso R, Bonet-Costa V, Gimeno-Mallench L, et al. Properties of resveratrol: *in vitro* and *in vivo* studies about metabolism, bioavailability, and biological effects in animal models and humans. Oxidative Med Cell Longev. 2015;doi:10.1155/2015/837042.
17. Schneider Y, Chabert P, Stutzmann J, Coelho D, Fougerousse A, Gossé F, et al. Resveratrol analog (Z)-3, 5, 4′-trimethoxystilbene is a potent anti-mitotic drug inhibiting tubulin polymerization. Int J Cancer. 2003;107(2):189–96.

18. Antus C, Radnai B, Dombovari P, Fonai F, Avar P, Matyus P, et al. Anti-inflammatory effects of a triple-bond resveratrol analog: structure and function relationship. Eur J Pharmacol. 2015;748:61–7.
19. Fulda S. Resveratrol and derivatives for the prevention and treatment of cancer. Drug Discov Today. 2010;15(17–18):757–65.
20. Madadi NR, Zong HL, Ketkar A, Zheng C, Penthala NR, Janganati V, et al. Synthesis and evaluation of a series of resveratrol analogues as potent anti-cancer agents that target tubulin. Medchemcomm. 2015;6(5):788–94.
21. Savio M, Ferraro D, Maccario C, Vaccarone R, Jensen LD, Corana F, et al. Resveratrol analogue 4,4 '-dihydroxy-trans-stilbene potently inhibits cancer invasion and metastasis. Sci Rep-Uk. 2016:6.
22. Wang Y, Bhuiya MW, Zhou R, Yu O. Pterostilbene production by microorganisms expressing resveratrol O-methyltransferase. Ann Microbiol. 2015;65(2):817–26.
23. Wang S, Zhang S, Xiao A, Rasmussen M, Skidmore C, Zhan J. Metabolic engineering of *Escherichia coli* for the biosynthesis of various phenylpropanoid derivatives. Metab Eng. 2015;29:153–9.
24. Lin Y, Jain R, Yan Y. Microbial production of antioxidant food ingredients via metabolic engineering. Curr Opin Biotechnol. 2014;26:71–8.
25. Mora-Pale M, Sanchez-Rodriguez SP, Linhardt RJ, Dordick JS, Koffas MA. Metabolic engineering and *in vitro* biosynthesis of phytochemicals and non-natural analogues. Plant Sci. 2013;210:10–24.
26. Delaunois B, Cordelier S, Conreux A, Clément C, Jeandet P. Molecular engineering of resveratrol in plants. Plant Biotechnol J. 2009;7(1):2–12.
27. Donnez D, Jeandet P, Clément C, Courot E. Bioproduction of resveratrol and stilbene derivatives by plant cells and microorganisms. Trends Biotechnol. 2009;27(12):706–13.
28. Almagro L, Belchí-Navarro S, Sabater-Jara AB, Vera-Urbina JC, Sellés-Marchart S, Bru R, et al. Bioproduction of trans-resveratrol from grapevine cell cultures. In: Natural Products. Berlin Heidelberg: Springer; 2013. p. 1683–713.
29. Giovinazzo G, Ingrosso I, Paradiso A, De Gara L, Santino A. Resveratrol biosynthesis: plant metabolic engineering for nutritional improvement of food. Plant Foods Hum Nutr. 2012;67(3):191–9.
30. Jeandet P, Clément C, Courot E. Resveratrol production at large scale using plant cell suspensions. Eng Life Sci. 2014;14(6):622–32.
31. Halls C, Yu O. Potential for metabolic engineering of resveratrol biosynthesis. Trends Biotechnol. 2008;26(2):77–81.
32. Li M, Kildegaard KR, Chen Y, Rodriguez A, Borodina I, Nielsen J. *De novo* production of resveratrol from glucose or ethanol by engineered *Saccharomyces cerevisiae*. Metab Eng. 2015;32:1–11.
33. Kang S-Y, Lee JK, Choi O, Kim CY, Jang J-H, Hwang BY, et al. Biosynthesis of methylated resveratrol analogs through the construction of an artificial biosynthetic pathway in *E. coli*. BMC Biotechnol. 2014;14(1):1.
34. Kallscheuer N, Vogt M, Stenzel A, Gätgens J, Bott M, Marienhagen J. Construction of a *Corynebacterium glutamicum* platform strain for the production of stilbenes and (2S)-flavanones. Metab Eng. 2016;38:47–55.
35. Liu X, Lin J, Hu H, Zhou B, Zhu B. *De novo* biosynthesis of resveratrol by site-specific integration of heterologous genes in *Escherichia coli*. FEMS Microbiol Lett. 2016;363(8):fnw061.
36. Krivoruchko A, Nielsen J. Production of natural products through metabolic engineering of *Saccharomyces cerevisiae*. Curr Opin Biotechnol. 2015;35:7–15.
37. Wang J, Guleria S, Koffas MA, Yan Y. Microbial production of value-added nutraceuticals. Curr Opin Biotechnol. 2016;37:97–104.
38. Beekwilder J, Wolswinkel R, Jonker H, Hall R, de Vos CR, Bovy A. Production of resveratrol in recombinant microorganisms. Appl Environ Microbiol. 2006;72(8):5670–2.
39. Sydor T, Schaffer S, Boles E. Considerable increase in resveratrol production by recombinant industrial yeast strains with use of rich medium. Appl Environ Microbiol. 2010;76(10):3361–3.

40. Shin S-Y, Han NS, Park Y-C, Kim M-D, Seo J-H. Production of resveratrol from p-coumaric acid in recombinant *Saccharomyces cerevisiae* expressing 4-coumarate: coenzyme a ligase and stilbene synthase genes. Enzym Microb Technol. 2011;48(1):48–53.
41. Wang Y, Halls C, Zhang J, Matsuno M, Zhang Y, Yu O. Stepwise increase of resveratrol biosynthesis in yeast *Saccharomyces cerevisiae* by metabolic engineering. Metab Eng. 2011;13(5):455–63.
42. Wang Y, Yu O. Synthetic scaffolds increased resveratrol biosynthesis in engineered yeast cells. J Biotechnol. 2012;157(1):258–60.
43. Watts KT, Lee PC, Schmidt-Dannert C. Biosynthesis of plant-specific stilbene polyketides in metabolically engineered *Escherichia coli*. BMC Biotechnol. 2006;6(1):22.
44. Katsuyama Y, Funa N, Miyahisa I, Horinouchi S. Synthesis of unnatural flavonoids and stilbenes by exploiting the plant biosynthetic pathway in *Escherichia coli*. Chem Biol. 2007;14(6):613–21.
45. Zhang E, Guo X, Meng Z, Wang J, Sun J, Yao X, et al. Construction, expression, and characterization of *Arabidopsis thaliana* 4CL and *Arachis hypogaea* RS fusion gene 4CL:: RS in *Escherichia coli*. World J Microbiol Biotechnol. 2015;31(9):1379–85.
46. Yang Y, Lin Y, Li L, Linhardt RJ, Yan Y. Regulating malonyl-CoA metabolism via synthetic antisense RNAs for enhanced biosynthesis of natural products. Metab Eng. 2015;29:217–26.
47. Lim CG, Fowler ZL, Hueller T, Schaffer S, Koffas MA. High-yield resveratrol production in engineered *Escherichia coli*. Appl Environ Microbiol. 2011;77(10):3451–60.
48. Bhan N, Xu P, Khalidi O, Koffas MA. Redirecting carbon flux into malonyl-CoA to improve resveratrol titers: proof of concept for genetic interventions predicted by OptForce computational framework. Chem Eng Sci. 2013;103:109–14.
49. Katsuyama Y, Funa N, Horinouchi S. Precursor-directed biosynthesis of stilbene methyl ethers in *Escherichia coli*. Biotechnol J. 2007;2(10):1286–93.
50. Wu J, Liu P, Fan Y, Bao H, Du G, Zhou J, et al. Multivariate modular metabolic engineering of *Escherichia coli* to produce resveratrol from L-tyrosine. J Biotechnol. 2013;167(4):404–11.
51. Choi O, C-Z W, Kang SY, Ahn JS, Uhm T-B, Hong Y-S. Biosynthesis of plant-specific phenylpropanoids by construction of an artificial biosynthetic pathway in *Escherichia coli*. J Ind Microbiol Biotechnol. 2011;38(10):1657–65.
52. Zhang Y, Li S-Z, Li J, Pan X, Cahoon RE, Jaworski JG, et al. Using unnatural protein fusions to engineer resveratrol biosynthesis in yeast and mammalian cells. J Am Chem Soc. 2006;128(40):13030–1.
53. Marienhagen J, Bott M. Metabolic engineering of microorganisms for the synthesis of plant natural products. J Biotechnol. 2013;163(2):166–78.
54. Huang Q, Lin Y, Yan Y. Caffeic acid production enhancement by engineering a phenylalanine over-producing *Escherichia coli* strain. Biotechnol Bioeng. 2013;110(12):3188–96.
55. van Summeren-Wesenhagen PV, Marienhagen J. Metabolic engineering of *Escherichia coli* for the synthesis of the plant polyphenol pinosylvin. Appl Environ Microbiol. 2015;81(3):840–9.
56. Liang JL, Guo LQ, Lin J-F, He Z-Q, Cai F-J, Chen J-F. A novel process for obtaining pinosylvin using combinatorial bioengineering in *Escherichia coli*. World J Microbiol Biotechnol. 2016;32(6):1–10.
57. Kim D-H, Ahn T, Jung H-C, Pan J-G, Yun C-H. Generation of the human metabolite piceatannol from the anticancer-preventive agent resveratrol by bacterial cytochrome P450 BM3. Drug Metab Dispos. 2009;37(5):932–6.
58. Lee N, Kim EJ, Kim B-G. Regioselective hydroxylation of trans-resveratrol via inhibition of tyrosinase from *Streptomyces avermitilis* MA4680. ACS Chem Biol. 2012;7(10):1687–92.
59. Furuya T, Kino K. Regioselective synthesis of piceatannol from resveratrol: catalysis by two-component flavin-dependent monooxygenase HpaBC in whole cells. Tetrahedron Lett. 2014;55(17):2853–5.
60. Jeong YJ, An CH, Woo SG, Jeong HJ, Kim Y-M, Park S-J, et al. Production of pinostilbene compounds by the expression of resveratrol O-methyltransferase genes in *Escherichia coli*. Enzym Microb Technol. 2014;54:8–14.

61. Baerson SR, Dayan FE, Rimando AM, Nanayakkara ND, Liu C-J, Schröder J, et al. A functional genomics investigation of allelochemical biosynthesis in Sorghum Bicolor root hairs. J Biol Chem. 2008;283(6):3231–47.
62. Rimando AM, Pan Z, Polashock JJ, Dayan FE, Mizuno CS, Snook ME, et al. In planta production of the highly potent resveratrol analogue pterostilbene via stilbene synthase and O-methyltransferase co-expression. Plant Biotechnol J. 2012;10(3):269–83.
63. Schmidlin L, Poutaraud A, Claudel P, Mestre P, Prado E, Santos-Rosa M, et al. A stress-inducible resveratrol O-methyltransferase involved in the biosynthesis of pterostilbene in grapevine. Plant Physiol. 2008;148(3):1630–9.
64. Ozaki S-i, Imai H, Iwakiri T, Sato T, Shimoda K, Nakayama T, et al. Regioselective glucosidation of trans-resveratrol in *Escherichia coli* expressing glucosyltransferase from *Phytolacca americana*. Biotechnol Lett. 2012;34(3):475–81.
65. Choi O, Lee JK, Kang S-Y, Pandey RP, Sohng J-K, Ahn JS, et al. Construction of artificial biosynthetic pathways for resveratrol glucoside derivatives. J Microbiol Biotechnol. 2014;24(5):614–8.
66. Rodriguez A, Kildegaard KR, Li M, Borodina I, Nielsen J. Establishment of a yeast platform strain for production of p-coumaric acid through metabolic engineering of aromatic amino acid biosynthesis. Metab Eng. 2015;31:181–8.
67. Juminaga D, Baidoo EE, Redding-Johanson AM, Batth TS, Burd H, Mukhopadhyay A, et al. Modular engineering of L-tyrosine production in *Escherichia coli*. Appl Environ Microbiol. 2012;78(1):89–98.
68. Zhang H, Stephanopoulos G. Engineering *E. coli* for caffeic acid biosynthesis from renewable sugars. Appl Microbiol Biotechnol. 2013;97(8):3333–41.
69. Lütke-Eversloh T, Stephanopoulos G. L-tyrosine production by deregulated strains of *Escherichia coli*. Appl Microbiol Biotechnol. 2007;75(1):103–10.
70. Chávez-Béjar MI, Lara AR, López H, Hernández-Chávez G, Martinez A, Ramírez OT, et al. Metabolic engineering of *Escherichia coli* for L-tyrosine production by expression of genes coding for the chorismate mutase domain of the native chorismate mutase-prephenate dehydratase and a cyclohexadienyl dehydrogenase from *Zymomonas mobilis*. Appl Environ Microbiol. 2008;74(10):3284–90.
71. Lütke-Eversloh T, Stephanopoulos G. Combinatorial pathway analysis for improved L-tyrosine production in *Escherichia coli*: identification of enzymatic bottlenecks by systematic gene overexpression. Metab Eng. 2008;10(2):69–77.
72. Jiang H, Wood KV, Morgan JA. Metabolic engineering of the phenylpropanoid pathway in *Saccharomyces cerevisiae*. Appl Environ Microbiol. 2005;71(6):2962–9.
73. Bulter T, Bernstein JR, Liao JC. A perspective of metabolic engineering strategies: moving up the systems hierarchy. Biotechnol Bioeng. 2003;84(7):815–21.
74. Na D, Yoo SM, Chung H, Park H, Park JH, Lee SY. Metabolic engineering of *Escherichia coli* using synthetic small regulatory RNAs. Nat Biotechnol. 2013;31(2):170–4.
75. Takamura Y, Nomura G. Changes in the intracellular concentration of acetyl-CoA and malonyl-CoA in relation to the carbon and energy metabolism of *Escherichia coli* K12. Microbiology. 1988;134(8):2249–53.
76. Zha W, Rubin-Pitel SB, Shao Z, Zhao H. Improving cellular malonyl-CoA level in *Escherichia coli* via metabolic engineering. Metab Eng. 2009;11(3):192–8.
77. Finzel K, Lee DJ, Burkart MD. Using modern tools to probe the structure–function relationship of fatty acid synthases. Chembiochem. 2015;16(4):528–47.
78. Lu Y, Shao D, Shi J, Huang Q, Yang H, Jin M. Strategies for enhancing resveratrol production and the expression of pathway enzymes. Appl Microbiol Biotechnol. 2016;100(17):7407–21.
79. Wu J, Yu O, Du G, Zhou J, Chen J. Fine-tuning of the fatty acid pathway by synthetic antisense RNA for enhanced (2S)-naringenin production from L-tyrosine in *Escherichia coli*. Appl Environ Microbiol. 2014;80(23):7283–92.
80. Wu J, Du G, Chen J, Zhou J. Enhancing flavonoid production by systematically tuning the central metabolic pathways based on a CRISPR interference system in *Escherichia coli*. Sci Rep. 2015;5

81. Wang Y, Yi H, Wang M, Yu O, Jez JM. Structural and kinetic analysis of the unnatural fusion protein 4-coumaroyl-CoA ligase:: stilbene synthase. J Am Chem Soc. 2011;133(51):20684–7.
82. Dueber JE, GC W, Malmirchegini GR, Moon TS, Petzold CJ, Ullal AV, et al. Synthetic protein scaffolds provide modular control over metabolic flux. Nat Biotechnol. 2009;27(8):753–9.
83. Bhan N, Cress BF, Linhardt RJ, Koffas M. Expanding the chemical space of polyketides through structure-guided mutagenesis of *Vitis vinifera* stilbene synthase. Biochimie. 2015;115:136–43.

Anthocyanin Production in Engineered Microorganisms

4

Jian Zha and Mattheos A.G. Koffas

Abbreviations

4CL	4-coumaroyl-CoA ligase
ANS	Anthocyanidin synthase
AOMT	Anthocyanin *O*-methyltransferase
CHI	Chalcone isomerase
CHS	Chalcone synthase
DFR	Dihydroflavonol 4-reductase
DSSC	Dye-sensitized solar cell
F3'5'H	Flavonoid 3′, 5′-hydroxylase
F3'H	Flavonoid 3′-hydroxylase
F3GT	Flavonoid 3-glucosyltransferase
F3H	Flavanone 3-hydroxylase
FGT	Flavonoid glucosyltransferase
SAM	*S*-adenosyl-L-methionine
UV	Ultraviolet

J. Zha
Department of Chemical and Biological Engineering, Rensselaer Polytechnic Institute, Center for Biotechnology and Interdisciplinary Studies, Troy, 12180 NY, USA

M.A.G. Koffas (✉)
Department of Chemical and Biological Engineering, Rensselaer Polytechnic Institute, Center for Biotechnology and Interdisciplinary Studies, Troy, 12180 NY, USA

Department of Biological Sciences, Rensselaer Polytechnic Institute, Center for Biotechnology and Interdisciplinary Studies, Troy, 12180 NY, USA
e-mail: koffam@rpi.edu

W. Schwab et al. (eds.), *Biotechnology of Natural Products*,
https://doi.org/10.1007/978-3-319-67903-7_4

4.1 Anthocyanins and Their Industrial Applications

Anthocyanins are ubiquitous pigments in many plants. Macroscopically, they endow flowers, leaves and fruits with diverse colors, which are important traits pursued by floriculture and horticulture. Microscopically, these chemicals protect plants from irradiation damage, oxidative stress, and pathogens [1–4]. As a group of colorful compounds with similar structures that belong to the flavonoid group of polyphenols (Fig. 4.1), anthocyanins have found their uses extended beyond their physiological roles in plants. With an increasing preference for natural food colorants, there is an ever-growing demand for anthocyanins as dietary supplements, colorants, and cosmetic additives [5–7]. Great effort has been invested towards improving the production of natural anthocyanins and developing new technologies for highly efficient and stable anthocyanin production [8].

4.1.1 Pharmaceutical Applications

Anthocyanins are promising drug candidates in preventing and treating diseases in animal models and in humans [9–12]. With the specific mechanisms remaining

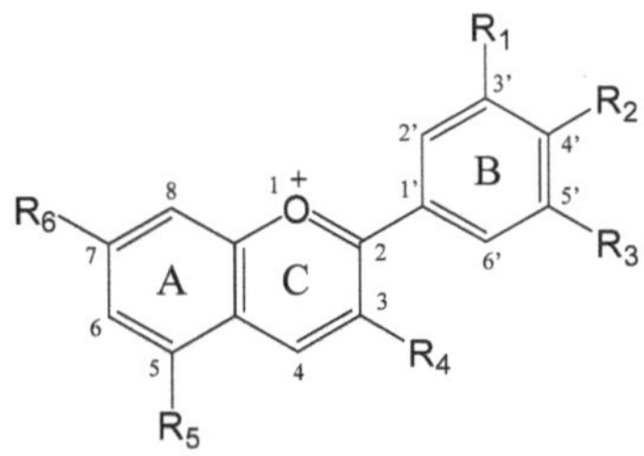

	Anthocyanidins	R_1	R_2	R_3	R_4	R_5	R_6
1	Pelargonidin	H	OH	H	OH	OH	OH
2	Cyanidin	OH	OH	H	OH	OH	OH
3	Delphinidin	OH	OH	OH	OH	OH	OH
4	Peonidin	OMe	OH	H	OH	OH	OH
5	Petunidin	OMe	OH	OH	OH	OH	OH
6	Malvidin	OMe	OH	OMe	OH	OH	OH

Fig. 4.1 The basic structure of natural anthocyanins. Different decorations including glycosylation, methylation and acylation can occur at C3' (R1), C4' (R2), C5' (R3), C3 (R4), C5 (R5), and C7 (R6), and over 600 anthocyanins with such a basic core structure have been identified. Most anthocyanins are distinguished into six anthocyanidins shown in the table. Me: a methyl group

elusive, there have been many studies probing the cellular and global response to anthocyanins, and *in vivo* anthocyanin metabolism [13]. Among other things, it has been shown that anthocyanins may inhibit body fat accumulation and obesity-induced inflammation in animal models, presumably by suppressing fat synthesis in the liver and white adipose tissue [10, 14], and by increasing glutathione peroxidase 3 expression while reducing the expression of inflammatory genes [15]. In treating diabetes in mice, cyanidin 3-*O*-glucoside was shown to downregulate retinol-binding protein 4 (RBP4) [16], while crude bilberry extract was shown to activate AMP-activated protein kinase (AMPK) [10], both ameliorating hyperglycemia symptoms. Anthocyanin extracts also help mitigate osteoclast-induced postmenopausal bone loss [17], lower blood pressure [18], and improve visual functions [10, 19]. In addition, anthocyanins block interleukin-1β, tumor necrosis factor-α, and nuclear factor (NF)-κβ in animal models, and therefore help with the suppression of neuroinflammation, neurodegradation, and brain aging [10]. These health benefits, though, are mostly observed with animal models and have yet to be verified in human clinical trials.

4.1.2 Food Colorants

Colorants are important additives to enhance the attractiveness of processed foods to consumers. Artificial colorants such as azo dyes dominated the market until a few decades ago, when consumers became more concerned about safety issues. The social tendency of "going natural" has stimulated the rapid increase in the use of pigments from natural sources as colorants owing to their specific characteristics, such as color variation at different pHs, pharmaceutical activities, biosafety, etc. [6].

Among natural pigments, anthocyanins, with potential health attributes and relatively low toxicity in animals and humans at high doses [20], are leading the market (together with carotenoids) [21–23]. In the US, four anthocyanin-based colorants are exempt from FDA certification [24]. In the European Union, anthocyanin-containing colorants are treated as natural colorants [25]. Besides anthocyanins, their acylated products are also widely used for improved color stability [26]. Nowadays, most anthocyanins are derived from grape pomace in winemaking processes [5], and grape extracts are widely used in coloring ice creams, dairy products, and sweets [21].

4.1.3 Cosmetic Industry

Besides the nutraceutical and food industries, anthocyanins also have potential applications in the cosmetic industry [27]. As effective antioxidants against reactive oxygen species, anthocyanins strongly absorb visible and ultraviolet (UV) light owing to the specific polyphenol structure [28], and protect skin from aging and UV-induced damage [29], such as inflammation and oxidative damage in the epidermis, dermis, and adnexal organs [27]. The underlying mechanisms have been demonstrated in several *in vitro* cellular and animal models, although detailed *in vivo* investigations are to be established [30, 31]. In general, anthocyanins reduce the UV-induced elevation of cyclooxygenase-2 and prostaglandin E2 through the NF-κβ-dependent pathways. Moreover, anthocyanins decrease apoptotic cell death by inhibiting caspase-3 activation and reduce the proapoptotic Bax protein levels [30, 32]. So far, no cosmetic products containing pure anthocyanins have been approved; however, there have been trials on the development of anthocyanin-colored lipsticks [33]. The incorporation of anthocyanins into cosmetics and skin care products may facilitate the alleviation of skin problems caused by direct contact with certain chemicals in these products, and may help to rejuvenate skin by reducing wrinkles, dark spots, redness, and other problems resulting from aging and skin damage [34, 35]. At present, a few companies have been trying to incorporate anthocyanins into cosmetics. Among them, Nutrasorb, LLC. developed engineered lettuce, whose extract is used both as a food supplement and a cosmetic additive. With more investigations into the skin protecting functions of anthocyanins and their decreased production costs, anthocyanin-based cosmetics may find their way into the market in the near future.

4.1.4 Other Fields

Beyond the applications in food, drugs and cosmetics that come into direct contact with humans and animals, anthocyanins have been exploited in dye sensitive solar cells (DSSCs) for the conversion of visible light to electricity. The key component in a DSSC is the sensitizer, which should have strong absorption in a wide spectrum and good adherence to the TiO_2 surface [36]. Traditionally, a transition metal coordination complex is used, making the synthesis expensive and complicated [37]. Natural dyes such as anthocyanins, however, are vastly available at much lower cost. Anthocyanins interact with TiO_2 through hydroxyl and carbonyl groups, allowing for electron transfer to the conducting band of TiO_2 films [38, 39]. The energy conversion efficiency is influenced by many factors such as the source of anthocyanins and the extraction method [37, 40], and efficiencies of up to 2% have been achieved with anthocyanins extracted from different plants [37, 41, 42]. Although the efficiency of anthocyanin-based DSSCs is substantially lower compared to those using synthetic sensitizers (~10% efficiency), it can be further enhanced by chemical modifications [43, 44]. Research is continuing in this field to improve the efficiency of DSSCs using anthocyanins alone or in combination with other natural dyes as photosensitizers.

4.2 Plant-Based Anthocyanin Production

Anthocyanins are synthesized via the general flavonoid pathway in plants, whereby three molecules of malonyl-CoA and one molecule of 4-coumaroyl-CoA derived from the general phenylpropanoid pathway are condensed to form naringenin chalcone by chalcone synthase (CHS) (Fig. 4.2). In the subsequent step, naringenin chalcone is converted to its isomer naringenin by chalcone isomerase (CHI). Next, naringenin is hydroxylated by enzymes such as flavanone 3-hydroxylase (F3H), flavonoid 3′-hydroxylase (F3'H) and flavonoid 3′, 5′-hydroxylase (F3'5'H), forming different dihydroflavonols. The dihydroflavonols are then reduced to the corresponding leucoanthocyanidins by dihydroflavonol 4-reductase (DFR), followed by the oxidation from anthocyanidin synthase (ANS) to generate the unstable precursor anthocyanidins. Anthocyanidins are flavylium cations that undergo glycosylation at C3 or other positions by flavonoid glucosyltransferases (FGTs), giving rise to anthocyanins (**see also Chap. 9 of this book**). The most common saccharide unit incorporated in this step is glucose, whereas galactose, xylose, and other sugar units are also found in natural anthocyanins [45]. Beyond glycosylation, other modifications, such as acylation, and methylation of the hydroxyl groups on B ring, have also been reported [46]. These modifications are performed in plants for improved stability or for specific physiological functions.

Anthocyanins in plants show different colors according to the pH in vacuoles [5, 45], which also affects anthocyanin stability. These compounds are quite labile at neutral and basic pH values. Structural modifications, lowered pH and co-pigmentation in vacuoles are all means adopted by plants to stabilize anthocyanins. The complexity of anthocyanin biosynthesis and their instability make their production in controlled systems a great challenge.

4.2.1 Extraction from Plants

So far the prevailing way of industrial anthocyanin production is by extraction from plants. For example, the anthocyanin supply for food colorants mainly comes from the waste products of the winemaking industry. The extraction is typically performed in solvents, of which the most commonly used is ethanol, because of its environmental friendliness, safety, and little interference with anthocyanin recovery [47]. Other extraction methods, such as pressurized liquid extraction, and novel extraction tools/agents, including ultrasound, subcritical water, and polymeric absorber resins, have also been reported to be effective in obtaining anthocyanins from crops and fruits [48–51]. These methods have different applications and should be selected with care. For example, water-extracted anthocyanins from plant flowers can be directly used for making DSSCs, whereas ethanol assisted extraction leads to photocatalytic decomposition of the extracted anthocyanins by TiO_2 in the solar cell, and hence lead to low efficiency of energy conversion [37].

4.2.2 Anthocyanin Production from Suspension Cell Culture

Plant suspension cell culture is a technology that introduces the bioreactor concept to cultivate plant cells for anthocyanin production with tight control over the fermentation processes [52, 53]. This technology involves development of suitable cell

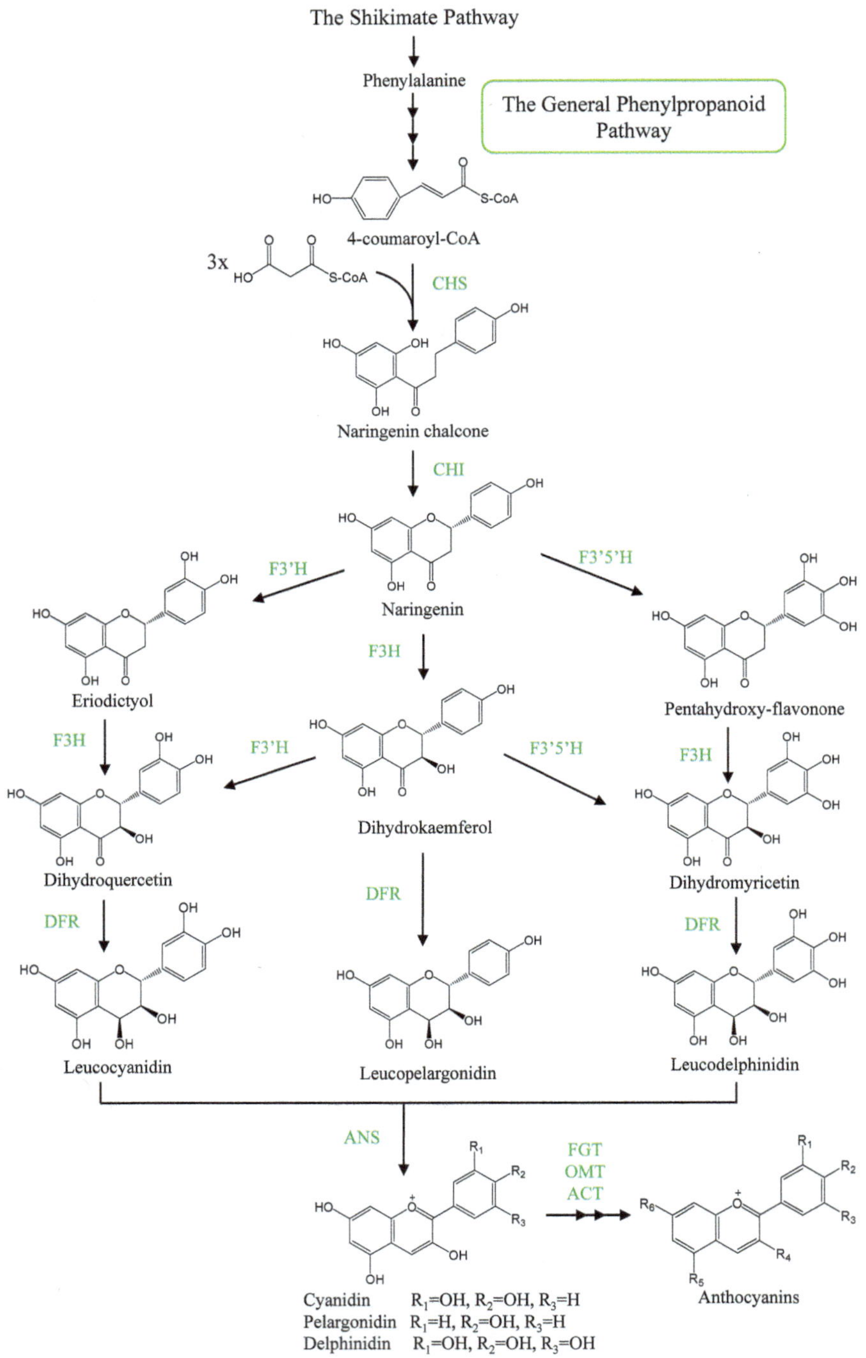

Fig. 4.2 The biosynthetic pathway of anthocyanins in plants. The general precursor phenylalanine, obtained from the shikimate pathway, enters the phenylpropanoid pathway to provide the intermediate coumaroyl-CoA for the production of flavonoids. Coumaroyl-CoA undergoes

lines, optimization of operating conditions of the bioreactors, and scaling up of fermentation (**see also Chap. 8 of this book**). For several decades, plants have been used to develop anthocyanin-producing cell lines, such as grapes, Cleome rose, sweet potatoes, aspen, wild carrots, etc. [54, 55]. However, currently there are no systems that are commercially feasible.

At present, the total anthocyanins obtained from plant suspension cell culture can reach up to 10% of the total dry weight of the producing cells, whereas the yields of specific types of anthocyanins are usually low [5, 56]. A major difficulty in suspension cell culture is the instability of cell lines, with anthocyanin production decreasing drastically over time [5]. Although the underlying mechanisms are not fully understood, one possible cause is the inadequate cell differentiation, since fast-growing, undifferentiated cells cannot produce anthocyanins. Besides the low anthocyanin yield in unstable and inefficient cell lines, other hurdles restricting the application of plant suspension cell culture in anthocyanin production include high production cost, low consumer acceptance, and strict biosafety regulations on cell line-derived compounds, especially those obtained from genetically modified cell lines [57, 58]. These limitations directly affect the economic competitiveness and attractiveness to investment and commercialization of anthocyanins. Subsequent metabolic engineering of cell lines and cell line selection are necessary for enhanced metabolic flux towards the anthocyanin pathway and improved accumulation of specific anthocyanins. In the future, the emerging global genome editing technologies, such as the clustered regularly interspaced short palindromic repeat (CRISPR)/CRISPR-associated protein 9 (Cas9) system, zinc finger nucleases (ZFNs), and transcription activator-like effector nucleases (TALENs), can be applied in developing engineered cell lines to support stable and efficient anthocyanin production [59]. Moreover, detailed investigations on fermentation design, in terms of medium composition, culture conditions, elicitation, and precursor feeding, are required to release the maximal potential of plant cell culture.

4.3 Anthocyanin Production in Microorganisms

As the most commonly used workhorse in metabolic engineering, *E. coli* has been engineered for the production of many flavonoids including naringenin, kaempferol, and catechin [46, 60–62]. Anthocyanins have also attracted much attention. In 2005, the genes of F3H and ANS from *Malus domestica*, DFR from *Anthurium andraeanum*, and flavonoid 3-glucosyltransferase (F3GT) from *Petunia hybrida* were successfully expressed in *E. coli* [63], and the recombinant strain could

Fig. 4.2 (continued) condensation with malonyl-CoA to form naringenin chalcone, which experiences various modifications to form diverse anthocyanin compounds. R1-R5 are functional groups involved in the modification of different carbons in anthocyanin molecules, such as glycosyl, acyl, methyl, and hydroxyl groups. Abbreviations of enzymes: *CHS* chalcone synthase, *CHI* chalcone isomerase, *F3H* flavanone 3-hydroxylase, *F3'H* flavonoid 3′-hydroxylase, *F3'5'H* flavonoid 3′, 5′-hydroxylase, *DFR* dihydroflavonol reductase, *ANS* anthocyanidin synthase, *FGT* flavonoid-glucosyltransferase, *OMT* *O*-methyltransferase, *ACT* acyltransferase

produce 6.0 μg/L of cyanidin 3-*O*-glucoside and 5.6 μg/L of pelargonidin 3-*O*-glucoside (**see also Chap. 9 of this book**) using naringenin and eriodictyol as the respective feeding precursors. Subsequent optimization of the enzyme source and the UDP-glucose pool, regulation of precursor uptake, and optimization of the production process greatly increased final product titers [64–66]. The highest production of cyanidin 3-*O*-glucoside and pelargonidin 3-*O*-glucoside was 350 mg/L and 113 mg/L using catechin and afzelechin as the respective precursors. These approaches have also been extended to the microbial biosynthesis of methylated anthocyanins. For example, the production of peonidin 3-*O*-glucoside (an *O*-methylated anthocyanin) from catechin was achieved in *E. coli* with the introduction of *P. hybrida* ANS, *Arabidopsis thaliana* F3GT, and *Vitis vinifera* anthocyanin *O*-methyltransferase (AOMT), and a final titer of 56 mg/L was reported upon pathway optimization [67]. To date, the reported microbial hosts of anthocyanin biosynthesis are still limited to *E. coli*, although the heterologous production of other flavonoids has been extended to *Saccharomyces cerevisiae* and *Streptomyces venezuelae* [46, 63], and the production of stilbenes and flavanones has been established in the amino acid-producing strain *Corynebacterium glutamicum* [68–70]. It will remain to be seen if other microbial hosts can be engineered for the production of anthocyanins.

4.3.1 Engineering of Pathway Enzymes

Engineering of the anthocyanin pathway involves coexpression of enzymes from plants. Heterologous expression of plant genes in prokaryotes is generally challenging, and typically, the genes/enzymes need to be modified prior to their functional expression (Fig. 4.3). For example, to achieve functional expression of a plant P450 F3'5'H from *Catharanthus roseus*, the four codons at the 5′-end of the gene were removed, and the fifth codon was replaced with ATG as the new start codon, while the sixth codon was changed from leucine to alanine [71]. The resulting new F3'5'H was fused to a shortened P450-reductase from *C. roseus* to form a chimeric protein, which catalyzed the formation of the flavonol quercetin by feeding coumaric acid.

Apart from modification of individual enzymes in the metabolic pathway, translational fusion of multiple enzymes in successive steps is another effective method of improving anthocyanin production (Fig. 4.3). Such fusions can maximize the local concentrations of substrates for each enzyme in the fusion system while minimizing the degradation of unstable intermediates, allowing multiple reactions to occur efficiently [72]. Using this strategy, it has been shown in *E. coli* that the translational fusion of F3GT from *Arabidopsis* to the N-terminus of ANS from *Petunia* could better convert catechin to cyanidin 3-*O*-glucoside compared with the tandem expression of ANS and F3GT [64]. In this case, the fused protein complex could catalyze the successive biochemical reactions 16.9% more efficiently than the uncoupled enzymes due to the faster conversion of the unstable intermediate anthocyanidin.

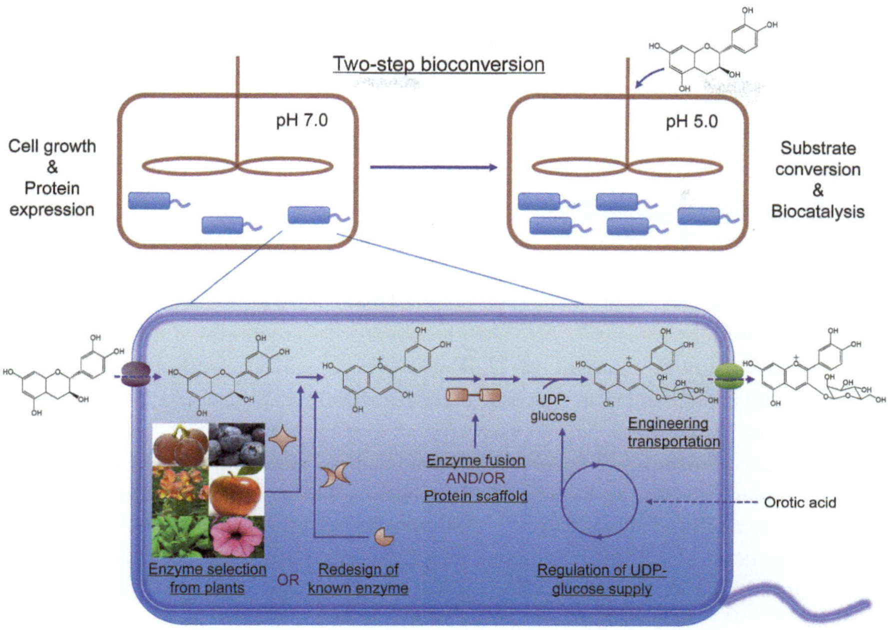

Fig. 4.3 The strategies applied in microbial production of anthocyanins. The whole strategies include the engineering of anthocyanin-producing strains and the optimization of the biocatalytic process. Strain modification focuses on screening and engineering of enzymes in the metabolic pathway, the transportation of the substrate and the product, and the supply of UDP-glucose. The biocatalysis is separated into two phases to maximize anthocyanin accumulation while maintaining normal cell growth. The content shown here is the example of cyanidin 3-*O*-glucoside production from catechin

Beyond direct enzyme engineering, selection of enzymes from diverse species is another way of improving the production of anthocyanins and other flavonoids (Fig. 4.3) [73]. The orthologous enzymes from different species that catalyze the same reactions usually exhibit diverse kinetic and thermodynamic properties, resulting in varied metabolic behaviors and different levels of production. In a study, the *in vivo* activities of ANS from four plants were compared, and the enzyme from *P. hybrida* produced 0.19- to 5.4-fold and 0.47- to 4.9-fold higher cyanidin and cyanidin 3-*O*-glucoside, respectively, in *E. coli* than the enzymes from *Antirrhinum majus*, *Gerbera hybrida*, and *M. domestica* [64]. In the production of peonidin 3-*O*-glucoside, five sources of AOMTs were compared and the one from *V. vinifera* led to the best substrate conversion with the lowest byproduct production [67]. Similarly, selection of DFR was conducted based on *in vitro* characterization of DFR orthologs from different plant sources during the *de novo* production of anthocyanins from flavonols [63, 74]. In another study, different combinations of three 4-coumaroyl CoA ligases, two CHSs and two CHIs, each enzyme having a distinct plant origin, resulted in a 3-fold increase in naringenin production in *E. coli* [75]. In an effort to synthesize resveratrol in recombinant *E. coli*, the *in vitro* kinetics of stilbene synthases from four plants were analyzed, and the data correlated well with the *in vivo* production [76] (**see also Chap. 3 of this book**).

4.3.2 Supply of Cofactors and Cosubstrates

Besides pathway enzymes, the biosynthesis of anthocyanins is also dependent on cofactors and cosubstrates that are involved in electron transfer, and enzyme activation or stabilization. For example, the enzyme ANS uses ferrous ions and sodium ascorbate as cofactors, and 2-oxoglutarate as a cosubstrate to conduct a two-electron oxidation of its substrates [64, 77]. The glycosylation of cyanidin at the C3 position requires an equimolar amount of UDP-glucose. Therefore, sufficient supply of cofactors and cosubstrates is a prerequisite for efficient, high-yield production of anthocyanins.

UDP-glucose is required for glycosylation in some anthocyanins (**see also Chap. 9 of this book**). As a valuable chemical that takes part in many cellular functions, from the generation of metabolic intermediates to the biosynthesis of cellular structural components, UDP-glucose must undergo global and elaborate regulation to reach a suitable level. In general, the regulation lies in altered expression of genes involved in UDP-glucose biosynthesis and/or its consumption. In an *E. coli* strain that produced cyanidin 3-*O*-glucoside, the abundance of UDP-glucose was increased by overexpressing one or more genes responsible for its biosynthesis from orotic acid (*pyrE*, *pyrR*, *cmk*, *ndk*, *pgm*, and *galU*) while blocking the competitive UDP-glucose consumption pathways. The resulting production of cyanidin 3-*O*-glucoside increased by 20-fold [64, 65]. Interestingly, even the overexpression of *pgm* and *galU* alone, under the control of independent T7 promoters on the same plasmid, led to a 57.8% increase in cyanidin 3-*O*-glucoside production [64]. These studies demonstrate that the supply of UDP-glucose is an important limiting factor for the overproduction of glycosylated anthocyanins. Considering the high cost of UDP-glucose and its precursor orotic acid, engineered intracellular biosynthesis of UDP-glucose from cheap nutrients would be useful for its supplementation.

Sodium ascorbate is another necessary ingredient to support the overproduction of anthocyanins. The addition of sodium ascorbate was found to significantly increase the consumption of the substrate catechin and the production of anthocyanin 3-*O*-glucoside in *E. coli*, whereas extra addition of the cosubstrate 2-oxoglutarate was unnecessary, probably because 2-oxoglutarate is the intermediate compound in the Krebs cycle and its supply is commonly abundant [64].

S-Adenosyl-L-methionine (SAM), a cosubstrate commonly involved in the transfer of methyl groups by methyltransferases, is generally required for the production of methylated anthocyanins. SAM supply can be increased by supplementing methionine and/or upregulating genes associated with SAM production. However, the generation of SAM undergoes feedback repression by methionine biosynthesis regulator *MetJ* based on the intracellular concentration of SAM [78], thus limiting the high-level accumulation of SAM and the rate of methylation of the target compounds. Recently, this difficulty was overcome by

CRISPRi-mediated deregulation of the methionine and SAM biosynthetic pathways through the silencing of *MetJ* in the production of peonidin 3-*O*-glucoside from catechin in *E. coli*, and a twofold increase in the production titer was achieved with such an approach [67].

4.3.3 Engineering Anthocyanin Secretion

Metabolic engineering has resulted in accomplishing the production of many compounds, natural or unnatural, in microorganisms. However, some of these compounds are toxic to cells by either directly reducing cell viability or indirectly interfering with cellular functions and metabolism, thus limiting their high-yield production. A feasible scheme is to pump out the products continuously during their biosynthesis to extracellular media, where their toxic effects are attenuated. To achieve this, identification of specific transporters is critical. In addition, the incorporation of transporters for enhanced substrate uptake also facilitates the production of the target chemicals. In an *E. coli* strain that converted catechin to cyanidin 3-*O*-glucoside, the overexpression of the product-associated efflux pump YadH increased the production by 15%, and the deletion of another efflux pump TolC, which was probably responsible for the secretion of catechin, enhanced production by 55%. The combined effect was a 63% promotion in cyanidin 3-*O*-glucoside production [66].

Anthocyanins in their natural plant hosts are transported to vacuoles after their synthesis, and this process requires both cytoplasmic transporters and transmembrane transporters. The most commonly studied plant-based transporters are glutathione S-transferase and ATP-binding cassette (ABC) transporters (**see also Chap. 9 of this book**). Since both plant tonoplasts (membranes surrounding the vacuoles) and microbial cell membranes comprise lipid bilayers, it may be useful to investigate the performance of engineered plant-based transporters in microorganisms for anthocyanin delivery across the cytoplasmic membrane and the outer membrane.

4.3.4 Optimization of the Production Process

The instability of anthocyanins is a major problem for accomplishing their efficient production in microorganisms. In plants, the naturally synthesized anthocyanins are stabilized in vacuoles through pH adjustment and co-pigmentation [45, 79]. However, in bacterial cells that are engineered as artificial producing hosts, there is a shortage of protection mechanisms for produced anthocyanins. The microbially synthesized anthocyanins are quite unstable inside or outside cells, considering that the intracellular and extracellular pH is around 7 for commonly used bacteria under their normal growth conditions. To solve this issue, a two-step biocatalysis strategy has been proposed [64]. During the first phase, cells are cultured in a medium at pH 7 to support normal growth and enzyme expression. In the second step, cells at a particular growth stage are transferred to fresh medium at pH 5.0 to facilitate anthocyanin production and accumulation (Fig. 4.3). Protective agents such as glutamate can be added to minimize acid-induced cell lysis. With such an approach, the production of cyanidin 3-*O*-glucoside in *E. coli* was ~15-fold higher than that from the traditional single-step production [64].

Concentration of dissolved oxygen is another parameter that shows great impact on anthocyanin biosynthesis and stability. Oxygen is critical for the maintenance of ANS functionality and the synthesis of anthocyanins; however, an excessive amount of dissolved oxygen may oxidize anthocyanins. Although the specific roles of oxygen in microbial production of anthocyanins are poorly understood, it is clear that an optimal supply of oxygen is important. In a study of catechin production from eriodictyol, increased concentration of dissolved oxygen led to enhanced production of catechin, which might be related to increased NADPH supply [73]. However, no such investigations have been reported for anthocyanin production.

Besides pH and oxygen, temperature and induction point also have remarkable effect on anthocyanin production [66, 75]. Temperature generally imposes direct impact on cell viability and protein expression or folding, and hence influences anthocyanin bioconversion indirectly. Induction time-points are correlated with growth stages and conditions of the producing cells, and differential enzyme expression at diverse growth phases can result in significantly different production efficiencies.

Pathway balancing should also be considered for efficient anthocyanin production. The aim is to reduce metabolic burden exerted on host cells during the overproduction of anthocyanins, and meanwhile, to maintain normal cell growth and metabolism to the most extent [80]. Many tools have been established to balance the metabolic pathways, such as the ePathBrick vectors, the ePathOptimize platform, and biosensor-based dynamic regulation in flavonoid biosynthesis [81–85]. Recently, a dCas9-based toolbox has been developed to orchestrate the expression of multiple genes simultaneously in *E. coli* [86, 87]. This strategy can be exploited for the identification of the potential regulation points relevant for anthocyanin production.

4.4 Conclusions and Future Perspectives

Anthocyanins are very useful flavonoids with applications as dietary supplements, food colorants, and cosmetic additives. The current supply is largely dependent on extraction from plant materials, while emerging technologies delve into sustainable production either in engineered plant cells or in recombinant microbial cells. In this chapter, we focused on metabolic engineering of anthocyanin production in microorganisms, especially in *E. coli*. We presented the strategies that have been applied in optimizing the biosynthetic pathways, the host strains, and the bioreaction processes. However, many issues still remain to be addressed, such as poor expression of anthocyanin biosynthetic genes, imbalance of genes in the pathway, and stabilization of the final product. With the elucidation of anthocyanin biosynthesis in plants, sophisticated redesign of related enzymes, and regulation of the constructed pathways based on metabolic models, it is expected that engineered microorganisms will become an important source of providing anthocyanins.

References

1. Gould KS, McKelvie J, Markham KR. Do anthocyanins function as antioxidants in leaves? Imaging of H_2O_2 in red and green leaves after mechanical injury. Plant Cell Environ. 2002;25(10):1261–9.
2. Winkel-Shirley B. Flavonoid biosynthesis: a colorful model for genetics, biochemistry, cell biology, and biotechnology. Plant Physiol. 2001;126(2):485–93.
3. Steyn WJ, Wand SJE, Holcroft DM, Jacobs G. Anthocyanins in vegetative tissues: a proposed unified function in photoprotection. New Phytol. 2002;155(3):349–61.
4. Lorenc-Kukuła K, Jafra S, Oszmiański J, Szopa J. Ectopic expression of anthocyanin 5-*O*-glucosyltransferase in potato tuber causes increased resistance to bacteria. J Agric Food Chem. 2005;53(2):272–81.
5. Ananga A, Georgiev V, Ochieng J, Phills B, Tsolova V. Production of anthocyanins in grape cell cultures: a potential source of raw material for pharmaceutical, food, and cosmetic industries. In: Poljuha D, Sladonja B, editors. The Mediterranean genetic code – Grapevine and Olive. Rijeka: InTech; 2013.
6. Aberoumand A. A review article on edible pigments properties and sources as natural biocolorants in foodstuff and food industry. World J Dairy Food Sci. 2011;6(1):71–8.
7. Tanaka Y, Ohmiya A. Seeing is believing: engineering anthocyanin and carotenoid biosynthetic pathways. Curr Opin Biotechnol. 2008;19(2):190–7.
8. Wang J, Guleria S, Koffas MAG, Yan Y. Microbial production of value-added nutraceuticals. Curr Opin Biotechnol. 2016;37:97–104.
9. Lila MA, Burton-Freeman B, Grace M, Kalt W. Unraveling anthocyanin bioavailability for human health. Annu Rev Food Sci Technol. 2016;7:375–93.
10. Tsuda T. Dietary anthocyanin-rich plants: biochemical basis and recent progress in health benefits studies. Mol Nutr Food Res. 2012;56(1):159–70.
11. de Pascual-Teresa S, Moreno DA, García-Viguera C. Flavanols and anthocyanins in cardiovascular health: a review of current evidence. Int J Mol Sci. 2010;11(4):1679–703.
12. Burton-Freeman B, Linares A, Hyson D, Kappagoda T. Strawberry modulates LDL oxidation and postprandial lipemia in response to high-fat meal in overweight hyperlipidemic men and women. J Am Coll Nutr. 2010;29(1):46–54.
13. Czank C, Cassidy A, Zhang Q, Morrison DJ, Preston T, Kroon PA, Botting NP, Kay CD. Human metabolism and elimination of the anthocyanin, cyanidin-3-glucoside: a ^{13}C-tracer study. Am J Clin Nutr. 2013;97(5):995–1003.
14. Tsuda T. Regulation of adipocyte function by anthocyanins: possibility of preventing the metabolic syndrome. J Agric Food Chem. 2008;56(3):642–6.
15. DeFuria J, Bennett G, Strissel KJ, Perfield JW, Milbury PE, Greenberg AS, Obin MS. Dietary blueberry attenuates whole-body insulin resistance in high fat-fed mice by reducing adipocyte death and its inflammatory sequelae. J Nutr. 2009;139(8):1510–6.
16. Sasaki R, Nishimura N, Hoshino H, Isa Y, Kadowaki M, Ichi T, Tanaka A, Nishiumi S, Fukuda I, Ashida H, Horio F, Tsuda T. Cyanidin 3-glucoside ameliorates hyperglycemia and insulin sensitivity due to downregulation of retinol binding protein 4 expression in diabetic mice. Biochem Pharmacol. 2007;74(11):1619–27.
17. Zheng X, Mun S, Lee SG, Vance TM, Hubert P, Koo SI, Lee S-K, Chun OK. Anthocyanin-rich blackcurrant extract attenuates ovariectomy-induced bone loss in mice. J Med Food. 2016;19(4):390–7.
18. Erlund I, Koli R, Alfthan G, Marniemi J, Puukka P, Mustonen P, Mattila P, Jula A. Favorable effects of berry consumption on platelet function, blood pressure, and HDL cholesterol. Am J Clin Nutr. 2008;87(2):323–31.
19. Zia Ul Haq M, Riaz M, Saad B. Anthocyanins and human health: biomolecular and therapeutic aspects. In: Book Anthocyanins and human health: biomolecular and therapeutic aspects. 1st ed. Cham: Springer; 2016.
20. Wallace TC, Giusti MM. Anthocyanins. Adv Nutr. 2015;6(5):620–2.

21. Mateus N, de Freitas V. Anthocyanins as food colorants. In: Winefield C, Davies K, Gould K, editors. Anthocyanins: biosynthesis, functions, and applications. New York: Springer; 2009.
22. Mora-Pale M, Sanchez-Rodriguez SP, Linhardt RJ, Dordick JS, Koffas MAG. Biochemical strategies for enhancing the *in vivo* production of natural products with pharmaceutical potential. Curr Opin Biotechnol. 2014;25:86–94.
23. Chin-Giaw L, Koffas MAG. Bioavailability and recent advances in the bioactivity of flavonoid and stilbene compounds. Curr Org Chem. 2010;14(16):1727–51.
24. Wrolstad RE. Anthocyanin pigments—bioactivity and coloring properties. J Food Sci. 2004;69(5):C419–25.
25. Socaciu C. Food colorants: chemical and functional properties. In: Food colorants: chemical and functional properties. Boca Raton: CRC Press; 2007.
26. Giusti MM, Wrolstad RE. Acylated anthocyanins from edible sources and their applications in food systems. Biochem Eng J. 2003;14(3):217–25.
27. Wallace TC, Giusti MM, Rojo LE, Roopchand DE, Graf B, Cheng DM, Ribnicky D, Fridlender B, Raskin I. Role of anthocyanins in skin aging and UV-induced skin damage. In: Anthocyanins in health and disease. Boca Raton: CRC Press; 2013.
28. Giusti MM, Wrolstad RE. Characterization and measurement of anthocyanins by UV-visible spectroscopy. In: Current protocols in food analytical chemistry. New York: Wiley; 2001.
29. Chan CF, Lien CY, Lai YC, Huang CL, Liao WC. Influence of purple sweet potato extracts on the UV absorption properties of a cosmetic cream. J Cosmet Sci. 2010;61(5):333–41.
30. Tsoyi K, Park HB, Kim YM, Chung JI, Shin SC, Shim HJ, Lee WS, Seo HG, Lee JH, Chang KC, Kim HJ. Protective effect of anthocyanins from black soybean seed coats on UVB-induced apoptotic cell death *in vitro* and *in vivo*. J Agric Food Chem. 2008;56(22):10600–5.
31. Tarozzi A, Marchesi A, Hrelia S, Angeloni C, Andrisano V, Fiori J, Cantelli-Forti G, Hrelia P. Protective effects of cyanidin-3-*O*-beta-glucopyranoside against UVA-induced oxidative stress in human keratinocytes. Photochem Photobiol. 2005;81(3):623–9.
32. Tsoyi K, Park HB, Kim YM, Chung JI, Shin SC, Lee WS, Seo HG, Lee JH, Chang KC, Kim HJ. Anthocyanins from black soybean seed coats inhibit UVB-induced inflammatory cylooxygenase-2 gene expression and PGE2 production through regulation of the nuclear factor-kappaB and phosphatidylinositol 3-kinase/Akt pathway. J Agric Food Chem. 2008;56(19):8969–74.
33. Westfall A. Evaluation of the efficacy of anthocyanins as biologically active ingredients in lipstick formulations. In: Food science and technology. Columbus: The Ohio State University; 2015.
34. Ghimeray AK, Jung US, Lee HY, Kim YH, Ryu EK, Chang MS. *In vitro* antioxidant, collagenase inhibition, and *in vivo* anti-wrinkle effects of combined formulation containing *Punica granatum*, *Ginkgo biloba*, *Ficus carica*, and *Morus alba* fruits extract. Clin Cosmet Invest Dermatol. 2015;8:389–96.
35. Mukherjee PK, Maity N, Nema NK, Sarkar BK. Bioactive compounds from natural resources against skin aging. Phytomedicine. 2011;19(1):64–73.
36. Tennakone K, Kumara GRRA, Kumarasinghe AR, Sirimanne PM, Wijayantha KGU. Efficient photosensitization of nanocrystalline TiO_2 films by tannins and related phenolic substances. J Photochem Photobiol A. 1996;94(2):217–20.
37. Wongcharee K, Meeyoo V, Chavadej S. Dye-sensitized solar cell using natural dyes extracted from rosella and blue pea flowers. Sol Energy Mater Sol Cells. 2007;91(7):566–71.
38. Hao S, Wu J, Huang Y, Lin J. Natural dyes as photosensitizers for dye-sensitized solar cell. Sol Energy. 2006;80(2):209–14.
39. Dai Q, Rabani J. Photosensitization of nanocrystalline TiO_2 films by pomegranate pigments with unusually high efficiency in aqueous medium. Chem Commun. 2001;20:2142–3.
40. Calogero G, Marco GD. Red Sicilian orange and purple eggplant fruits as natural sensitizers for dye-sensitized solar cells. Sol Energy Mater Sol Cells. 2008;92(11):1341–6.
41. Calogero G, Yum J-H, Sinopoli A, Di Marco G, Grätzel M, Nazeeruddin MK. Anthocyanins and betalains as light-harvesting pigments for dye-sensitized solar cells. Sol Energy. 2012;86(5):1563–75.
42. Ramamoorthy R, Radha N, Maheswari G, Anandan S, Manoharan S, Victor Williams R. Betalain and anthocyanin dye-sensitized solar cells. J Appl Electrochem. 2016;46(9):929–41.

43. Amao Y, Yamada Y, Aoki K. Preparation and properties of dye-sensitized solar cell using chlorophyll derivative immobilized TiO_2 film electrode. J Photochem Photobiol A. 2004;164(1–3):47–51.
44. Calogero G, Sinopoli A, Citro I, Di Marco G, Petrov V, Diniz AM, Parola AJ, Pina F. Synthetic analogues of anthocyanins as sensitizers for dye-sensitized solar cells. Photochem Photobiol Sci. 2013;12(5):883–94.
45. Zhang Y, Butelli E, Martin C. Engineering anthocyanin biosynthesis in plants. Curr Opin Plant Biol. 2014;19:81–90.
46. Pandey RP, Parajuli P, Koffas MAG, Sohng JK. Microbial production of natural and non-natural flavonoids: pathway engineering, directed evolution and systems/synthetic biology. Biotechnol Adv. 2016;34(5):634–62.
47. Delgado-Vargas F, Paredes-Lopez O. Natural colorants for food and nutraceutical uses. In: Natural colorants for food and nutraceutical uses. Boca Raton: CRC Press; 2002.
48. Corrales M, Toepfl S, Butz P, Knorr D, Tauscher B. Extraction of anthocyanins from grape by-products assisted by ultrasonics, high hydrostatic pressure or pulsed electric fields: a comparison. Innovative Food Sci Emerg Technol. 2008;9(1):85–91.
49. Ju Z, Howard LR. Subcritical water and sulfured water extraction of anthocyanins and other phenolics from dried red grape skin. J Food Sci. 2005;70(4):S270–6.
50. Kammerer D, Gajdos Kljusuric J, Carle R, Schieber A. Recovery of anthocyanins from grape pomace extracts (*Vitis vinifera L. cv. Cabernet Mitos*) using a polymeric adsorber resin. Eur Food Res Technol. 2005;220(3):431–7.
51. ZY J, Howard LR. Effects of solvent and temperature on pressurized liquid extraction of anthocyanins and total phenolics from dried red grape skin. J Agric Food Chem. 2003;51(18):5207–13.
52. Matkowski A. Plant *in vitro* culture for the production of antioxidants — a review. Biotechnol Adv. 2008;26(6):548–60.
53. Deroles S. Anthocyanin biosynthesis in plant cell cultures: a potential souce of natural colorants. In: Anthocyanins: Springer; 2008.
54. Simões C, Albarello N, de Castro TC, Mansur E. Production of anthocyanins by plant cell and tissue culture strategies. In: Biotechnological production of plant secondary metabolites. Dubai: Bentham Science Publishers; 2012.
55. Rao SR, Ravishankar GA. Plant cell cultures: chemical factories of secondary metabolites. Biotechnol Adv. 2002;20(2):101–53.
56. Yue W, Ming QL, Lin B, Rahman K, Zheng CJ, Han T, Qin LP. Medicinal plant cell suspension cultures: pharmaceutical applications and high-yielding strategies for the desired secondary metabolites. Crit Rev Biotechnol. 2016;36(2):215–32.
57. Davies KM, Deroles SC. Prospects for the use of plant cell cultures in food biotechnology. Curr Opin Biotechnol. 2014;26:133–40.
58. Angelov A, Gotcheva V. Safety assessment and regulations for food ingredients derived from plant *in vitro* systems. In: Pavlov A, Bley T, editors. Bioprocessing of plant *in vitro* systems. Cham: Springer; 2017.
59. Bortesi L, Fischer R. The CRISPR/Cas9 system for plant genome editing and beyond. Biotechnol Adv. 2015;33(1):41–52.
60. Chemler JA, Leonard E, Koffas MAG. Flavonoid biotransformations in microorganisms. In: Winefield C, Davies K, Gould KS, editors. Anthocyanins: biosynthesis, funcitons, and applications: Springer; 2009.
61. Trantas EA, Koffas MA, Xu P, Ververidis F. When plants produce not enough or at all: metabolic engineering of flavonoids in microbial hosts. Front Plant Sci. 2015;6:7.
62. Chemler JA, Koffas MAG. Metabolic engineering for plant natural product biosynthesis in microbes. Curr Opin Biotechnol. 2008;19(6):597–605.
63. Yan Y, Chemler J, Huang L, Martens S, Koffas MAG. Metabolic engineering of anthocyanin biosynthesis in *Escherichia coli*. Appl Environ Microbiol. 2005;71(7):3617–23.
64. Yan Y, Li Z, Koffas MA. High-yield anthocyanin biosynthesis in engineered *Escherichia coli*. Biotechnol Bioeng. 2008;100(1):126–40.

65. Leonard E, Yan Y, Fowler ZL, Li Z, Lim C-G, Lim K-H, Koffas MAG. Strain improvement of recombinant *Escherichia coli* for efficient production of plant flavonoids. Mol Pharm. 2008;5(2):257–65.
66. Lim CG, Wong L, Bhan N, Dvora H, Xu P, Venkiteswaran S, Koffas MAG. Development of a recombinant *Escherichia coli* strain for overproduction of plant pigment, anthocyanin. Appl Environ Microbiol. 2015;81(18):6276–84.
67. Cress BF, Leitz QD, Kim DC, Amore TD, Suzuki JY, Linhardt RJ, Koffas MA. CRISPRi-mediated metabolic engineering of *E. coli* for *O*-methylated anthocyanin production. Microb Cell Factories. 2017;16(1):10.
68. Kallscheuer N, Vogt M, Stenzel A, Gatgens J, Bott M, Marienhagen J. Construction of a *Corynebacterium glutamicum* platform strain for the production of stilbenes and (2*S*)-flavanones. Metab Eng. 2016;38:47–55.
69. Kallscheuer N, Vogt M, Marienhagen J. A novel synthetic pathway enables microbial production of polyphenols independent from the endogenous aromatic amino acid metabolism. ACS Synth Biol. 2017;6(3):410–5.
70. van Summeren-Wesenhagen PV, Marienhagen J. Putting bugs to the blush: metabolic engineering for phenylpropanoid-derived products in microorganisms. Bioengineered. 2013;4(6):355–62.
71. Leonard E, Yan Y, Koffas MAG. Functional expression of a P450 flavonoid hydroxylase for the biosynthesis of plant-specific hydroxylated flavonols in *Escherichia coli*. Metab Eng. 2006;8(2):172–81.
72. Springob K, Nakajima J, Yamazaki M, Saito K. Recent advances in the biosynthesis and accumulation of anthocyanins. Nat Prod Rep. 2003;20(3):288–303.
73. Zhao S, Jones JA, Lachance DM, Bhan N, Khalidi O, Venkataraman S, Wang Z, Koffas MAG. Improvement of catechin production in *Escherichia coli* through combinatorial metabolic engineering. Metab Eng. 2015;28:43–53.
74. Leonard E, Yan Y, Chemler J, Matern U, Martens S, Koffas MAG. Characterization of dihydroflavonol 4-reductases for recombinant plant pigment biosynthesis applications. Biocatal Biotransfor. 2008;26(3):243–51.
75. Jones JA, Vernacchio VR, Sinkoe AL, Collins SM, Ibrahim MHA, Lachance DM, Hahn J, Koffas MAG. Experimental and computational optimization of an *Escherichia coli* co-culture for the efficient production of flavonoids. Metab Eng. 2016;35:55–63.
76. Lim CG, Fowler ZL, Hueller T, Schaffer S, Koffas MAG. High-yield resveratrol production in engineered *Escherichia coli*. Appl Environ Microbiol. 2011;77(10):3451–60.
77. Turnbull JJ, Nakajima J-i, Welford RWD, Yamazaki M, Saito K, Schofield CJ. Mechanistic studies on three 2-oxoglutarate-dependent oxygenases of flavonoid biosynthesis: anthocyanidin synthase, flavonol synthase, and flavanone 3β-hydroxylase. J Biol Chem. 2004;279(2):1206–16.
78. Martí-Arbona R, Teshima M, Anderson PS, Nowak-Lovato KL, Hong-Geller E, Unkefer CJ, Unkefer PJ. Identification of new ligands for the methionine biosynthesis transcriptional regulator (MetJ) by FAC-MS. J Mol Microbiol Biotechnol. 2012;22(4):205–14.
79. Passeri V, Koes R, Quattrocchio FM. New challenges for the design of high value plant products: stabilization of anthocyanins in plant vacuoles. Front Plant Sci. 2016;7:153.
80. Wu G, Yan Q, Jones JA, Tang YJ, Fong SS, Koffas MA. Metabolic burden: cornerstones in synthetic biology and metabolic engineering applications. Trends Biotechnol. 2016;34(8):652–64.
81. Xu P, Vansiri A, Bhan N, Koffas MAG. ePathBrick: a synthetic biology platform for engineering metabolic pathways in *E. coli*. ACS Synth Biol. 2012;1(7):256–66.
82. Jones JA, Vernacchio VR, Lachance DM, Lebovich M, Fu L, Shirke AN, Schultz VL, Cress B, Linhardt RJ, Koffas MAG. ePathOptimize: a combinatorial approach for transcriptional balancing of metabolic pathways. Sci Rep. 2015;5:11301.
83. Xu P, Li L, Zhang F, Stephanopoulos G, Koffas M. Improving fatty acids production by engineering dynamic pathway regulation and metabolic control. Proc Natl Acad Sci. 2014;111(31):11299–304.

84. Skjoedt ML, Snoek T, Kildegaard KR, Arsovska D, Eichenberger M, Goedecke TJ, Rajkumar AS, Zhang J, Kristensen M, Lehka BJ, Siedler S, Borodina I, Jensen MK, Keasling JD. Engineering prokaryotic transcriptional activators as metabolite biosensors in yeast. Nat Chem Biol. 2016;12(11):951–8.
85. Cress BF, Trantas EA, Ververidis F, Linhardt RJ, Koffas MAG. Sensitive cells: enabling tools for static and dynamic control of microbial metabolic pathways. Curr Opin Biotechnol. 2015;36:205–14.
86. Cress BF, Jones JA, Kim DC, Leitz QD, Englaender JA, Collins SM, Linhardt RJ, Koffas MAG. Rapid generation of CRISPR/dCas9-regulated, orthogonally repressible hybrid T7-lac promoters for modular, tuneable control of metabolic pathway fluxes in *Escherichia coli*. Nucleic Acids Res. 2016;44(9):4472–85.
87. Cress BF, Toparlak ÖD, Guleria S, Lebovich M, Stieglitz JT, Englaender JA, Jones JA, Linhardt RJ, Koffas MAG. CRISPathBrick: modular combinatorial assembly of type II-A CRISPR arrays for dCas9-mediated multiplex transcriptional repression in *E. coli*. ACS Synth Biol. 2015;4(9):987–1000.

Microbial Synthesis of Plant Alkaloids 5

Michael E. Pyne, Lauren Narcross, Cecilie I.C. Hansen, Alex Campbell, Meghan E. Davies, Leanne Bourgeois, Elena Fossati, and Vincent J.J. Martin

Abbreviations

10HC	10-hydroxycitronellol
10HG	10-hydroxygeraniol
10HGO	10-hydroxygeraniol oxidoreductase
3,4-dHPAA	3,4-dihydroxyphenylacetaldehyde
4′OMT	3′-hydroxy-*N*-methylcoclaurine 4′-*O*-methyltransferase
4-HPAA	4-hydroxyphenylacetaldehyde
4-HPP	4-hydroxyphenylpyruvate
6OMT	Norcoclaurine 6-*O*-methyltransferase
7-DLGA	7-deoxyloganic acid
7DLH	7-deoxyloganic acid hydroxylase
AAAT	Aromatic amino acid transferase
AADC	Amino acid decarboxylase
ACAT	Acetoacetyl-CoA thiolase
ARS	Autonomously replicating sequence
AT	Acetyl transferase
BBE	Berberine bridge enzyme
BIA	Benzylisoquinoline alkaloid
CAS	Canadine synthase
CFS	Cheilanthifoline synthase

M.E. Pyne • L. Narcross • A. Campbell • M.E. Davies
L. Bourgeois • E. Fossati • V.J.J. Martin (✉)
Department of Biology, Centre of Applied Synthetic Biology, Centre for Structural and Functional Genomics, Concordia University, Montréal, QC, Canada
e-mail: vincent.martin@concordia.ca

C.I.C. Hansen
Plant Biochemistry Laboratory, Department of Plant and Environmental Sciences, University of Copenhagen, Copenhagen, Denmark

W. Schwab et al. (eds.), *Biotechnology of Natural Products*,
https://doi.org/10.1007/978-3-319-67903-7_5

CNMT	Coclaurine *N*-methyltransferase
CODM	Codeine demethylase
COR	Codeinone reductase
CPR	Cytochrome P450 reductase
CXE	Carboxylesterase
CYB5	Cytochrome *b*5
CYP	Cytochrome P450
DMAPP	Dimethylallyl pyrophosphate
DODC	DOPA decarboxylase
DOPA	3,4-dihydroxyphenylalanine
DXP	Deoxyxylulose 5-phosphate
ER	Endoplasmic reticulum
FBR	Feedback resistant
FPP	Farnesyl pyrophosphate
G10H	Geraniol 10-hydroxylase
GES	Geraniol synthase
GPP	Geranyl pyrophosphate
HMG-CoA	3-hydroxy-3-methylglutaryl CoA
HMGR	HMG-CoA reductase
HMGS	HMG-CoA synthase
IO	Iridoid oxidase
IPP	Isopentenyl pyrophosphate
IS	Iridoid synthase
MAO	Monoamine oxidase
MIA	Monoterpene indole alkaloid
MSH	(*S*)-*N*-methylstylopine 14-hydroxylase
MVD	Mevalonate pyrophosphate decarboxylase
MVK	Mevalonate kinase
NCS	Norcoclaurine synthase
NMCH	*N*-methylcoclaurine hydroxylase
P6H	Protopine 6-hydroxylase
PDH	Prephenate dehydrogenase
PMVK	Phosphomevalonate kinase
Prx1	Class III peroxidase
REPI	Reticuline epimerase
ROS	Reactive oxygen species
S9OMT	Scoulerine 9-*O*-methyltransferase
SAR	Salutaridine reductase
SAS	Salutaridine synthase
SAT	Salutaridinol acetyltransferase
SDR	Short-chain dehydrogenase/reductase
SLS	Secologanin synthase
SPS	Stylopine synthase
STS	Strictosidine synthase
T6ODM	Thebaine 6-*O*-demethylase

TNMT	Tetrahydroprotoberberine *N*-methyltransferase
TYR	Tyrosine hydroxylase

5.1 Introduction

Microbial synthesis has emerged as an attractive and sustainable alternative for the production of high-value plant natural products [1]. Within the alkaloid family of metabolites, the monoterpene indole alkaloids (MIAs) and benzylisoquinoline alkaloids (BIAs) are particularly noteworthy for their potent pharmaceutical activities and broad structural diversity (Fig. 5.1). More than 5500 known MIAs and BIAs are found in nature, several of which exhibit analgesic, antimicrobial, anticancer, antispasmodic, and antitussive activities [2–5]. Notable compounds include the anticancer MIAs vinblastine and vincristine and the BIA analgesics morphine and codeine, all of which are included in the World Health Organization's Model List of Essential Medicines [6]. The biosynthetic pathways mediating formation of MIAs and BIAs share many general features. For example, both pathways branch from the aromatic amino acid pathway and employ a key Pictet-Spengler condensation responsible for the core structures from which all members are derived. Strictosidine is the condensation product in the MIA pathway (Fig. 5.1a), while norcoclaurine is generated in the analogous BIA reaction (Fig. 5.1b). Downstream metabolites from both metabolite classes are then diversified through a network of complex enzymatic and spontaneous rearrangements that generate the immense diversity characterized by plant natural products.

Presently, the principal source of most alkaloid-derived pharmaceuticals remains direct extraction from source plants and plant cell cultures (**see also Chap. 8 of this book**). Owing to its comparatively low structural complexity and lack of a chiral center, papaverine is the only naturally-occurring BIA pharmaceutical produced through complete chemical synthesis [2]. Selective breeding and mutagenesis cycles have generated cultivars of opium poppy (*Papaver somniferum*) for supply of morphinans (thebaine, codeine, and morphine) and noscapine BIAs, which together with papaverine constitute the most abundant alkaloids in the latex of commercially cultivated opium poppy [2, 7]. Conversely, the antimicrobial BIA berberine and the MIA precursor geraniol are produced by means of plant cell cultures [8] (**see also Chap. 8 of this book**). While the supply of these select metabolites and their derivatives currently meet market demands [7], crop-based manufacturing of less abundant alkaloids is generally unfeasible. The MIAs vinblastine and vincristine, for example, are harvested from the leaves of mature periwinkle (*Catharanthus roseus*) at very low yields and highly variable concentrations [9]. Indeed, of the thousands of known MIA and BIA structures, only a few select compounds accumulate to sufficient quantities in plant tissues to justify large-scale production and extraction. In this regard, the overwhelming majority of MIA and BIA activities remains untapped, resulting in an attractive pool of bioactive candidates awaiting drug discovery. In an effort to tap into this pharmaceutical potential, microbial synthesis has emerged as a promising alternative to natural product extraction and total chemical synthesis.

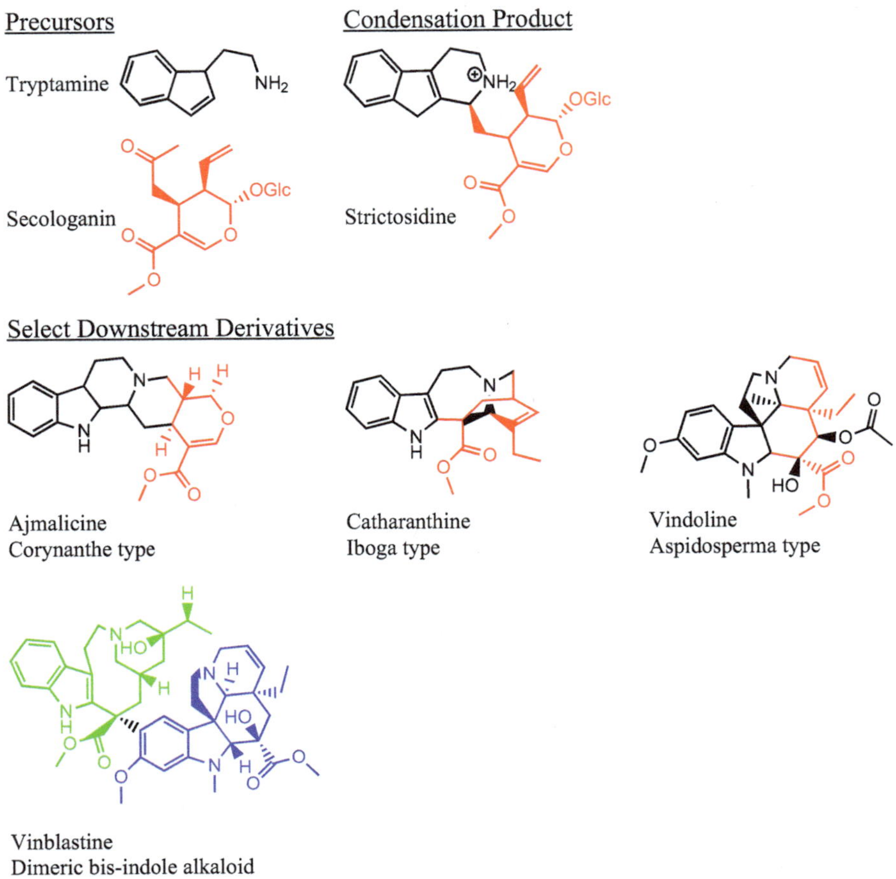

Fig. 5.1 Chemical structures of major MIA and BIA classes. (**a**) Chemical structures of notable MIA pathway intermediates and structural classes. Black and red portions highlight backbones derived from tryptamine and secologanin, respectively. Green and blue regions of vinblastine derive from the MIAs catharanthine and vindoline, respectively. (**b**) Chemical structures of key BIA pathway intermediates and structural classes. Black and red portions highlight backbones derived from dopamine and 3,4-dihydroxyphenylacetaldehyde (3,4-dHPAA) or 4-hydroxyphenylacetaldehyde (4-HPAA), respectively. Downstream derivatives are color-coordinated according to derivation from (*S*)-norcoclaurine

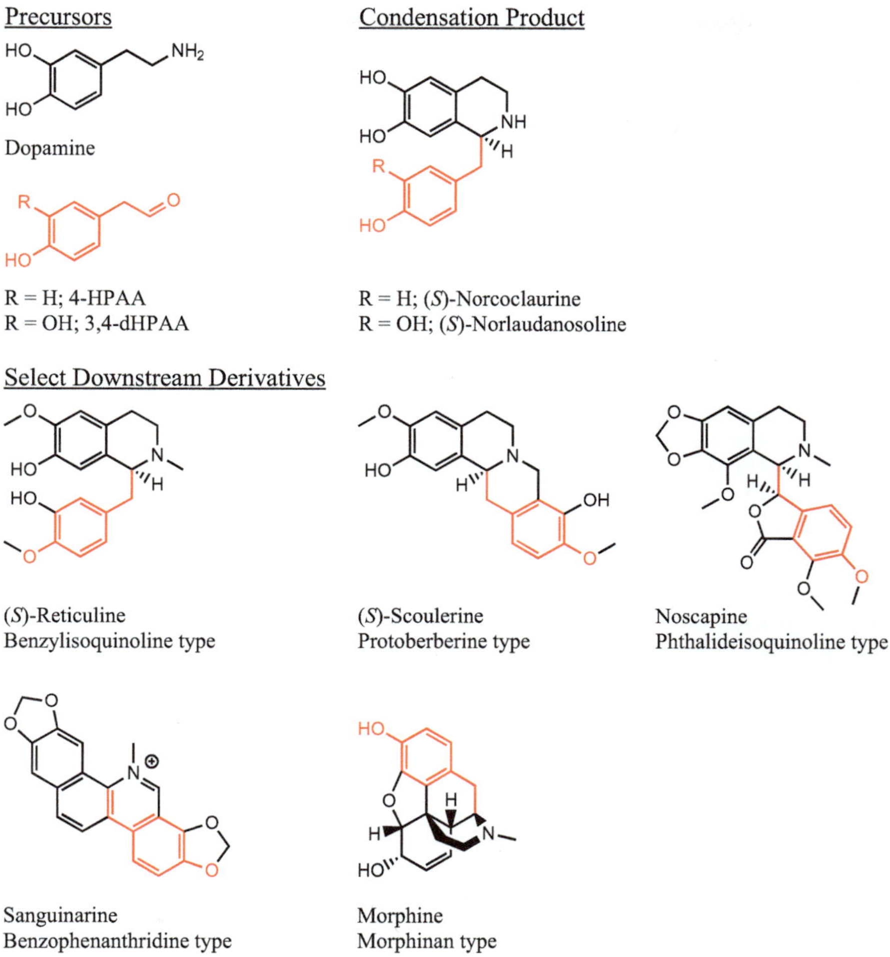

Fig. 5.1 (continued)

Microbial systems exploit the exceptional regio- and stereo-selectivity of enzymes and offer fast growth rates, well-developed genetic engineering technologies, and relatively simple product purification schemes. Although host systems based on both bacteria (*Escherichia coli*) and yeast (*Saccharomyces cerevisiae*) have been envisioned for MIA and BIA biosynthesis, assembling plant secondary metabolic pathways in prokaryotes is challenging, as bacteria lack organelles required to support functionality of complex plant enzymes. Instead, the yeast *S. cerevisiae* is often regarded as the favorable microbial host for producing high-value plant alkaloids. The most notable and abundant membrane-associated enzymes present in alkaloid biosynthesis are cytochromes P450 (CYPs), which catalyze a

Table 5.1 Production of key MIA and BIA precursors and upstream intermediates in microbial hosts

	Maximum titer (g/L)		
Metabolite	*E. coli*	*S. cerevisiae*	Key references
MIA biosynthesis			
Mevalonate	47 17.6 (Cell-free)[a]	41 (Downstream)[b]	[1, 23, 24]
Geraniol	1.119	0.036	[12, 25, 26]
BIA biosynthesis			
Tyrosine	55	0.0847 (Precursor)[c] 0.0898 (Downstream)[d]	[18, 19, 27, 28]
L-DOPA	8.67	0.00364	[20, 21]
Dopamine	2.15	0.0238	[21, 29, 30]

[a]Cell-free lysates derived from *E. coli* hosts were employed
[b]Titer reported for the downstream product amorphadiene
[c]Titer reported for the upstream precursor 4-hydroxyphenylpyruvate (4-HPP)
[d]Titer reported for the downstream metabolite 4-hydroxybenzoic acid

range of complex biochemical transformations, particularly in downstream MIA and BIA pathways (Sect. 5.4.1). CYPs require pairing with a suitable membrane-bound cytochrome P450 reductase (CPR) to mediate electron shuttling from NADPH. While recent successes in the synthesis of downstream alkaloids have been reported using *E. coli* (2.1 mg/L of thebaine) [10], pathways abundant in CYPs, such as the sanguinarine and noscapine BIA branches, in addition to the key branch point MIA strictosidine (0.5 mg/L), have only been successfully reconstructed in *S. cerevisiae* [5, 11–16]. On the other hand, current titers of BIA precursors and upstream intermediates, namely tyrosine [17–19], L-DOPA [20], and dopamine [21, 22], are substantially higher in bacterial systems (55, 8.67, and 2.15 g/L, respectively) compared to yeast (roughly 0.09, 0.00364, and 0.0238 g/L, respectively) (Table 5.1). While both yeast and *E. coli* have been engineered for production of the MIA precursor mevalonate with comparable titers (41–47 g/L) [1, 23, 24], production of the intermediate geraniol appears to be favorable in *E. coli* (1.119 g/L compared to 0.036 g/L) [12, 25, 26]. Advances in the application of cell-free systems have also generated impressive mevalonate titers (17.6 g/L) that could be exploited for MIA biosynthesis [31]. Given the infancy of microbial alkaloid production, host systems based on both bacteria and yeast warrant further investigation.

De novo synthesis of alkaloids in microbial hosts involves connecting plant secondary pathways with host central metabolism. However, several enzymes involved in MIA and BIA synthesis have only been recently identified due to advances in sequencing technologies and gene silencing techniques. Together these efforts form the basis of microbial synthesis of plant natural products. For example, the final missing steps of the *C. roseus* pathway leading to production of the key branching MIA strictosidine were only unveiled in 2014 [32–35]. More recently, the elusive

Table 5.2 *De novo* synthesis of downstream MIA and BIA products in microbial hosts

Metabolite	Titer (mg/L) *E. coli*	*S. cerevisiae*	Key references
MIA biosynthesis			
Strictosidine	–	0.5	[11]
BIA biosynthesis			
(*S*)-reticuline	40.5	0.082 0.0806	[21, 22, 38]
Thebaine	2.1[a]	0.0064	[10, 38]
Hydrocodone	0.36[a]	0.0003	[10, 38]

[a]Stepwise fermentation involving four engineered *E. coli* strains

enzyme involved in the stereochemical inversion of (*S*)- to (*R*)-reticuline in the morphinan node of the BIA pathway was identified [36, 37]. *De novo* production of strictosidine in yeast (0.5 mg/L) [11] and thebaine in both *E. coli* (2.1 mg/L) [10] and yeast (0.0064 mg/L) [38] soon followed (Table 5.2), thus solidifying microbial synthesis as a potential source of high-value plant alkaloids. Prior to these monumental studies, *de novo* production of plant alkaloids was hindered by gaps in our understanding of MIA and BIA biosynthetic pathways. To overcome these obstacles, early efforts focused on supplementation approaches, in which key MIA and BIA pathway intermediates were fed to cells expressing partial alkaloid pathways. While total *de novo* production of the BIAs sanguinarine and noscapine has yet to be demonstrated, the corresponding pathway branches have been elucidated very recently and promising partial pathway reconstructions have been achieved [13, 15, 16, 39]. Given the considerable effort required to engineer strictosidine biosynthesis in yeast [11, 12], it is unsurprising that presently no downstream MIAs have been produced in a microbial host.

In this chapter we describe recent successes in the microbial synthesis of MIAs and BIAs (Sects. 5.2 and 5.3, respectively), highlight underlying challenges gleaned from these studies (Sect. 5.4), and discuss integrated engineering strategies for optimizing microbial production (Sect. 5.5). While present titers are still several orders of magnitude below levels required for large-scale production, it is evident that microbial systems possess exceptional potential for synthesizing natural products, as well as manufacturing novel structures and activities. Indeed, the number of proof-of-concept syntheses reported in 2015 and 2016 [10, 11, 21, 38] lays a promising foundation for the development of industrial bioprocesses based on microbial production of plant metabolites. We anticipate that this chapter will serve as an informative and comprehensive introduction to the MIA and BIA metabolite classes and provide unique insights into the production of plant natural products using microbial species. Readers are directed to recent reviews covering a range of detailed topics spanning MIA and BIA biosynthesis, regulation, and gene discovery [2, 9, 40–43], as well as engineering strategies for microbial production of these promising compounds [5, 44–47].

5.2 Reconstitution of MIA Biosynthetic Pathways

Biosynthesis of MIAs is extremely complex and spans 37–40 enzymatic steps, in which mevalonate-derived secologanin and the tryptophan derivative tryptamine are condensed to yield strictosidine, the MIA structural scaffold. Strictosidine is subsequently derivatized to generate an array of dimeric bis-indole products, such as vinblastine and vincristine (Fig. 5.1a). Compared to BIAs, downstream MIA pathways are poorly characterized, as many biochemical steps remain to be elucidated. In this section we provide an overview of MIA biosynthesis and survey microbial engineering strategies based on both *S. cerevisiae* and *E. coli*. For simplicity, we partition the MIA biosynthetic pathway into three sections consisting of precursor pathways, strictosidine formation, and downstream derivatization (Fig. 5.2).

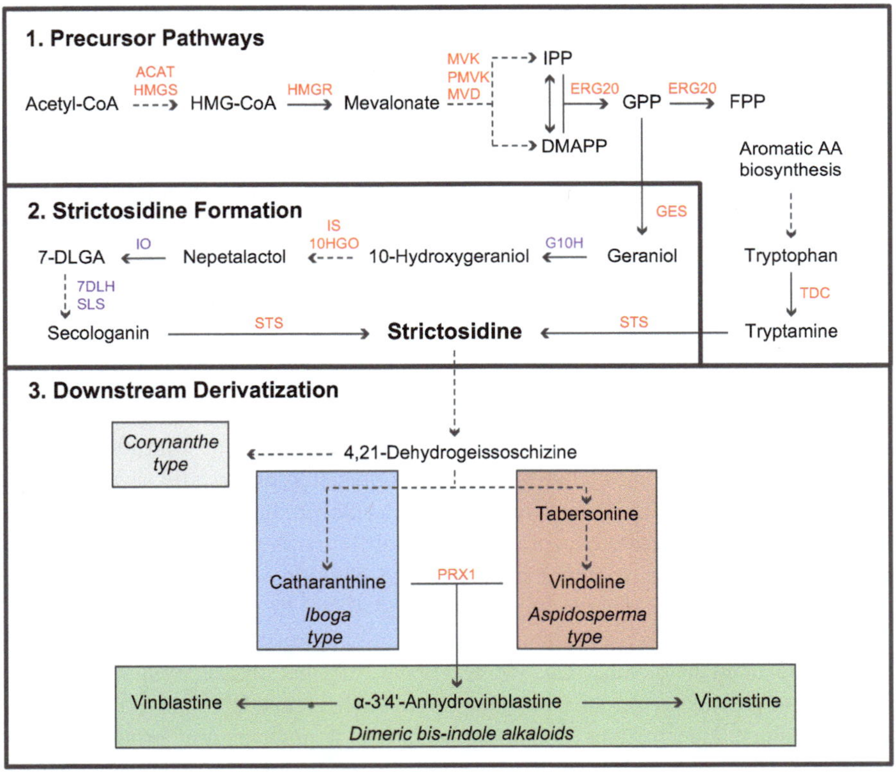

Fig. 5.2 Overview of major MIA biosynthetic pathways. The MIA network is divided into three portions consisting of precursor pathways, strictosidine formation, and downstream derivatization. Solid and dashed arrows represent single and multi-enzyme conversions, respectively. Major MIAs corresponding to corynanthe (grey), iboga (blue), aspidosperma (red), and bis-indole (green) types are boxed. Transformations catalyzed by cytochrome P450 enzymes (CYPs) are depicted in violet font. All other enzymes are shown in red font. Refer to text for abbreviations

5.2.1 Precursor Pathways

The mevalonate pathway is responsible for initiating synthesis of the secologanin precursor that, when condensed with tryptamine, constitutes the structural backbone of all MIAs [48]. Whereas tryptamine is derived from tryptophan by tryptophan decarboxylase (TDC), secologanin is produced in a complex series of transformations that begins with acetyl-CoA (Fig. 5.2). Three molecules of acetyl CoA are first condensed by acetoacetyl-CoA thiolase (ACAT) and 3-hydroxy-3-methylglutaryl CoA (HMG-CoA) synthase (HMGS) to yield the intermediate HMG-CoA, which is subsequently reduced to mevalonate by the rate-controlling enzyme HMG-CoA reductase (HMGR) in the first committed step of the pathway [49, 50]. HMGR ameliorates the toxicity caused by HMG-CoA [51] and is regulated through a multivalent feedback mechanism mediated by downstream sterol and non-sterol metabolites [52, 53]. Mevalonate is subsequently converted to mevalonate-5-phosphate by mevalonate kinase (MVK), which is feedback regulated by the downstream isoprenoids geranyl pyrophosphate (GPP) and farnesyl pyrophosphate (FPP) [54]. Mevalonate 5-phosphate is further phosphorylated and decarboxylated by phosphomevalonate kinase (PMVK) and mevalonate pyrophosphate decarboxylase (MVD), respectively, to produce the universal isoprenoid building blocks isopentenyl pyrophosphate (IPP) and dimethylallyl pyrophosphate (DMAPP) [52, 55]. IPP and DMAPP exemplify the final products of the mevalonate pathway and undergo condensation to form the base monoterpene GPP by the action of farnesyl diphosphate synthase (Erg20p). Erg20p also catalyzes an additional condensation of DMAPP with GPP to form FPP, representing the precursor to many important sterols and forming the basis of sesquiterpene biosynthesis. This mevalonate-dependent pathway prevails in higher eukaryotes (including plants) and archaea, while an alternative mevalonate-independent route known as the deoxyxylulose 5-phosphate (DXP) pathway is employed by most bacteria, as well as plants. This pathway generates the essential IPP and DMAPP isoprenoid precursors through condensation of pyruvate and glyceraldehyde 3-phosphate [56, 57]. Detailed knowledge of the biochemistry of both the mevalonate and DXP pathways has enabled unique opportunities for optimizing isoprenoid production in a number of organisms [58, 59].

The widespread use of isoprenoids in pharmaceuticals, fragrances, and flavorings has stimulated extensive metabolic engineering strategies to enhance production titers [11, 12, 57, 60]. The prevailing strategy involves reinforcing the mevalonate or DXP pathway to increase the pool of IPP and DMAPP precursors. In this regard, Martin et al. [57] engineered *E. coli* to express the heterologous *S. cerevisiae* mevalonate pathway, in addition to its native DXP pathway, which resulted in titers of over 100 mg/L of the sesquiterpene olefin amorphadiene. Expression of the *S. cerevisiae* pathway in *E. coli* circumvents host control mechanisms that would hinder isoprenoid production using the native DXP route [57]. Similar approaches were employed in *S. cerevisiae* by overexpressing eight genes of the mevalonate pathway, including two additional copies of a truncated feedback-resistant form of HMGR [61], resulting in the production of over

1.2 g/L of amorphadiene. Recent literature has described additional methods of optimizing monoterpene biosynthesis in *S. cerevisiae*, such as mutagenesis of the *ERG20* gene to reduce off-target activity [62, 63]. A K197E amino acid substitution was found to increase the available pool of GPP by minimizing its conversion to FPP [63]. These strategies provide rationale for producing highly complex MIAs in microbial hosts by providing an abundant supply of monoterpene precursors. Indeed, coupling expression of appropriate plant terpene synthases with high level monoterpene formation has enabled the production of various non-native products in *S. cerevisiae*, including limonene (0.61 mg/L) [64], pinene (32 mg/L) [65], citronellol (1180 mg/L) [66], and geraniol (36.04 mg/L) [25].

5.2.2 Strictosidine Formation

The seco-iridoid strictosidine has been identified as the core intermediate in the derivatization of all MIAs across various plant families [48]. Strictosidine biosynthesis is a nine-step process beginning with conversion of GPP to geraniol by geraniol synthase (GES) [11] (Fig. 5.2). The hydroxylation of geraniol to 10-hydroxygeraniol (10HG) is the first committed step of MIA biosynthesis, and is catalyzed by geraniol 10-hydroxylase (G10H), one of four CYPs present in the pathway [48]. 10HG is converted to nepetalactol via reduction by 10HG oxidoreductase (10HGO) and subsequent cyclization by iridoid synthase (IS), an NADPH-dependent reductive cyclase. A complex series of reactions involving three CYPs [iridoid oxidase (IO; CYP76A26), 7-deoxyloganic acid hydroxylase (7DLH; CYP72A224), and secologanin synthase (SLS; CYP72A1)] converts nepetalactol to secologanin, via oxidative ring cleavage. It has been suggested that conversion of loganin to secologanin by SLS is the rate limiting step in strictosidine formation [48]. In the final step of this pathway, secologanin is condensed with tryptamine by strictosidine synthase (STS), resulting in the synthesis of strictosidine [11, 67]. Elucidation of the enzymes and corresponding genes involved in strictosidine biosynthesis by *C. roseus* [33] has unlocked the possibility of producing this complex molecule in microbial hosts. Following genomic integration of all required biosynthetic genes from *C. roseus*, Brown et al. [11] succeeded in engineering *S. cerevisiae* to produce strictosidine. Trace amounts of the compound were generated following resolution of a major bottleneck in geraniol hydroxylation due to poor activity of the G10H enzyme. Enhancing gene dosage through integration of four additional copies of the G10H gene led to an increase in strictosidine production from 0.03 to 0.5 mg/L [11]. Campbell et al. [12] subsequently demonstrated that G10H and IS exhibit promiscuous activities that limit flux through the target pathway. While tremendous efforts will be required to overcome such issues as enzyme promiscuity (Sect. 5.4.2) and low titers (Sect. 5.5), the development of a strictosidine-producing yeast paves the way for microbial synthesis of even higher complexity downstream MIAs.

5.2.3 Downstream Derivatization

Based on alkaloids naturally produced by *C. roseus*, downstream MIAs are divided into four major subtypes: the corynanthe, iboga, and aspidosperma types, as well as the dimeric bis-indole alkaloids [48, 68, 69] (Fig. 5.1a). The first three types are classified based on the structural orientation of the monoterpene group, while the dimeric bis-indole alkaloids are more loosely defined as dimers of strictosidine derivatives. There are roughly 3000 examples of such molecules, all of which are based on the strictosidine backbone [11]. The first downstream transformation involves deglycosylation of strictosidine by strictosidine glucosidase [70], which forms an unstable hemi-acetal that triggers a series of spontaneous reactions and yields 4,21-dehydrogeissoschizine [71]. 4,21-Dehydrogeissoschizine exists in equilibrium with cathenamine [72], a precursor to the corynanthe alkaloids, including ajmalicine, serpentine, and tetrahydroalsonine. Alternatively, 4,21-dehydrogeissoschizine also feeds into the aspidosperma or iboga type MIA branches through the intermediate preakuammicine. Although several mechanisms have been proposed to clarify the conversion of 4,21-dehydrogeissoschizine to preakuammicine, the precise mechanism remains unknown [73]. Preakuammicine is then reduced to stemmadenine, which acts as a precursor for iboga- and aspidosperma-type alkaloids when supplemented to *C. roseus* suspension cultures [74]. Seven steps catalyze the formation of vindoline from tabersonine, embodying one of the only downstream pathways that has been reconstructed in microbes. Supplementation of tabersonine to yeast expressing the reconstituted pathway produced vindoline with a yield of 7.5% [75]. The class III peroxidase Prx1 has been demonstrated to couple vindoline and catharanthine *in vitro*, producing an unstable iminium dimer that spontaneously rearranges to α-3′,4′-anhydrovinblastine [76]. Hydroxylation of the α-3′,4′-anhydrovinblastine double bond generates vinblastine, while formylation of the *N*-methyl group yields vincristine.

The production of downstream MIAs in microbes remains a significant challenge due to gaps in our understanding of MIA biosynthesis in native plant species. The corynanthe-type family of MIAs is presently the most accessible, as the reductases involved in the formation of tetrahydroalsonine and ajmalicine from cathenamine have been identified [77]. Theoretically, these enzymes could be expressed in yeast, given a base strain engineered for strictosidine biosynthesis has been constructed [11]. Unfortunately, synthesis of aspidosperma- and iboga-type alkaloids is not possible at the present time, as the enzymatic steps involved in conversion of 4,21-dehydrogeissoschizine to preakuammicine, as well as formation of catharanthine and tabersonine from stemmadenine, remain unknown. In this context, it is clear that discovery of elusive pathway enzymes is the chief factor impeding our capacity to produce valuable MIAs in heterologous hosts. It is anticipated that the recent success in engineering yeast for strictosidine biosynthesis will serve as a catalyst for the identification of new pathway genes, leading to new microbial engineering opportunities and contributing to a deeper understanding of MIA biosynthesis.

5.3 Reconstitution of BIA Biosynthetic Pathways

While enzyme discovery remains a bottleneck for the microbial synthesis of key MIAs, the BIA family has recently experienced an intense period of pathway elucidation and enzyme discovery. Although morphine was first isolated from opium poppy in 1806, the elucidation of its biosynthetic pathway was only completed in 2015 [36, 37]. Analogously, the chemical structure of noscapine was determined more than a century ago (Fig. 5.1b), yet the majority of the enzymatic steps from the branch point (*S*)-canadine were characterized very recently [78–83]. Nevertheless, formation of most BIA structures proceeds through the branch point intermediate (*S*)-reticuline, itself derived from (*S*)-norcoclaurine (Fig. 5.3). The protoberberine (*S*)-scoulerine and the aporphine (*S*)-corytuberine are produced from (*S*)-reticuline by the berberine bridge enzyme (BBE) and corytuberine synthase (CYP80G2), while the morphinan class of BIAs are produced from (*R*)-reticuline via stereochemical inversion of (*S*)-reticuline by a fusion enzyme composed of aldo-keto reductase and CYP domains. The protoberberine (*S*)-canadine is another intermediate common to the synthesis of both berberine and phthalideisoquinoline BIAs. Because many downstream pathways share common intermediates, BIAs are prime candidates for microbial synthesis whereby, an optimized host producing high titers of a common branch point metabolite serves as a platform for the synthesis of different classes of downstream compounds [84]. In this section we present an overview of such BIA formation pathways and summarize the recent flurry of activity directed at engineering microbial hosts for the synthesis of valuable BIA pharmaceuticals.

5.3.1 Upstream BIA Pathways

Prior to the elucidation of all biosynthetic steps involved in a specific pathway, heterologous BIA production strategies were limited to the assembly of partial pathways in microbial hosts. Such approaches rely on the availability of costly BIA pathway intermediates and contrast *de novo* production strategies, in which glucose acts as substrate for both cell growth and BIA biosynthesis. In this regard, BIA feeding schemes have been employed to reconstruct both mid- and down-stream pathways in yeast for the production of high-value BIAs, including (*R,S*)-reticuline, sanguinarine, (*S*)-canadine and noscapine from (*R,S*)-norlaudanosoline [13, 39], as well as an array of morphinan alkaloids from supplemented thebaine or (*R*)-reticuline [14, 85]. With the elucidation of entire metabolic networks in native BIA-producing plants (Fig. 5.3), *de novo* production was first demonstrated for the upstream precursor dopamine in yeast (23.8 mg/L) and subsequently extended to (*S*)-reticuline (0.0806 mg/L) [21]. *De novo* production strategies entail linking heterologous plant biosynthetic pathways with endogenous host metabolism. Since MIAs and BIAs are derived from tryptophan and tyrosine, respectively, the aromatic amino acid pathway acts as the key starting point for many plant alkaloid engineering approaches. The norcoclaurine structural backbone of BIAs is formed through

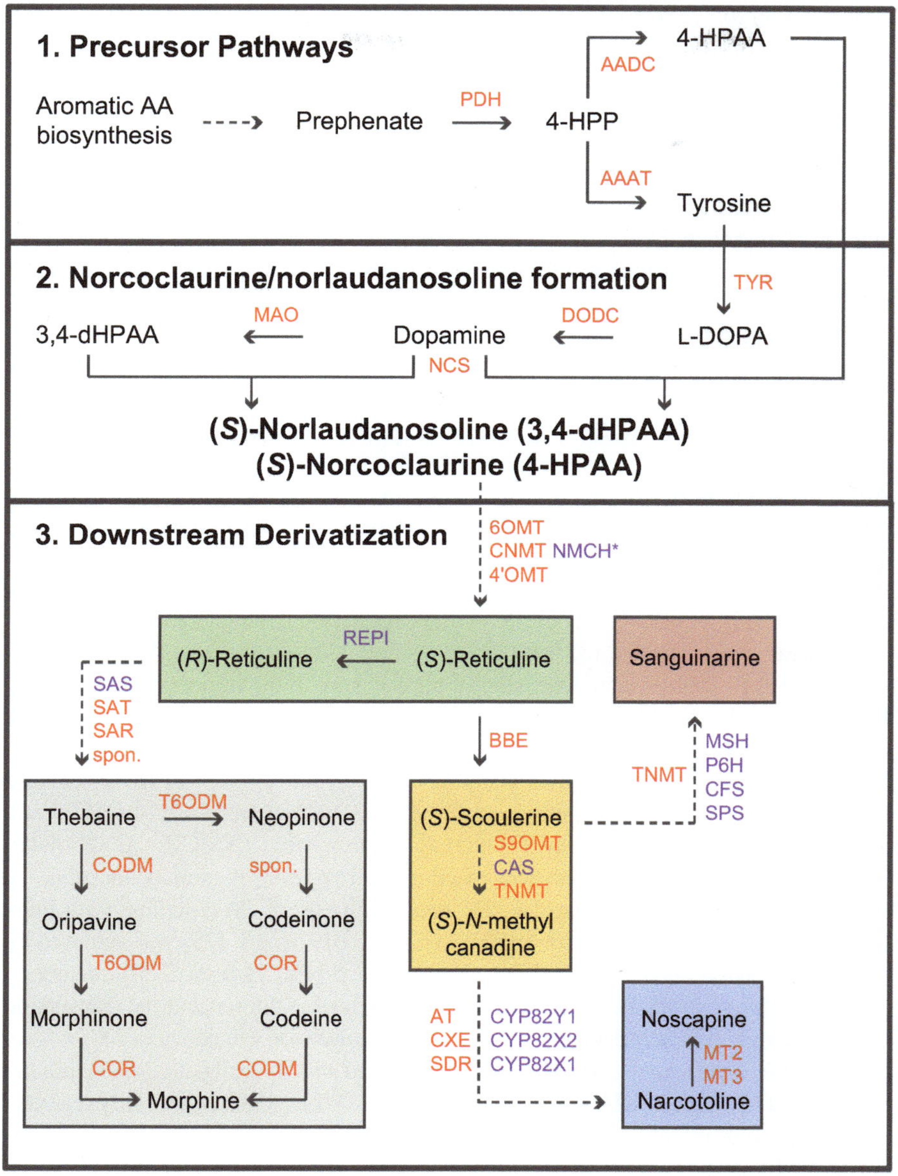

Fig. 5.3 Overview of major BIA biosynthetic pathways. The BIA network is divided into three portions consisting of precursor pathways, (*S*)-norcoclaurine/(*S*)-norlaudanosoline formation, and downstream derivatization. Solid and dashed arrows represent single and multi-enzyme conversions, respectively. BIAs corresponding to the benzylisoquinoline (green), protoberberine (orange), phthalideisoquinoline (blue), benzophenanthridine (red), and morphinan (grey) types are boxed. Transformations catalyzed by cytochrome P450 enzymes (CYPs) are depicted in violet font. All other enzymes are shown in red font. An asterisk denotes that NMCH is not required when norlaudanosoline is utilized as substrate. The AADC reaction is not utilized in *E. coli* BIA production schemes. Refer to text for abbreviations

condensation of dopamine and 4-hydroxyphenylacetaldehyde (4-HPAA), both derivatives of tyrosine. Alternatively, dopamine can be condensed with 3,4-dihydroxyphenylacetaldehyde (3,4-dHPAA), itself a derivative of dopamine through the action of monoamine oxidase (MAO). Although yeast naturally produces 4-HPAA, the organism lacks a source of dopamine, necessitating the introduction of a heterologous pathway to enable conversion of tyrosine to dopamine via L-DOPA. Using a novel enzyme biosensor to monitor pathway fluxes, DeLoache et al. [21] engineered yeast to produce 23.8 mg/L dopamine from glucose through the combined expression of an engineered tyrosine hydroxylase from sugar beet ($CYP76AD1^{W13L,F309L}$) and a bacterial DOPA decarboxylase (DODC). High-level dopamine production was also dependent on expression of a feedback resistant (FBR) mutant of Aro4p ($Aro4^{FBR}$) to enhance supply of the tyrosine precursor [86]. Condensation of the resulting dopamine pool with endogenous 4-HPAA using norcoclaurine synthase (NCS) from opium poppy yielded 0.1046 mg/L (*S*)-norcoclaurine, a conversion efficiency of only 0.25% based on dopamine levels. Analysis of the norcoclaurine-producing strain revealed the accumulation of (*S*)-norcoclaurine, as well as substantial levels of dopamine, in the extracellular medium [21]. Secretion of intermediates from non-optimized pathways is a common challenge of yeast BIA production [5] and, therefore, preventing efflux of BIA pathway intermediates is a critical facet of yeast BIA production strategies (Sect. 5.4.3).

Whereas production of (*S*)-norcoclaurine from glucose in yeast (0.1046 mg/L) requires further optimization, extension of the pathway to the important branch point BIA (*S*)-reticuline (0.0806 mg/L) has proven to be relatively efficient. (*S*)-Reticuline is produced from (*S*)-norcoclaurine in a four-step conversion involving three methyltransferases [norcoclaurine 6-*O*-methyltransferase (6OMT), coclaurine *N*-methyltransferase (CNMT), and 3′-hydroxy-*N*-methylcoclaurine 4′-*O*-methyltransferase (4′OMT)] and a CYP hydroxylase [*N*-methylcoclaurine hydroxylase (NMCH; CYP80B1)]. Alternatively, the NMCH-catalyzed reaction is bypassed if norlaudanosoline is employed in place of norcoclaurine. *De novo* synthesis of (*S*)-reticuline with titers reaching 0.082 mg/L [38], 0.0806 mg/L [21], and 0.0192 mg/L [29] was achieved by expressing methyltransferases from opium poppy and NMCH from California poppy (*Eschscholzia californica*). Based on measured levels of (*S*)-norcoclaurine, conversion to (*S*)-reticuline was surprisingly efficient (molar conversion efficiency of 63.5%), yet less than 3% of the total product was found associated with the cell pellet, again indicating substantial efflux of BIA pathway intermediates. While yeast is generally regarded as the preferred microbial host for the expression of complex downstream BIA pathways, the highest *de novo* (*S*)-reticuline production titers to date have been achieved using bacteria. Nakagawa et al. [22] engineered *E. coli* to produce 46 mg/L (*S*)-reticuline from glycerol by rewiring host tyrosine metabolism and engineering an efficient (*S*)-reticuline pathway composed of a diversity of variants from bacteria and plants. Regardless of the microbial host selected for BIA biosynthesis, *de novo* production of (*S*)-reticuline remains an impressive feat, as the pathway can be extended to yield an assortment of high-value BIA pharmaceuticals, including compounds from the morphinan, sanguinarine, and noscapine nodes.

5.3.2 Microbial Production of Morphinan Alkaloids

The reticuline epimerase (REPI) is responsible for the stereochemical inversion of (*S*)- to (*R*)-reticuline [36, 37]. (*R*)-Reticuline is then converted into thebaine in a linear four-step conversion involving salutaridine synthase (CYP719B1; SAS), salutaridine reductase (SAR), salutaridinol acetyltransferase (SAT), and a spontaneous rearrangement. Thebaine represents an important morphinan building block, acting as a precursor to morphine, codeine, and the semi-synthetic oxycodone. Two convergent pathways have been shown to mediate production of morphine. The prevalent pathway in opium poppy proceeds through neopinone and codeinone via the enzymes thebaine 6-*O*-demethylase (T6ODM) and codeinone reductase (COR) [87]. Codeine is then demethylated to morphine by codeine demethylase (CODM). The alternative route to morphine utilizes the same enzyme set and produces oripavine in a codeine-independent pathway. Like most plant MIA and BIA biosynthetic pathways, the morphinan node is punctuated with spontaneous chemical rearrangements and highly promiscuous enzymes (Sect. 5.4.2) [14]. In this context, downstream BIA metabolic networks resemble complex web-like architectures rather than simple linear pathways. For example, the enzyme COR catalyzes the reduction of codeinone to codeine, yet is also responsible for the formation of a number of off-target products, such as neopine and 14-hydroxycodeine [85]. Moreover, the product of the SAT reaction, salutaridinol-7-*O*-acetate, is spontaneously rearranged to thebaine, the preferred product, or dibenz[*d,f*]azonine, a side product. Culture pH was shown to be the chief driving force, whereby thebaine formation is favored at an elevated pH of 8–9 [14], a recurring theme in microbial BIA production approaches (Sect. 5.5.2) [5, 46]. For these reasons, microbial morphinan production is highly complex and difficult to harness, contributing to the currently low product titers and yields [10, 14, 38, 85].

Following identification of the REPI enzyme, *de novo* production of thebaine and hydrocodone was first demonstrated in yeast [38], followed by *E. coli* [10] (Table 5.2). While thebaine production titers are more than 300-fold higher in *E. coli* compared to *S. cerevisiae* (2.1 and 0.0064 mg/L, respectively), it is noteworthy that a stepwise fermentation approach involving four engineered strains of *E. coli* was utilized to attain titers in the mg/L range [10]. In contrast, a complete thebaine pathway consisting of 21 non-native enzymatic activities was introduced into a single yeast host [38]. Until most BIA production issues are resolved in the respective microbial hosts, present studies point to a two-stage approach, in which bacteria synthesize pathway precursors and early BIA intermediates, which are then derivatized in complex downstream reactions by engineered yeast. Examples of this approach are the synthesis of 8.3 mg/L (*S*)-scoulerine or 7.2 mg/L magnoflorine from 766 mg/L supplemented dopamine using co-cultures of *E. coli* expressing the dopamine-to-(*S*)-reticuline biosynthetic pathway and a *S. cerevisiae* strain expressing either BBE or CYP80G2 along with CNMT [88].

5.3.3 Microbial Production of Protoberberine, Benzophenanthridine, and Phthalideisoquinoline Alkaloids

Berberine (protoberberine type), sanguinarine (benzophenanthridine type), and noscapine (phthalideisoquinoline type) biosynthetic pathways all proceed through the common intermediate (*S*)-scoulerine, synthesized from (*S*)-reticuline by the enantioselective BBE. (*S*)-Canadine is the following branch point intermediate shared by protoberberine and phthalideisoquinoline alkaloids. In 2008, Minami et al. [88] reported the production of (*S*)-scoulerine from exogenously supplied dopamine using a co-culture of an *E. coli* strain producing (*S*)-reticuline and an *S. cerevisiae* strain expressing the BBE from *Coptis japonica*. In the same year Hawkins and Smolke [39] produced (*S*)-canadine from (*R,S*)-norlaudanosoline by expressing seven heterologous genes from a range of plant species. The pathway was further optimized to sustain production of the downstream alkaloids berberine and noscapine several years later [15, 89]. A total of 1.8, 0.621, and 0.700 mg/L (*S*)-canadine were obtained in shake flask, batch, and fed-batch conditions, respectively. Optimization consisted of tuning expression levels, screening enzyme orthologs involved in the conversion of (*S*)-scoulerine to (*S*)-canadine, and using buffered media to maintain neutral pH. Berberine was also detected as a spontaneous oxidation product from (*S*)-canadine. The tetrahydroberberine oxidase from *Berberis wilsoniae*, which was reported to oxidize (*S*)-canadine to berberine [90], didn't increase berberine production compared to the spontaneous reaction in *S. cerevisiae*.

5.3.3.1 Synthesis of Dihydrosanguinarine and Sanguinarine in Yeast

Sanguinarine is an orange-red benzophenanthridine alkaloid produced by certain members of the Papaveraceae family, such as bloodroot (*Sanguinaria canadensis*) and Mexican poppy (*Argemone mexicana*). Like most major BIA pathway nodes, the sanguinarine branch was fully elucidated following a series of recent studies [91–93]. Sanguinarine is produced from (*S*)-reticuline in a seven-step process that proceeds through the key intermediate (*S*)-scoulerine. This transformation involves four plant CYPs [cheilanthifoline synthase (CFS; CYP710A25), stylopine synthase (SPS; CYP719A20), (*S*)-*N*-methylstylopine 14-hydroxylase (MSH; CYP82N4), and protopine 6-hydroxylase (P6H; CYP82N2v2)], in addition to BBE and tetrahydroprotoberberine *N*-methyltransferase (TNMT). Following elucidation of the sanguinarine biosynthetic pathway, Fossati et al. [13] devised a feeding strategy to reconstitute the pathway from (*R,S*)-norlaudanosoline by partitioning the pathway into three enzyme blocks. Pathway flux through each module could then be monitored and quantified by feeding with different pathway intermediates, namely (*R,S*)-norlaudanosoline, (*S*)-reticuline, (*S*)-scoulerine, or (*S*)-stylopine. Whereas molar conversion for the full pathway from (*R,S*)-norlaudanosoline to dihydrosanguinarine was 1.5%, yields for enzyme blocks 2 + 3 and block 3 alone were 4.4% and 37%, respectively, demonstrating significant losses at each portion of the pathway. Promiscuity of TNMT and CYP719, as well as efflux of (*S*)-*N*-methylstylopine, significantly hindered efficient conversion. The former issue was comprehensively resolved in an ensuing effort [94] (Sect. 5.4.2), leading to a boost in molar conversion to 10% from fed (*R,S*)-norlaudanosoline without production of off-target products. Unsurprisingly, however, resolution of the previous bottleneck spawned a new downstream target for optimization based on accumulation and efflux of (*S*)-*N*-methylstylopine. The importance of the CFS reaction was further emphasized by a subsequent sanguinarine biosynthesis study, in which CFS variants were combinatorially expressed with various plant CPRs [16]. Although all tested CFS variants exhibited activity, the authors demonstrated that pairing with an optimal CPR was critical for achieving high flux through (*S*)-cheilanthifoline (Sect. 5.4.1). A final molar conversion of 0.012% was achieved from (*R,S*)-norlaudanosoline to sanguinarine, which contrasts the 10% yield achieved by the previous study through extensive optimization of the conversion of (*S*)-scoulerine to (*S*)-*N*-methylstylopine [94]. Despite these partial pathway reconstructions, *de novo* production of dihydrosanguinarine or sanguinarine has yet to be demonstrated.

5.3.3.2 Noscapine Synthesis in Yeast

In a manner similar to the morphine biosynthesis pathway, the noscapine branch has remained elusive until very recently despite the fact that its structure was characterized more than a century ago. In 2012 Winzer et al. [95]. identified a 10-gene cluster tightly linked to the formation of noscapine in select opium poppy variants. The cluster was found to encode three *O*-methyltransferases, four CYPs, an acetyltransferase, a carboxylesterase, and a short chain dehydrogenase/reductase (SDR). Detailed biochemical characterization of most of the enzymatic activities associated with the cluster was completed within the following 3 years [78–83]. Together these efforts led to an in-depth understanding of the noscapine biosynthetic pathway employed by opium poppy and paved the way to its reconstitution in yeast. As with other key BIA pathway elucidations, assembly of the noscapine pathway in *S. cerevisiae* was reported soon after its discovery [15]. Heterologous production of noscapine required expression of 16 plant enzymes together catalyzing 14 biosynthetic transformations from fed (*R,S*)-norlaudanosoline leading to the production of approximately 0.678 mg/L noscapine. The authors were forced to tweak the overall noscapine biosynthetic schema, resulting in characterization of the last missing catalytic step: *O*-methylation of the 4′ hydroxyl group of narcotoline to generate noscapine. The reaction was found to be catalyzed by a heterodimer composed of two previously uncharacterized *O*-methyltransferases, MT2 and MT3, expressed from the noscapine cluster in *P. somniferum*. Due to the complexity of noscapine biosynthesis, the pathway was found to exhibit all major challenges associated with the production of alkaloids in yeast. For example, the pathway expresses four complex plant CYPs (Sect. 5.4.1), several promiscuous enzymatic activities were detected (Sect. 5.4.2), almost all pathway intermediates were exported from the cell (Sect. 5.4.3), and molar conversion was only 0.082% from a supplemented substrate, as conversion was too low to support *de novo* biosynthesis [15]. Hence, significant improvements will be required to link the existing noscapine pathway to yeast central metabolism.

5.4 Microbial Engineering Challenges

Building highly complex plant secondary metabolic pathways in microbes has proven to be an immensely arduous challenge. A number of recurring metabolic engineering challenges undermine attempts to reconstruct both MIA and BIA pathways in *E. coli* and *S. cerevisiae*. In this section we focus on the most pertinent engineering hurdles: functionally expressing complex plant CYPs, taming enzyme promiscuity, and limiting efflux of intermediates. We briefly outline specific approaches that have proven successful at overcoming such challenges in the context of microbial synthesis of MIAs and BIAs. In Sect. 5.5, we highlight broader genetic and bioprocessing strategies to enhance flux through target alkaloid pathways.

5.4.1 Functional Expression of Cytochromes P450 (CYPs)

CYPs are abundant in plant specialized metabolism and contribute to the vast structural diversity of MIAs and BIAs by adding hydroxyl functional groups and catalyzing oxidation, methylene-dioxy bridge formation, and C-O and C-C phenol coupling reactions [2, 96]. CYPs require a source of electrons for catalysis, which is typically provided by CPRs following oxidation of NADPH [97]. Plant CYPs and CPRs localize to the endoplasmic reticulum (ER), where N-terminal membrane-binding domains anchor them. The presence of endomembrane structures affords a significant advantage to eukaryotes as hosts for alkaloid production. Indeed, most efforts to assemble CYP-containing pathways have been carried out in yeast [5]. The strictosidine, morphinan, noscapine, and sanguinarine branches are examples of complex multi-CYP pathways [11, 13–16, 38]. Surprisingly, *E. coli* has also proven to be a suitable host for *de novo* production of thebaine, albeit through partitioning of the pathway into four engineered strains [10]. The authors employed a truncated SAS (CYP719B1) and co-expression of truncated CPR2 from *Arabidopsis thaliana* (ATR2) to generate salutaridine from (*R*)-reticuline. In contrast, functional expression of REPI in *E. coli* was unsuccessful, leading the authors to opt for an alternative route to (*R*)-reticuline [10].

As a result of their inherent complexity, CYP-catalyzed steps often produce a bottleneck in microbial alkaloid production. Overall, heterologous CYPs suffer from poor expression and activity, resulting in low titers of target compounds. Improvements in the functional expression of CYPs have been made, yet catalytic efficiency often remains suboptimal. No strategy alone seems to be superior, pointing to an integrated strategy for improving CYP activity. In this regard, high throughput assays greatly assist efforts to enhance CYP expression and activity, such as the biosensor screen employed to boost activity of a tyrosine hydroxylase (CYP76AD1) by a factor of 2.8 [21]. Unfortunately, such high throughput screens are rarely available to facilitate protein engineering and mutagenesis. An effective alternative entails engineering the N-terminal membrane anchor to increase CYP stability. Fossati et al. [13] swapped the N-terminus of SPS from *P. somniferum* with the membrane domain from lettuce germacrene A oxidase, thus enabling functional expression in *S. cerevisiae*. In another study, N-terminal engineering was used to improve SAS (CYP719B1) activity [38]. Western blotting detected glycosylation of SAS in yeast, resulting in misprocessing of the protein. Subsequent removal of the glycosylation sites by site-directed mutagenesis prevented glycosylation but reduced the activity of the enzyme. Instead, replacing the N-terminus with that from an unglycosylated CYP (CFS; CYP719A5) increased the activity of SAS. Other strategies to enhance expression and activity of CYPs include broadly applicable genetic strategies, such as promoter swapping and boosting gene copy number (Sect. 5.5.1.3).

Although *S. cerevisiae* expresses a native CPR, designated Ncp1p, co-expression of an ancillary plant CPR is usually necessary to assist activity of plant CYPs [98]. The choice of plant CPR varies amongst studies, where typically only one plant CPR is assessed, even in cases where a diversity of CYPs are co-expressed. Trenchard and Smolke [16] compared functional expression of three distinct CPRs, in addition to the native yeast CPR, to support activity of CFS from *E. californica*. Co-expression of either CPR1 from *A. thaliana* (ATR1) or CPR from *P. somniferum* resulted in similar production of (*S*)-cheilanthifoline, while co-expression of CPR from *E. californica* led to lower titers of (*S*)-cheilanthifoline comparable to those attained using the endogenous yeast CPR. Efficient electron transfer between CYP and CPR is paramount, as de-coupling leads to formation of toxic reactive oxygen species (ROS) [99]. ROS inhibit CYP activity and are deleterious to yeast growth [1, 99]. Collectively, these studies suggest that an optimal ratio of CYP:CPR exists to optimize electron shuttling and limit formation of ROS. While few alkaloid studies have investigated tuning of CYP expression [15, 16, 89], the optimal ratio of CYP:CPR has not been directly investigated. The cytochrome *b*5 (CYB5) family of proteins are also known to support electron transfer to some CYPs, in combination with or independent of a CPR partner [100, 101]. CYB5 has been used as redox support for heterologously expressed CYPs in the strictosidine pathway [11], although the direct contribution of co-expressing CYB5 was not investigated. However, high-level production of artemisinic acid in *S. cerevisiae* was found to require CYB5 together with CPR1 from *Artemisia annua* [1]. Finally, overexpression of CYPs induces proliferation of ER in *S. cerevisiae* [102], presumably to condition membrane capacity to high amounts of embedded membrane proteins. In a localization study, cells expressing a CFS-GFP fusion protein from a high-copy plasmid showed patches of concentrated fluorescence associated with reduced CFS activity compared to the same fusion expressed from a low copy plasmid, which displayed evenly distributed fluorescence [16]. Similar observations were made for canadine synthase when expression from a low-copy plasmid was compared to that from a high-copy plasmid [89].

5.4.2 Taming Enzyme Promiscuity

Enzyme promiscuity, defined here as an unusually broad substrate range for a given enzyme (thereby releasing multiple products), has proven to be an underlying theme and challenge of microbial production of MIAs and BIAs. While many metabolic networks employ distinct pathways to synthesize a common metabolite, the transport and accumulation of pathway intermediates in different plant tissues provides a means for regulating secondary metabolite production [103]. Removing these spatially-separated enzymes from their native context and combining them within the yeast cytosol often leads to unpredictable outcomes in the form of off-target reactions. While in some instances enzyme promiscuity can be harnessed to yield "new-to-nature" structures, most often undesirable side products result. In the former case, non-canonical enzyme activities tend to exhibit poor catalytic efficiencies, necessitating improvement through protein engineering if pathway flux is to be harnessed and diverted to non-native routes [5].

Within the MIA network, both G10H and IS have been shown to divert pathway flux to off-target routes. In addition to its reductive cyclase activity, IS was found to reduce both geraniol and 10GH without cyclization, yielding the side-products citronellol (200 mg/L) and 10-hydroxycitronellol (10HC) (5.6 mg/L), respectively [12]. To further convolute this pathway, G10H was shown to hydroxylate citronellol to 10HC. In this case, levels of the citronellol side-product dramatically outweighed that of the target 10HG metabolite, underscoring the complex pathway challenges that arise from enzyme promiscuity. Since flux through the citronellol pathway was found to be highly efficient, pathway diversion to reconnect off-target reactions to the trunk pathway is ostensibly the most promising troubleshooting strategy [5]. This approach is dependent on the ability to reductively cyclize 10HC, necessitating the identification or engineering of such an enzymatic activity. An example of this approach was reported by Narcross et al. [94], in which a side reaction of the sanguinarine pathway was diverted and reconnected to the target pathway. Promiscuous activities of CYP719s (CFS and SPS) and TNMT from the sanguinarine branch initially led to the accumulation of numerous side products, including (*S*)-nandinine, (*S*)-*N*-methylscoulerine, and (*S*)-*N*-methylcheilanthifoline, placing constraints on the formation of dihydrosanguinarine and sanguinarine. A comprehensive library of 15 methyltransferases and 54 CYP719s were screened to identify variant combinations leading to reduced off-target products. Using this strategy, novel CYP719s possessing activity on the side product (*S*)-nandinine were identified such that off-target flux could be efficiently diverted back to the trunk pathway through conversion to (*S*)-stylopine. Most importantly, several methyltransferase-CYP719 variant pairs were identified that completely abolished the formation of side products, effectively increasing the yield of dihydrosanguinarine. Other microbial MIA and BIA production strategies that have effectively decreased off-target activities include the identification of a tyrosine hydroxylase with reduced DOPA oxidase activity [21], mutagenesis of the ERG20 gene to diminish FPP synthase activity [12], and the use of bacterial variants in place of promiscuous plant enzymes for production of natural and semi-synthetic opioids [85]. Perhaps the most promiscuous class of enzymes involved in alkaloid biosynthesis are methyltransferases. Examples include the TNMT enzyme discussed above [94], MT2 and MT3 of the noscapine branch [15], and all three methyltransferases involved in formation of (*S*)-reticuline from norcoclaurine or norlaudanosoline [104–107]. In addition to pathway redesign and enzyme bioprospecting, enzyme promiscuity is overcome by tuning enzyme levels or engineering synthetic protein scaffolds to better balance metabolic flux, or through spatial or temporal separation of promiscuous enzymes from non-target substrates [5, 108]. To prevent formation of neomorphine in yeast, Thodey et al. [85] localized COR to the ER, resulting in decreased production of neomorphine and slightly higher titers of morphine. Spatial sequestration allowed for the spontaneous conversion of neopinone to codeinone to proceed before COR could divert available neopinone to neomorphine. Because *E. coli* lacks membrane structures for subcellular localization, Nakagawa et al. [30] sequestered promiscuous enzymes between strains rather than organelles by employing a stepwise fermentation approach.

5.4.3 Limiting Efflux of Intermediates

The passage of molecules across cell walls and phospholipid bilayers has become an increasingly common target for metabolic engineering. Evidence suggests that the majority of small molecule passage across membranes is facilitated by transporters [109]. Molecules carried by microbial transporters include endogenous metabolites as well as toxins and heterologous products [110]. Almost every step in the production of MIAs and BIAs from simple sugars is accompanied by secretion of intermediates: synthesis of aromatic amino acids [27, 111], precursors of the initial alkaloid [21, 38], as well as derivatization of the downstream alkaloids [13, 15, 112]. Overexpression of efflux transporters reduces the intracellular concentration of end-products, minimizing the effects of feedback-inhibition [113, 114] as well as product toxicity [115–117]. Final recovery of end-products from supernatant is also easier than recovery from cell pellets [118]. Thus, a great deal of effort has been dedicated to identifying and improving transporters capable of secreting molecules of interest [119–122]. However, when pathway intermediates are secreted, total yield is reduced [123]. This is especially true when a pathway is not well-balanced, leading to intermediate accumulation [124], such as current endeavors to build MIA and BIA pathways in yeast. In such cases, enhancing fluxes through target pathways should reduce efflux of intermediates. Finally, a few studies have unveiled a significant association between cultivation pH and both the uptake and efflux of BIA intermediates (Sect. 5.5.2).

5.5 Optimization Strategies

5.5.1 Genetic and Pathway Engineering Techniques

Reconstitution of MIA and BIA biosynthetic networks in microbes entails more than simply expressing plant enzymes in a suitable host, as troubleshooting and optimization are essential facets of metabolic engineering. This process demands a suite of genetic and metabolic engineering strategies involving the use of combinatorial enzyme libraries, diverse pathway assembly tools, and a genetic tuning toolkit [125]. In this section we provide a broad overview of such pathway engineering techniques to aid efforts directed at enhancing microbial production of MIA and BIA alkaloids.

5.5.1.1 Building Combinatorial Enzyme Libraries

One of the most effective methods to improve metabolic flux and troubleshoot pathway bottlenecks is to survey nature's vast biodiversity by screening enzyme variants for improved catalytic function [46]. Major advances in next generation sequencing technology have led to an explosion of sequencing data deposited in publically accessible online databases. The 1000 Plants Project (http://www.onekp.com) and PhytoMetaSyn (http://www.phytometasyn.ca) are plant transcriptome initiatives providing a valuable resource for the identification of enzymes involved in biosynthesis of plant natural products [84, 126]. These and other databases are queried, often using characterized enzymes from source plants, to identify orthologs possessing significant sequence similarity [46]. Building combinatorial enzyme libraries in this manner has become more accessible with the decreasing cost of DNA synthesis, effectively abolishing reliance on plant cDNAs [94]. Libraries of synthetic coding sequences are codon optimized and cloned into custom expression vectors to optimize translation efficiency and gene expression within a target microbial host. Comparison of three codon optimized orthologs each of CFS and SPS allowed improvement of (*S*)-stylopine production within the sanguinarine pathway [16]. The same approach was used to identify the best available CAS variant for production of (*S*)-canadine [89]. Unlike other pathway engineering strategies, such as protein engineering and directed evolution, screening using combinatorial libraries often provides a more efficient and comprehensive approach for pathway optimization. Rather than improving single enzymatic conversions, combinatorial libraries can reveal enzyme synergies through parallel screening at multiple enzymatic steps. As outlined previously (Sect. 5.4.2), Narcross et al. [94] exploited this approach by sampling the sequence diversity of transcriptome databases to combinatorially screen a set of 15 methyltransferase and 54 CYP719 candidates for optimized production of (*S*)-stylopine from (*S*)-scoulerine.

5.5.1.2 Pathway Assembly

The reconstitution of alkaloid metabolic pathways can be simplified by dividing large pathways into smaller manageable portions through grouping of enzymes into blocks and assembling corresponding genes into multigene expression cassettes [13]. This strategy reduces the number of manipulations required to assemble full metabolic pathways in a target host. Enzyme blocks can be screened in parallel by feeding supplemented pathway intermediates and can also be further improved prior to combining into a single strain. To assess activity of candidate genes, gene cassettes are often first assembled and expressed from replicating plasmids because they are easily manipulated and efficiently co-transformed. It is important to consider the copy number of plasmids selected for gene expression, as expression profiles are typically dependent on the origin of replication employed. High copy plasmids based on the 2μ replicon (>50 copies per cell) and low copy plasmids based on an Autonomously Replicating Sequence (ARS; 1–2 copies per cell) exhibit vastly different effects on gene expression. Despite a reduction in gene dosage, low copy plasmids have the advantage of minimizing metabolic stress inflicted on the host cell. Still, the use of plasmids presents several challenges. For example, low copy plasmids show varying levels of gene expression between cells in the same population, leading to inconsistencies in pathway productivity and culture performance [127]. Because of these challenges and following advancements in the use of chromosome-encoded expression systems, it is now generally preferred to integrate pathways directly into host chromosome(s), thus ensuring consistent cell-to-cell gene expression and long-term stability of biosynthetic pathways without the need for antibiotics or auxotrophies for plasmid maintenance. Multiple promoters, genes, and terminators can be shuffled, assembled, and integrated into any target genomic locus by harnessing the efficient homologous recombination capability of yeast [128]. Homologous recombination in *S. cerevisiae* affords an additional benefit to the use of this organism as a host for the heterologous synthesis of plant natural products, as even phage-assisted recombination in *E. coli* is less efficient than the native yeast system [129, 130]. Until recently, targeted chromosomal integration in yeast was only possible through homologous recombination with co-integration of a selectable marker. The use of chromosomal expression platforms have become widespread following the discovery and exploitation of CRISPR-Cas systems for genome editing [131]. When coupled with yeast homologous recombination, the CRISPR-Cas9 system enables highly efficient marker-free gene integration at virtually any chromosomal locus [132]. This technology has the capacity to greatly increase the efficiency of pathway assembly and optimization by allowing multiplexed integration of very large DNA fragments comprised of multi-gene pathways [133–135].

5.5.1.3 Tuning Gene Expression

The most widespread approaches for tuning gene expression for pathway optimization involve swapping genetic regulatory elements and modulating gene copy number. Extensive yeast promoter and transcriptional terminator libraries, including native, synthetic, and hybrid elements, have been characterized in yeast for applications in metabolic engineering [136–139]. Despite these efforts, typically only a few select regulatory elements are routinely employed for yeast metabolic engineering, including the *TEF1*, *TDH3*, *TPI1*, and *PGK1* promoters and the *PGI1*, *ADH1*, *CYC1*, and *TGI1* terminators. Tuning gene expression by swapping regulatory elements has been shown to lead to a six-fold difference in enzyme activity between promoters ($N = 14$) [139] and a 70-fold difference between terminators ($N = 5302$) [140]. Another approach to optimize enzyme activity and alleviate pathway bottlenecks is to modulate gene copy number. Pathway assembly begins by first integrating a single copy of each pathway gene into the chromosome and subsequently probing pathway flux for inefficient enzymatic conversions. Catalytic efficiency of enzyme bottlenecks is subsequently improved through iterative rounds of gene integration until sufficient chemical conversion is attained or no further increase in product yield is observed. For example, a total of four copies of the gene encoding G10H were required to partially alleviate a key bottleneck in the formation of strictosidine by *S. cerevisiae* [11].

5.5.2 Cultivation Methods

The importance of bioprocess engineering and strain cultivation cannot be overstated, as substantial improvements in titer and productivity are achieved by optimizing growth conditions. The decade-long quest to engineer artemisinin production in microbes serves as a benchmark for all subsequent microbial synthesis efforts. Interestingly, strain engineering techniques, such as deletion of native *S. cerevisiae* genes and overexpression of a codon-optimized mevalonate pathway from strong inducible promoters, led to only modest production of amorphadiene (2 g/L) [11]. Through extensive engineering strategies, including phosphate limitation, the use of ethanol as carbon source, and elimination of catabolite repression of galactose, the amorphadiene titer was elevated to more than 40 g/L [24]. Nonetheless, the synthesis of the target downstream product artemisinic acid was hardly improved relative to the preexisting strain despite tremendous increases in the formation of amorphadiene [24], again highlighting the challenges of producing highly functionalized compounds using complex CYP enzymes. While no studies have extensively investigated bioprocessing techniques to optimize microbial alkaloid synthesis, here we briefly recap strain cultivation conditions that have been employed to date in the context of MIA and BIA biosynthesis.

Several studies suggest that an intimate relationship exists between pH, alkaloid yield, and efflux of pathway intermediates [14, 89, 141]. An inherent incompatibility was identified, wherein the ideal pH for cultivation of *E. coli* and *S. cerevisiae* (pH 5–6) contrasts the alkaline conditions required to retain pathway intermediates and support activity of plant enzymes (pH 7.5–9) [89]. To further complicate these findings, however, alkaline conditions favor the non-productive oxidation of several upstream intermediates, particularly L-DOPA, dopamine, and norlaudanosoline [5]. Kim et al. [142] employed a pH of 6 to balance *E. coli* cell growth and synthesis of (*S*)-reticuline from dopamine. In the case of yeast hosts, most early BIA feeding strategies utilized a two-stage approach, wherein cells are grown to a target OD under slightly acidic conditions, which was followed by a BIA production stage buffered at pH 8 [14, 16]. Such a strategy is not as straightforward in *de novo* systems, as BIA production and cell growth are expected to proceed concomitantly. In this regard, pathway balancing is paramount to ensure intermediates are transformed enzymatically before spontaneous oxidation or efflux can occur. Nevertheless, the impact of pH on microbial alkaloid synthesis across different hosts and pathways demands careful investigation and optimization of this critical cultivation parameter. Similarly, selection of the growth substrate affects several facets of a bioprocess, including precursor supply, redox considerations, and production costs. As mentioned above, the switch from glucose to ethanol and subsequent optimization of the feeding regimen led to an astounding 20-fold increase in amorphadiene titer from 2 to 41 g/L [24]. Since amorphadiene and strictosidine are both derived from acetyl-CoA and the mevalonate pathway, such a strategy could be adopted for enhancing MIA production. In the case of bacterial systems, *de novo* BIA production has been shown to be more effective using glycerol rather than glucose as growth substrate [10], which is likely related to the effectiveness of this carbon source for producing compounds derived from the shikimate pathway [143]. Curiously, the sugar galactose appears to boost yeast production of (*S*)-stylopine, (*S*)-canadine, and berberine from fed (*R,S*)-norlaudanosoline by a factor of 2-to-3 compared to glucose [16, 89]. The authors did not provide a hypothesis for the increase in alkaloid yield using galactose, although a global effect on yeast metabolism was acknowledged. From these studies, it is clear that much remains to be known regarding the effect of carbon source on both *de novo* and supplemented alkaloid production schemes. Lastly, the temperature employed to cultivate engineered strains for alkaloid synthesis can have a minor effect on production of a few select BIAs, such as (*S*)-stylopine, which was elevated more than 3-fold by cultivating *S. cerevisiae* at a reduced temperature of 25 °C [16]. While this strategy was not effective for enhancing synthesis of (*S*)-canadine or berberine [89], it has been demonstrated that the activity and folding of CYPs in *S. cerevisiae* is improved through cultivation at reduced temperatures [16, 144]. Since CFS and SPS are the only enzymes unique to the (*S*)-stylopine formation branch, it is apparent that one or both of these CYPs is at least partially unstable at typical cultivation temperatures of 30 °C. This finding again highlights the complexities associated with functionally expressing plant CYPs in microbial hosts such as *S. cerevisiae* (Sect. 5.4.1).

5.6 Conclusions and Future Directions

With the establishment of microbial synthesis as a promising platform for plant alkaloid production, considerable effort will be required to move this technology beyond the proof of concept stage. Elevating current MIA and BIA titers to levels warranting scale up is the most critical and challenging endeavor. Future efforts will be wise to draw from prior successes of microbial synthesis, such as the production of amorphadiene, which was initiated at 24 μg/ml in *E. coli* [57] and meticulously elevated to 41 g/L in *S. cerevisiae* over a 10-year span [24]. In this context, present yeast titers of (*S*)-reticuline mirror early outputs of amorphadiene, forecasting that the coming years will witness tremendous increases in the production of key MIA and BIA intermediates. Enhanced alkaloid pathway fluxes will also facilitate assembly of longer and more complex *de novo* pathways, such as the production of noscapine, which currently relies on supplementation of (*R,S*)-norlaudanosoline due to poor overall conversion efficiency [15]. The second area of focus encompasses expanding alkaloid pathways to produce non-native scaffolds and tailor-made pharmacological activities [141]. Harnessing enzyme promiscuity, as detailed in this chapter, will play a key role in exploiting microbial systems for drug discovery and engineering "new-to-nature" structures. The third research direction involves elucidating the highly complex downstream MIA pathways and corresponding enzymes from *C. roseus* that have thus far eluded discovery. While MIA pathway elucidation currently lags behind that of BIAs, it is noteworthy that yeast production of the key branching MIA strictosidine (0.5 mg/L) is roughly 6-fold higher than present levels of the corresponding BIA intermediate, (*S*)-reticuline (0.082 mg/L) (Table 5.2). We anticipate that this discrepancy will act as a catalyst for the unveiling of downstream MIA pathways and enable *de novo* production of highly derivatized MIAs.

Finally, it must be noted that several members of the morphinan branch of BIAs are controlled substances in many countries. While these compounds embody some of the most important medicines in the world, they are also some of the most abused and addictive substances known to mankind. For example, the naturally occurring morphinans thebaine, morphine, and codeine can be chemically converted into more potent semisynthetic opioids, including oxycodone, hydrocodone, and heroin. Since *de novo* production of thebaine and hydrocodone has been validated in two highly tractable microbes, albeit at low levels (Table 5.2), the threat of "home brewed opioids" has become a reality. Therefore, ethical considerations and policies surrounding the construction of morphinan-producing microorganisms, including the genetic safeguarding of such strains against release or misuse, must be firmly established prior to the deployment of large-scale processes based on microbial synthesis of opioids.

References

1. Paddon CJ, Westfall PJ, Pitera D, Benjamin K, Fisher K, McPhee D, et al. High-level semi-synthetic production of the potent antimalarial artemisinin. Nature. 2013;496(7446):528–32.
2. Hagel JM, Facchini PJ. Benzylisoquinoline alkaloid metabolism–a century of discovery and a brave new world. Plant Cell Physiol. 2013;54(5):647–72.
3. Kutchan TM. Alkaloid biosynthesis [mdash] the basis for metabolic engineering of medicinal plants. Plant Cell. 1995;7(7):1059.
4. Liscombe DK, Facchini PJ. Evolutionary and cellular webs in benzylisoquinoline alkaloid biosynthesis. Curr Opin Biotechnol. 2008;19(2):173–80.
5. Narcross L, Fossati E, Bourgeois L, Dueber JE, Martin VJ. Microbial factories for the production of benzylisoquinoline alkaloids. Trends Biotechnol. 2016;34(3):228–41.
6. Organization WH. WHO model list of essential medicines: 18th list, April 2013. 2013.
7. Narcotic drugs technical report: Estimated world requirements for 2016. United Nations, New York. 2016 [cited 2016 September 28]; Available from: http://www.incb.org/incb/en/narcotic-drugs/Technical_Reports/narcotic_drugs_reports.html.
8. McCoy E, O'Connor SE. Natural products from plant cell cultures, Natural compounds as drugs, vol. I. Basel: Springer; 2008. p. 329–70.
9. Glenn WS, Runguphan W, O'Connor SE. Recent progress in the metabolic engineering of alkaloids in plant systems. Curr Opin Biotechnol. 2013;24(2):354–65.
10. Nakagawa A, Matsumura E, Koyanagi T, Katayama T, Kawano N, Yoshimatsu K, et al. Total biosynthesis of opiates by stepwise fermentation using engineered *Escherichia coli*. Nat Commun. 2016;7:10390.
11. Brown S, Clastre M, Courdavault V, O'Connor SE. *De novo* production of the plant-derived alkaloid strictosidine in yeast. PNAS. 2015;112(11):3205–10.
12. Campbell A, Bauchart P, Gold ND, Zhu Y, De Luca V, Martin VJ. Engineering of a nepetalactol-producing platform strain of *Saccharomyces cerevisiae* for the production of plant seco-iridoids. ACS Synth Biol. 2016;5(5):405–14.

13. Fossati E, Ekins A, Narcross L, Zhu Y, Falgueyret J-P, Beaudoin GA, et al. Reconstitution of a 10-gene pathway for synthesis of the plant alkaloid dihydrosanguinarine in *Saccharomyces cerevisiae*. Nat Commun. 2014;5:3283.
14. Fossati E, Narcross L, Ekins A, Falgueyret J-P, Martin VJ. Synthesis of morphinan alkaloids in *Saccharomyces cerevisiae*. PLoS One. 2015;10:e0124459.
15. Li Y, Smolke CD. Engineering biosynthesis of the anticancer alkaloid noscapine in yeast. Nat Commun. 2016;7:12137.
16. Trenchard IJ, Smolke CD. Engineering strategies for the fermentative production of plant alkaloids in yeast. Metab Eng. 2015;30(7):96–104.
17. Gold ND, Gowen CM, Lussier F-X, Cautha SC, Mahadevan R, Martin VJ. Metabolic engineering of a tyrosine-overproducing yeast platform using targeted metabolomics. Microb Cell Factories. 2015;14(1):1.
18. Patnaik R, Zolandz RR, Green DA, Kraynie DF. L-tyrosine production by recombinant *Escherichia coli*: fermentation optimization and recovery. Biotechnol Bioeng. 2008;99(4):741–52.
19. Santos CNS, Xiao W, Stephanopoulos G. Rational, combinatorial, and genomic approaches for engineering L-tyrosine production in *Escherichia coli*. PNAS. 2012;109(34):13538–43.
20. Wei T, Cheng B-Y, Liu J-Z. Genome engineering *Escherichia coli* for L-DOPA overproduction from glucose. Sci Rep. 2016;6. 30080; doi: 10.1038/srep30080
21. DeLoache WC, Russ ZN, Narcross L, Gonzales AM, Martin VJ, Dueber JE. An enzyme-coupled biosensor enables (*S*)-reticuline production in yeast from glucose. Nat Chem Biol. 2015;11:465–71.
22. Nakagawa A, Minami H, Kim J-S, Koyanagi T, Katayama T, Sato F, et al. A bacterial platform for fermentative production of plant alkaloids. Nat Commun. 2011;2:326.
23. Tabata K, Hashimoto S-I. Production of mevalonate by a metabolically-engineered *Escherichia coli*. Biotechnol Lett. 2004;26(19):1487–91.
24. Westfall PJ, Pitera DJ, Lenihan JR, Eng D, Woolard FX, Regentin R, et al. Production of amorphadiene in yeast, and its conversion to dihydroartemisinic acid, precursor to the antimalarial agent artemisinin. PNAS. 2012;109(3):E111–E8.
25. Liu J, Zhang W, Du G, Chen J, Zhou J. Overproduction of geraniol by enhanced precursor supply in *Saccharomyces cerevisiae*. J Biotechnol. 2013;168(4):446–51.
26. Zhou J, Wang C, Yang L, Choi E-S, Kim S-W. Geranyl diphosphate synthase: an important regulation point in balancing a recombinant monoterpene pathway in *Escherichia coli*. Enzym Microb Technol. 2015;68:50–5.
27. Luttik M, Vuralhan Z, Suir E, Braus G, Pronk J, Daran J. Alleviation of feedback inhibition in *Saccharomyces cerevisiae* aromatic amino acid biosynthesis: quantification of metabolic impact. Metab Eng. 2008;10(3):141–53.
28. Krömer JO, Nunez-Bernal D, Averesch NJ, Hampe J, Varela J, Varela C. Production of aromatics in *Saccharomyces cerevisiae*—a feasibility study. J Biotechnol. 2013;163(2):184–93.
29. Trenchard IJ, Siddiqui MS, Thodey K, Smolke CD. *De novo* production of the key branch point benzylisoquinoline alkaloid reticuline in yeast. Metab Eng. 2015;31:74–83.
30. Nakagawa A, Matsuzaki C, Matsumura E, Koyanagi T, Katayama T, Yamamoto K, et al. (*R*, *S*)-tetrahydropapaveroline production by stepwise fermentation using engineered *Escherichia coli*. Sci Rep. 2014;4:6695.
31. Dudley QM, Anderson KC, Jewett MC. Cell-free mixing of *Escherichia coli* crude extracts to prototype and rationally engineer high-titer mevalonate synthesis. ACS Synth Biol. 2016;5(12):1578–88.
32. Asada K, Salim V, Masada-Atsumi S, Edmunds E, Nagatoshi M, Terasaka K, et al. A 7-deoxyloganetic acid glucosyltransferase contributes a key step in secologanin biosynthesis in Madagascar periwinkle. Plant Cell. 2013;25(10):4123–34.
33. Miettinen K, Dong L, Navrot N, Schneider T, Burlat V, Pollier J, et al. The seco-iridoid pathway from *Catharanthus roseus*. Nat Commun. 2014;5:3606.
34. Salim V, Wiens B, Masada-Atsumi S, Yu F, De Luca V. 7-Deoxyloganetic acid synthase catalyzes a key 3 step oxidation to form 7-deoxyloganetic acid in *Catharanthus roseus* iridoid biosynthesis. Phytochemistry. 2014;101:23–31.

35. Salim V, Yu F, Altarejos J, Luca V. Virus-induced gene silencing identifies *Catharanthus roseus* 7-deoxyloganic acid 7-hydroxylase, a step in iridoid and monoterpene indole alkaloid biosynthesis. Plant J. 2013;76(5):754–65.
36. Farrow SC, Hagel JM, Beaudoin GA, Burns DC, Facchini PJ. Stereochemical inversion of (*S*)-reticuline by a cytochrome P450 fusion in opium poppy. Nat Chem Biol. 2015;11(9):728–32.
37. Winzer T, Kern M, King AJ, Larson TR, Teodor RI, Donninger SL, et al. Morphinan biosynthesis in opium poppy requires a P450-oxidoreductase fusion protein. Science. 2015;349(6245):309–12.
38. Galanie S, Thodey K, Trenchard IJ, Interrante MF, Smolke CD. Complete biosynthesis of opioids in yeast. Science. 2015;349(6252):1095–100.
39. Hawkins KM, Smolke CD. Production of benzylisoquinoline alkaloids in *Saccharomyces cerevisiae*. Nat Chem Biol. 2008;4(9):564–73.
40. Beaudoin GA, Facchini PJ. Benzylisoquinoline alkaloid biosynthesis in opium poppy. Planta. 2014;240(1):19–32.
41. Kries H, O'Connor SE. Biocatalysts from alkaloid producing plants. Curr Opin Chem Biol. 2016;31:22–30.
42. Pan Q, Mustafa NR, Tang K, Choi YH, Verpoorte R. Monoterpenoid indole alkaloids biosynthesis and its regulation in *Catharanthus roseus*: a literature review from genes to metabolites. Phytochem Rev. 2016;15(2):221–50.
43. Thamm AM, Qu Y, De Luca V. Discovery and metabolic engineering of iridoid/secoiridoid and monoterpenoid indole alkaloid biosynthesis. Phytochem Rev. 2016;15(3):339–61.
44. Kishimoto S, Sato M, Tsunematsu Y, Watanabe K. Evaluation of biosynthetic pathway and engineered biosynthesis of alkaloids. Molecules. 2016;21(8):1078.
45. O'Connor SE. Engineering of secondary metabolism. Annu Rev Genet. 2015;49:71–94.
46. Pyne M, Narcross L, Fossati E, Bourgeois L, Burton E, Gold N, et al. Reconstituting plant secondary metabolism in *Saccharomyces cerevisiae* for production of high-value benzylisoquinoline alkaloids. Methods Enzymol. 2016;575:195–224.
47. Schläger S, Dräger B. Exploiting plant alkaloids. Curr Opin Biotechnol. 2016;37:155–64.
48. O'Connor SE, Maresh JJ. Chemistry and biology of monoterpene indole alkaloid biosynthesis. Nat Prod Rep. 2006;23(4):532–47.
49. Lange BM, Rujan T, Martin W, Croteau R. Isoprenoid biosynthesis: the evolution of two ancient and distinct pathways across genomes. PNAS. 2000;97(24):13172–7.
50. Rohmer M. The discovery of a mevalonate-independent pathway for isoprenoid biosynthesis in bacteria, algae and higher plants†. Nat Prod Rep. 1999;16(5):565–74.
51. Kizer L, Pitera DJ, Pfleger BF, Keasling JD. Application of functional genomics to pathway optimization for increased isoprenoid production. Appl Environ Microbiol. 2008;74(10):3229–41.
52. Brown MS, Goldstein JL. Multivalent feedback regulation of HMG CoA reductase, a control mechanism coordinating isoprenoid synthesis and cell growth. J Lipid Res. 1980;21(5):505–17.
53. Siperstein MD, Guest MJ. Studies on the site of the feedback control of cholesterol synthesis. J Clin Invest. 1960;39(4):642.
54. Dorsey JK, Porter JW. The inhibition of mevalonic kinase by geranyl and farnesyl pyrophosphates. J Biol Chem. 1968;243(18):4667–70.
55. Brown MS, Dana SE, Goldstein JL. Regulation of 3-hydroxy-3-methylglutaryl coenzyme a reductase activity in human fibroblasts by lipoproteins. PNAS. 1973;70(7):2162–6.
56. Hunter WN. The non-mevalonate pathway of isoprenoid precursor biosynthesis. J Biol Chem. 2007;282(30):21573–7.
57. Martin VJ, Pitera DJ, Withers ST, Newman JD, Keasling JD. Engineering a mevalonate pathway in *Escherichia coli* for production of terpenoids. Nat Biotechnol. 2003;21(7):796–802.
58. Kirby J, Keasling JD. Metabolic engineering of microorganisms for isoprenoid production. Nat Prod Rep. 2008;25(4):656–61.
59. Kirby J, Keasling JD. Biosynthesis of plant isoprenoids: perspectives for microbial engineering. Annu Rev Plant Biol. 2009;60:335–55.

60. Dinda B, Chowdhury DR, Mohanta BC. Naturally occurring iridoids, secoiridoids and their bioactivity. An updated review, part 3. Chem Pharm Bull. 2009;57(8):765–96.
61. Donald K, Hampton RY, Fritz IB. Effects of overproduction of the catalytic domain of 3-hydroxy-3-methylglutaryl coenzyme A reductase on squalene synthesis in *Saccharomyces cerevisiae*. Appl Environ Microbiol. 1997;63(9):3341–4.
62. Blanchard L, Karst F. Characterization of a lysine-to-glutamic acid mutation in a conservative sequence of farnesyl diphosphate synthase from *Saccharomyces cerevisiae*. Gene. 1993;125(2):185–9.
63. Fischer MJ, Meyer S, Claudel P, Bergdoll M, Karst F. Metabolic engineering of monoterpene synthesis in yeast. Biotechnol Bioeng. 2011;108(8):1883–92.
64. Jongedijk E, Cankar K, Ranzijn J, Krol S, Bouwmeester H, Beekwilder J. Capturing of the monoterpene olefin limonene produced in *Saccharomyces cerevisiae*. Yeast. 2015;32(1):159–71.
65. Sarria S, Wong B, Martín HG, Keasling JD, Peralta-Yahya P. Microbial synthesis of pinene. ACS Synth Biol. 2014;3(7):466–75.
66. Arifin AA, Don MM, Uzir MH. The feasibility of growing cells of *Saccharomyces cerevisiae* for citronellol production in a continuous-closed-gas-loop bioreactor (CCGLB). Bioresour Technol. 2011;102(19):9318–20.
67. Courdavault V, Papon N, Clastre M, Giglioli-Guivarc'h N, St-Pierre B, Burlat V. A look inside an alkaloid multisite plant: the *Catharanthus* logistics. Curr Opin Plant Biol. 2014;19:43–50.
68. Le Men J, Taylor WI. A uniform numbering system for indole alkaloids. Experientia. 1965;21(9):508–10.
69. Szabó LF. Rigorous biogenetic network for a group of indole alkaloids derived from strictosidine. Molecules. 2008;13(8):1875–96.
70. Geerlings A, Ibañez MM-L, Memelink J, van der Heijden R, Verpoorte R. Molecular cloning and analysis of strictosidine β-D-glucosidase, an enzyme in terpenoid indole alkaloid biosynthesis in *Catharanthus roseus*. J Biol Chem. 2000;275(5):3051–6.
71. Gerasimenko I, Sheludko Y, Ma X, Stöckigt J. Heterologous expression of a *Rauvolfia* cDNA encoding strictosidine glucosidase, a biosynthetic key to over 2000 monoterpenoid indole alkaloids. Eur J Biochem. 2002;269(8):2204–13.
72. Heinstein P, Höfle G, Stöckigt J. Involvement of cathenamine in the formation of *N*–analogues of indole alkaloids. Planta Med. 1979;37(12):349–57.
73. Benayad S, Ahamada K, Lewin G, Evanno L, Poupon E. Preakuammicine: a long-awaited missing link in the biosynthesis of monoterpene indole alkaloids. Eur J Org Chem. 2016;8:1494–99.
74. El-Sayed M, Choi YH, Frederich M, Roytrakul S, Verpoorte R. Alkaloid accumulation in *Catharanthus roseus* cell suspension cultures fed with stemmadenine. Biotechnol Lett. 2004;26(10):793–8.
75. Qu Y, Easson ML, Froese J, Simionescu R, Hudlicky T, De Luca V. Completion of the seven-step pathway from tabersonine to the anticancer drug precursor vindoline and its assembly in yeast. PNAS. 2015;112(19):6224–9.
76. Costa MMR, Hilliou F, Duarte P, Pereira LG, Almeida I, Leech M, et al. Molecular cloning and characterization of a vacuolar class III peroxidase involved in the metabolism of anticancer alkaloids in *Catharanthus roseus*. Plant Physiol. 2008;146(2):403–17.
77. Hemscheidt T, Zenk M. Partial purification and characterization of a NADPH dependent tetrahydroalstonine synthase from *Catharanthus roseus* cell suspension cultures. Plant Cell Rep. 1985;4(4):216–9.
78. Chen X, Dang T-TT, Facchini PJ. Noscapine comes of age. Phytochemistry. 2015;111:7–13.
79. Chen X, Facchini PJ. Short-chain dehydrogenase/reductase catalyzing the final step of noscapine biosynthesis is localized to laticifers in opium poppy. Plant J. 2014;77(2):173–84.
80. Dang T-TT, Chen X, Facchini PJ. Acetylation serves as a protective group in noscapine biosynthesis in opium poppy. Nat Chem Biol. 2015;11(2):104–6.
81. Dang T-TT, Facchini PJ. Characterization of three *O*-methyltransferases involved in noscapine biosynthesis in opium poppy. Plant Physiol. 2012;159(2):618–31.

82. Dang T-TT, Facchini PJ. CYP82Y1 is *N*-methylcanadine 1-hydroxylase, a key noscapine biosynthetic enzyme in opium poppy. J Biol Chem. 2014;289(4):2013–26.
83. Dang T-TT, Facchini PJ. Cloning and characterization of canadine synthase involved in noscapine biosynthesis in opium poppy. FEBS Lett. 2014;588(1):198–204.
84. Facchini PJ, Bohlmann J, Covello PS, De Luca V, Mahadevan R, Page JE, et al. Synthetic biosystems for the production of high-value plant metabolites. Trends Biotechnol. 2012;30(3):127–31.
85. Thodey K, Galanie S, Smolke CD. A microbial biomanufacturing platform for natural and semisynthetic opioids. Nat Chem Biol. 2014;10(10):837–44.
86. Fukuda K, Watanabe M, Asano K, Ouchi K, Takasawa S. A mutated *ARO4* gene for feedback-resistant DAHP synthase which causes both *o*-fluoro-dl-phenylalamine resistance and β-phenethyl-alcohol overproduction in *Saccharomyces cerevisiae*. Curr Genet. 1991;20(6):453–6.
87. Hagel JM, Facchini PJ. Dioxygenases catalyze the O-demethylation steps of morphine biosynthesis in opium poppy. Nat Chem Biol. 2010;6(4):273–5.
88. Minami H, Kim J-S, Ikezawa N, Takemura T, Katayama T, Kumagai H, et al. Microbial production of plant benzylisoquinoline alkaloids. PNAS. 2008;105(21):7393–8.
89. Galanie S, Smolke CD. Optimization of yeast-based production of medicinal protoberberine alkaloids. Microb Cell Factories. 2015;14(1):1–13.
90. Gesell A, Chávez MLD, Kramell R, Piotrowski M, Macheroux P, Kutchan TM. Heterologous expression of two FAD-dependent oxidases with (*S*)-tetrahydroprotoberberine oxidase activity from *Argemone mexicana* and *Berberis wilsoniae* in insect cells. Planta. 2011;233(6):1185–97.
91. Beaudoin GA, Facchini PJ. Isolation and characterization of a cDNA encoding (*S*)-*cis*-*N*-methylstylopine 14-hydroxylase from opium poppy, a key enzyme in sanguinarine biosynthesis. Biochem Biophys Res Commun. 2013;431(3):597–603.
92. Hagel JM, Beaudoin GA, Fossati E, Ekins A, Martin VJ, Facchini PJ. Characterization of a flavoprotein oxidase from opium poppy catalyzing the final steps in sanguinarine and papaverine biosynthesis. J Biol Chem. 2012;287(51):42972–83.
93. Takemura T, Ikezawa N, Iwasa K, Sato F. Molecular cloning and characterization of a cytochrome P450 in sanguinarine biosynthesis from *Eschscholzia californica* cells. Phytochemistry. 2013;91:100–8.
94. Narcross L, Bourgeois L, Fossati E, Burton E, Martin VJ. Mining enzyme diversity of transcriptome libraries through DNA synthesis for benzylisoquinoline alkaloid pathway optimization in yeast. ACS Synth Biol. 2016;5(12):1505–18.
95. Winzer T, Gazda V, He Z, Kaminski F, Kern M, Larson TR, et al. A Papaver somniferum 10-gene cluster for synthesis of the anticancer alkaloid noscapine. Science. 2012;336(6089):1704–8.
96. De Luca V, Salim V, Thamm A, Masada SA, Yu F. Making iridoids/secoiridoids and monoterpenoid indole alkaloids: progress on pathway elucidation. Curr Opin Plant Biol. 2014;19:35–42.
97. Meunier B, De Visser SP, Shaik S. Mechanism of oxidation reactions catalyzed by cytochrome P450 enzymes. Chem Rev. 2004;104(9):3947–80.
98. Pompon D, Louerat B, Bronine A, Urban P. Yeast expression of animal and plant P450s in optimized redox environments. Methods Enzymol. 1996;272:51–64.
99. Zangar RC, Davydov DR, Verma S. Mechanisms that regulate production of reactive oxygen species by cytochrome P450. Toxicol Appl Pharmacol. 2004;199(3):316–31.
100. Hatakeyama M, Kitaoka T, Ichinose H. Heterologous expression of fungal cytochromes P450 (CYP5136A1 and CYP5136A3) from the white-rot basidiomycete *Phanerochaete chrysosporium*: functionalization with cytochrome *b*5 in *Escherichia coli*. Enzym Microb Technol. 2016;89:7–14.
101. Schenkman JB, Jansson I. The many roles of cytochrome *b*5. Pharmacol Ther. 2003;97(2):139–52.
102. Zimmer T, Ogura A, Takewaka T, Zimmer R-M, Ohta A, Takagi M. Gene regulation in response to overexpression of cytochrome P450 and proliferation of the endoplasmic reticulum in *Saccharomyces cerevisiae*. Biosci Biotechnol Biochem. 2000;64(9):1930–6.

103. Yazaki K. Transporters of secondary metabolites. Curr Opin Plant Biol. 2005;8(3):301–7.
104. Choi K-B, Morishige T, Shitan N, Yazaki K, Sato F. Molecular cloning and characterization of coclaurine *N*-methyltransferase from cultured cells of *Coptis japonica*. J Biol Chem. 2002;277(1):830–5.
105. Desgagné-Penix I, Facchini PJ. Systematic silencing of benzylisoquinoline alkaloid biosynthetic genes reveals the major route to papaverine in opium poppy. Plant J. 2012;72(2):331–44.
106. Ounaroon A, Decker G, Schmidt J, Lottspeich F, Kutchan TM. (*R, S*)-Reticuline 7-*O*-methyltransferase and (*R, S*)-norcoclaurine 6-*O*-methyltransferase of *Papaver somniferum*–cDNA cloning and characterization of methyl transfer enzymes of alkaloid biosynthesis in opium poppy. Plant J. 2003;36(6):808–19.
107. Sato F, Tsujita T, Katagiri Y, Yoshida S, Yamada Y. Purification and characterization of *S*-adenosyl-L-methionine: norcoclaurine 6-*O*-methyltransferase from cultured *Coptis japonica* cells. Eur J Biochem. 1994;225(1):125–31.
108. Dueber JE, GC W, Malmirchegini GR, Moon TS, Petzold CJ, Ullal AV, et al. Synthetic protein scaffolds provide modular control over metabolic flux. Nat Biotechnol. 2009;27(8):753–9.
109. Kell DB, Swainston N, Pir P, Oliver SG. Membrane transporter engineering in industrial biotechnology and whole cell biocatalysis. Trends Biotechnol. 2015;33(4):237–46.
110. Diallinas G. Understanding transporter specificity and the discrete appearance of channel-like gating domains in transporters. Front Pharmacol. 2014;5:207.
111. Doroshenko V, Airich L, Vitushkina M, Kolokolova A, Livshits V, Mashko S. YddG from *Escherichia coli* promotes export of aromatic amino acids. FEMS Microbiol Lett. 2007;275(2):312–8.
112. Yu F, De Luca V. ATP-binding cassette transporter controls leaf surface secretion of anticancer drug components in *Catharanthus roseus*. PNAS. 2013;110(39):15830–5.
113. Jones CM, Lozada NJH, Pfleger BF. Efflux systems in bacteria and their metabolic engineering applications. Appl Microbiol Biotechnol. 2015;99(22):9381–93.
114. Liu Q, Cheng Y, Xie X, Xu Q, Chen N. Modification of tryptophan transport system and its impact on production of L-tryptophan in *Escherichia coli*. Bioresour Technol. 2012;114:549–54.
115. Boyarskiy S, Tullman-Ercek D. Getting pumped: membrane efflux transporters for enhanced biomolecule production. Curr Opin Chem Biol. 2015;28:15–9.
116. Mukhopadhyay A. Tolerance engineering in bacteria for the production of advanced biofuels and chemicals. Trends Microbiol. 2015;23(8):498–508.
117. Zhou G, Pereira JF, Delhaize E, Zhou M, Magalhaes JV, Ryan PR. Enhancing the aluminium tolerance of barley by expressing the citrate transporter genes *SbMATE* and *FRD3*. J Exp Bot. 2014;65(9):2381–90.
118. Doshi R, Nguyen T, Chang G. Transporter-mediated biofuel secretion. PNAS. 2013;110(19):7642–7.
119. Boyarskiy S, López SD, Kong N, Tullman-Ercek D. Transcriptional feedback regulation of efflux protein expression for increased tolerance to and production of *n*-butanol. Metab Eng. 2016;33:130–7.
120. Dunlop MJ, Dossani ZY, Szmidt HL, Chu HC, Lee TS, Keasling JD, et al. Engineering microbial biofuel tolerance and export using efflux pumps. Mol Syst Biol. 2011;7(1):487.
121. Yang N, Driessen AJ. The *saci_2123* gene of the hyperthermoacidophile *Sulfolobus acidocaldarius* encodes an ATP-binding cassette multidrug transporter. Extremophiles. 2015;19(1):101–8.
122. Zhang C, Chen X, Stephanopoulos G, Too HP. Efflux transporter engineering markedly improves amorphadiene production in *Escherichia coli*. Biotechnol Bioeng. 2016;113(8):1755–63.
123. Milne N, Wahl S, van Maris A, Pronk J, Daran J. Excessive by-product formation: a key contributor to low isobutanol yields of engineered *Saccharomyces cerevisiae* strains. Metab Eng Commun. 2016;3:39–51.
124. Kind S, Jeong WK, Schröder H, Zelder O, Wittmann C. Identification and elimination of the competing *N*-acetyldiaminopentane pathway for improved production of diaminopentane by *Corynebacterium glutamicum*. Appl Environ Microbiol. 2010;76(15):5175–80.

125. Lechner A, Brunk E, Keasling JD. The need for integrated approaches in metabolic engineering. Cold Spring Harb Perspect Biol. 2016:a023903.
126. Matasci N, Hung L-H, Yan Z, Carpenter EJ, Wickett NJ, Mirarab S, et al. Data access for the 1,000 plants (1KP) project. GigaScience. 2014;3(1):1–10.
127. Ryan OW, Skerker JM, Maurer MJ, Li X, Tsai JC, Poddar S, et al. Selection of chromosomal DNA libraries using a multiplex CRISPR system. eLife. 2014;3:e03703.
128. Shao Z, Zhao H, Zhao H. DNA assembler, an *in vivo* genetic method for rapid construction of biochemical pathways. Nucleic Acids Res. 2009;37(2):e16–e.
129. Datsenko KA, Wanner BL. One-step inactivation of chromosomal genes in *Escherichia coli* K-12 using PCR products. PNAS. 2000;97(12):6640–5.
130. Yu D, Ellis HM, Lee E-C, Jenkins NA, Copeland NG. An efficient recombination system for chromosome engineering in *Escherichia coli.* PNAS. 2000;97(11):5978–83.
131. Jinek M, Chylinski K, Fonfara I, Hauer M, Doudna JA, Charpentier E. A programmable dual-RNA–guided DNA endonuclease in adaptive bacterial immunity. Science. 2012;337(6096):816–21.
132. DiCarlo JE, Norville JE, Mali P, Rios X, Aach J, Church GM. Genome engineering in *Saccharomyces cerevisiae* using CRISPR-Cas systems. Nucleic Acids Res. 2013;41:4336–43.
133. Horwitz AA, Walter JM, Schubert MG, Kung SH, Hawkins K, Platt DM, et al. Efficient multiplexed integration of synergistic alleles and metabolic pathways in yeasts via CRISPR-Cas. Cell Syst. 2015;1(1):88–96.
134. Jakočiūnas T, Bonde I, Herrgård M, Harrison SJ, Kristensen M, Pedersen LE, et al. Multiplex metabolic pathway engineering using CRISPR/Cas9 in *Saccharomyces cerevisiae*. Metab Eng. 2015;28:213–22.
135. Shi S, Liang Y, Zhang MM, Ang EL, Zhao H. A highly efficient single-step, markerless strategy for multi-copy chromosomal integration of large biochemical pathways in *Saccharomyces cerevisiae*. Metab Eng. 2016;33:19–27.
136. Alper H, Fischer C, Nevoigt E, Stephanopoulos G. Tuning genetic control through promoter engineering. Proc Natl Acad Sci U S A. 2005;102(36):12678–83.
137. Curran KA, Karim AS, Gupta A, Alper HS. Use of expression-enhancing terminators in *Saccharomyces cerevisiae* to increase mRNA half-life and improve gene expression control for metabolic engineering applications. Metab Eng. 2013;19:88–97.
138. Curran KA, Morse NJ, Markham KA, Wagman AM, Gupta A, Alper HS. Short synthetic terminators for improved heterologous gene expression in yeast. ACS Synth Biol. 2015;4(7):824–32.
139. Sun J, Shao Z, Zhao H, Nair N, Wen F, JH X, et al. Cloning and characterization of a panel of constitutive promoters for applications in pathway engineering in *Saccharomyces cerevisiae.* Biotechnol Bioeng. 2012;109(8):2082–92.
140. Yamanishi M, Ito Y, Kintaka R, Imamura C, Katahira S, Ikeuchi A, et al. A genome-wide activity assessment of terminator regions in *Saccharomyces cerevisiae* provides a "terminatome" toolbox. ACS Synth Biol. 2013;2(6):337–47.
141. Ehrenworth A, Sarria S, Peralta-Yahya P. Pterin-dependent mono-oxidation for the microbial synthesis of a modified monoterpene indole alkaloid. ACS Synth Biol. 2015;4(12):1295–307.
142. Kim J-S, Nakagawa A, Yamazaki Y, Matsumura E, Koyanagi T, Minami H, et al. Improvement of reticuline productivity from dopamine by using engineered *Escherichia coli.* Biosci Biotechnol Biochem. 2013;77(10):2166–8.
143. Ahn JO, Lee HW, Saha R, Park MS, Jung J-K, Lee D-Y. Exploring the effects of carbon sources on the metabolic capacity for shikimic acid production in *Escherichia coli* using *in silico* metabolic predictions. J Microbiol Biotechnol. 2008;18(11):1773–84.
144. Zimmer T, Vogel F, Ohta A, Takagi M, Schunck W-H. Protein quality—a determinant of the intracellular fate of membrane-bound cytochromes P450 in yeast. DNA Cell Biol. 1997;16(4):501–14.

Caffeine

6

Misako Kato and Fumiyo Nakayama

6.1 Introduction

Caffeine was first isolated from tea and coffee in the early 1820s; however, the main biosynthetic pathway of this alkaloid has not been elucidated until relatively recently. In this chapter, the current knowledge about the distribution and the biosynthetic pathway of caffeine are summarized. In addition, the relationship between *N*-methyltransferases involved in caffeine biosynthesis and the motif B′ methyltransferase family is discussed.

6.2 Distribution of Caffeine in Plants

Caffeine (1,3,7-trimethylxanthine) is one of the well-known purine alkaloids (Fig. 6.1). Occurrence of theobromine (3,7-dimethylxanthine), which is the precursor of caffeine, has also been demonstrated in caffeine-containing plant species. Caffeine is widely distributed in plants compared with the limited distribution of some other alkaloids, such as nicotine and morphine. The major genera of caffeine-containing plants are *Camellia*, *Coffea*, *Theobroma, Paullinia*, *Ilex*, and *Cola* [2, 3, 6, 8]. The distribution of purine alkaloids in *Camellia*, *Coffea*, and *Theobroma,* has

M. Kato (✉)
Graduate School of Humanities and Sciences, Ochanomizu University, Otsuka, Bunkyo-ku, Tokyo 112-8610, Japan

Institute for Human Life Innovation, Ochanomizu University, Otsuka, Bunkyo-ku, Tokyo 112-8610, Japan
e-mail: kato.misako@ocha.ac.jp

F. Nakayama
Graduate School of Humanities and Sciences, Ochanomizu University, Otsuka, Bunkyo-ku, Tokyo 112-8610, Japan

W. Schwab et al. (eds.), *Biotechnology of Natural Products*,
https://doi.org/10.1007/978-3-319-67903-7_6

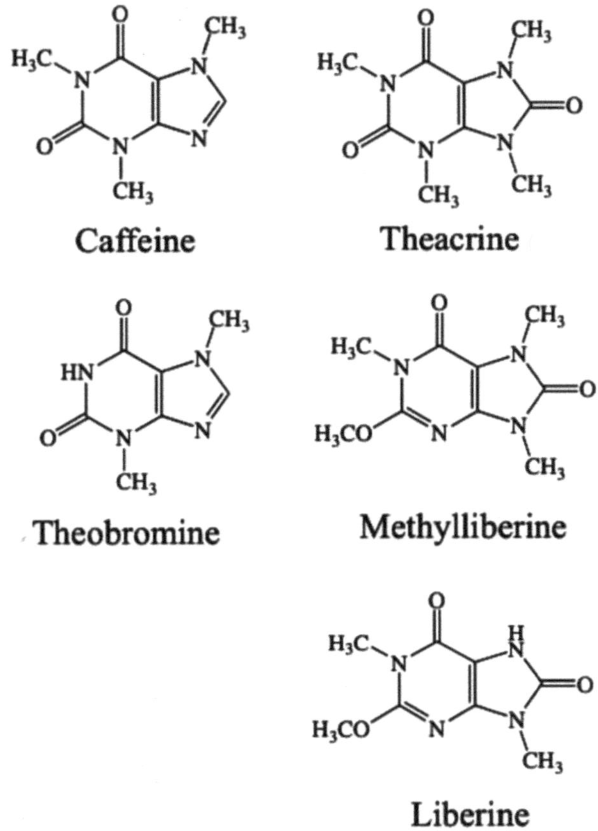

Fig. 6.1 Structure of purine alkaloids in plants

been well studied; however, there has been limited study on their occurrence in the remaining taxa.

Caffeine content in the young leaves of *Camellia sinensis* var. sinensis, *C. sinensis* var. *assamica*, and *C. taliensis* has been reported to be 2–3% of the dry weight. In contrast, no caffeine has been detected in *C. irrawadiensis* [39], where the predominant purine alkaloid is theobromine [40]. *C. ptilophylla* has also been investigated as one of the theobromine-accumulating *Camellia* species [4]. In addition to caffeine, theacrine (1,3,7,9-tetramethyluric acid) is a major purine alkaloid in *C. assamica* var. *kucha* [63]. *Camellia,* which consists of four subgenera, is the largest genus among the Teaceae, and all species containing purine alkaloids belong to section *Thea.* No purine alkaloid biosynthetic activity has been detected by radiolabeled tracer experiments in the leaves of species from subgenera *Protocamellia, Camellia,* and *Metacamellia* [5, 15, 20].

The caffeine content of seeds from *Coffea* species has been reported to be 0.4–2.4% of the dry weight [32]. According to these authors, the content of caffeine in *Coffea arabica,* which is the most widely cultivated species, is 0.6–1.2%. Caffeine and theobromine also occur in the leaves of *Coffea arabica,* and the maximum content of purine alkaloids in the leaves is 4% of the dry weight [11, 16]. However, no caffeine has been detected in roots or the brown parts of shoots [62]. In addition to its occurrence in *Camellia* species, theacrine has also been detected in some *Coffea* species. The presence of liberine (*O*(2),1,9-trimethyluric acid) and methylliberine (*O*(2),1,7,9-tetramethyluric acid) has also been demonstrated in *Coffea liberica*, *Coffea dewevrei*, and *Coffea abeokutae* [45, 46, 57].

In *Theobroma cacao,* the major purine alkaloid is theobromine [18, 50]. The theobromine and caffeine contents in seeds from three genotypes of *Theobroma cacao* have been shown to be 0.9–3.3% and 0.08–1.2%, respectively [18]. The leaves of *Theobroma cacao* also contain theobromine and caffeine [18, 28]. Theacrine also occurs in *Theobroma* spp., although not in *Theobroma cacao* [18].

Leaf extracts of *Ilex paraguariensis,* which in South America are commonly used to prepare maté, contain 0.8–0.9% caffeine and 0.08–0.16% theobromine [31]. *Paullinia cupana*, which is renowned as the source of Amazonian guarana fruits, contains 4.3% caffeine in the seed kernel [10]. Based on a screening of 34 species of *Paullinia*, three species (*P. cupana, P. yoco and P. panchycarpa*) were found to contain purine alkaloids [58]. The theobromine content was shown to be higher than that of caffeine in the leaves of five cultivars of *P. cupana* [49]. Moreover, significant amounts of caffeine and related purine alkaloids have been detected in *Citrus* flowers [29]. The content of purine alkaloids in selected plant species has previously been summarized by Ashihara et al. [6].

6.3 Biosynthesis of Caffeine from Xanthosine

The major route of caffeine biosynthesis from xanthosine is illustrated in Fig. 6.2. The xanthine skeleton of caffeine is derived from purine nucleotides. This pathway in *Camellia sinensis* and *Coffea arabica* has been verified by the substrate specificity of native and recombinant *N*-methyltransferases [23, 25, 26, 37, 38, 42, 55], and also by experiments using radiolabeled precursors [3, 5, 53]. The pathway through paraxanthine is one of the minor pathways operating in *C. sinensis* [23]. The pathway of caffeine biosynthesis, as mentioned above for *Camellia* and *Coffea*, is essentially the same in *Ilex paraguariensis* [1] and *Theobroma cacao* [28, 60].

Caffeine biosynthesis from xanthosine involves three methylation steps. In all these steps, the purine-base methyl donor is *S*-adenosyl-L-methionine (SAM) [53]. The first methylation involves the production of 7-methylxanthosine from xanthosine and is catalyzed by 7-methylxanthosine synthase (EC2.1.1.158). The only substrate that is available for 7-methylxanthosine synthase is xanthosine [37, 55]. The monoanionic form of xanthosine rather than the neutral form as widely adopted is the substrate for 7-methylxanthosine synthase [47]. Xanthine monophosphate (XMP) has been proposed as the substrate for the precursor of caffeine [52]; however, XMP does not function as a substrate of 7-methylxanthosine synthase [37, 55]. *CmXRS1* (AB034699), *CaXMT1* (AB048793), and *CaXMT2* (JX978522) from *Coffea arabica* and *CcXMT1* (JX978518) from *Coffea canephora* have been reported as 7-methylxanthosine synthases [33, 37, 55] (Table 6.1). To date, no

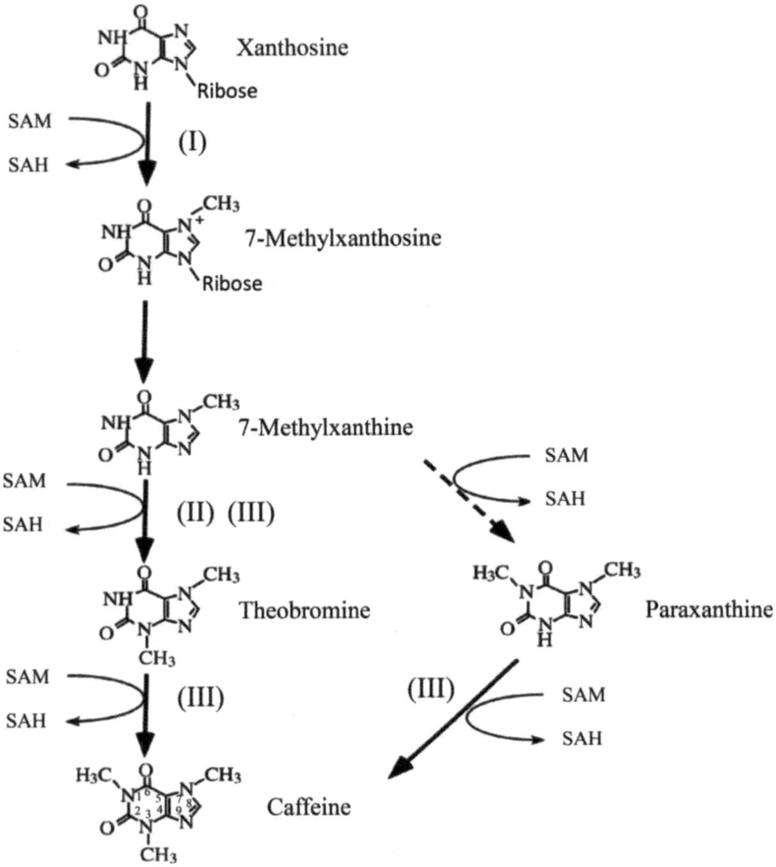

Fig. 6.2 Pathway for the biosynthesis of theobromine and caffeine in tea and coffee. Solid arrows indicate the major biosynthesis route. 7-Methylxanthosine synthase and theobromine synthase have high substrate specificity and only catalyze the conversion of xanthosine to 7-methylxanthosine and 7-methylxanthine to theobromine, respectively. Caffeine synthase has a broad substrate specificity. 7-Methylxanthosine synthase (I), theobromine synthase (II), caffeine synthase (III)

7-methylxanthosine synthase activity has been detected in recombinant enzyme studies of other species.

7-Methylxanthosine is converted to 7-methylxanthine by *N*-methylnucleosidase (EC3.2.2.25). Studies on the chemical structure of 7-methylxanthosine synthase from *C. canephora* have suggested that this enzyme has the dual functions of methyl transfer and nucleoside cleavage. The purine ring of 7-methylxanthosine has a partial positive charge that induce the flow of electrons from the ribose moiety. A nucleophilic attack by water on the oxocarbenium intermediate occur and the ribose moiety of 7-methylxanthosine is released [33].

The second and third methylations are catalyzed by caffeine synthase (EC2.1.1.160). Caffeine synthase was the first *N*-methyltransferase from

Camellia sinensis to be biochemically characterized and cloned, and this breakthrough led to the subsequent isolation of other *N*-methyltransferase genes involved in caffeine biosynthesis [25, 26]. *TCS1* (AB031280), which was isolated from the cDNA of *C. sinensis* as the caffeine synthase gene, consists of 1438 bp and encodes a protein of 369 amino acids. The apparent molecular mass of the native caffeine synthase from *Camellia sinensis* is 61 kDa, as estimated by gel-filtration chromatography, and 41 kDa, as estimated by sodium dodecyl sulfate-polyacrylamide gel electrophoresis [25]. Most members of the caffeine synthase family are believed to function in a dimeric form *in vivo* [33, 64]. *CCS1* (AB086414), *CtCS7* (AB086415), and *CaDXMT1* (AB084125) from *C. arabica* and *CcDXMT* (JX978516) from *C. canephora* have been identified as caffeine synthases from *Coffea* [33, 37, 55]. Caffeine synthase catalyzes 1-*N*- and 3-*N*-methylation reactions of mono- or di-methylxanthines. When dimethylxanthines are used as substrates for caffeine synthase, paraxanthine is the best methyl acceptor, followed by theobromine. 7-Methylxanthine is the best substrate of the three monomethylxanthines. The order of the methylation of purine bases by caffeine synthase from tea and coffee is *N*-3 > *N*-1. The amount of paraxanthine in

Table 6.1 *N*-methyltransferases involved in the biosynthesis of caffeine and the related compounds

Family	Genus	Nomen	Major alkaloids	Gene name	Substrate
Theales	*Camellia*	*Camellia sinensis*	Cf	*TCS1* (AB031280)	7mX/Tb
				TCS2 (AB031281)	–
		Camellia irrawadiensis	Tb	*ICS1* (AB056108)	7mX
				ICS2 (AB207816)	–
		Camellia ptilophylla	Tb	*PCS1* (AB207817)	7mX
				PCS2 (AB207818)	–
		Camellia japonica	–	*CjCS1* (AB297451)	7mX
		Camellia granthamiana	–	*CgCS1* (AB362882)	7mX
				CgCS2 (AB362883)	7mX
		Camellia lutchuensis	–	*ClCS1* (AB362885)	7mX
		Camellia kissi	–	*CkCS1* (AB362884)	7mX
		Camellia chrysantha	–	*CcCS2* (AB362886)	–

(continued)

Table 6.1 (continued)

Family	Genus	Nomen	Major alkaloids	Gene name	Substrate
Rubiales	*Coffea*	*Coffea arabica*	Cf, Tg	*CmXRS1* (AB034699)	XR
				CaXMT1 (AB048793)	XR
				CaXMT2 (JX978522)	XR
				CTS1 (AB034700)	7mX
				CaMXMT1 (AB048794)	7mX
				CaMXMT2 (AB084126)	7mX
				CCS1 (AB086414)	7mX/Tb
				CaDXMT1 (AB084125)	7mX/Tb
				CTgS1 (AB058482)	NA
				CTgS2 (AB054843)	NA
		Coffea canephora	Cf	*CcXMT1* (JX978518)	XR
				CcMXMT1 (JX978517)	7mX
				CcDXMT (JX978516)	7mX/Tb
Malvales	*Theobroma*	*Theobroma cacao*	Tb	*BTS1* (AB096699)	7mX
Sapindales	*Paullinia*	*Paullinia cupana*	Cf	*PcCS* (BK008796)	7mX/Tb

7mX 7-methylxanthine, *XR* xanthosine, *Cf* caffeine, *Tg* Trigonelline, *NA* nicotinic acid, *Tb* theobromine, – not detected

plant tissues is very low, suggesting that *N*-1-methylation of 7-methylxanthine is very slow [7]. In contrast, the best substrate of recombinant *PcCS* (BK008796) protein, which is a caffeine synthase from *Paullinia cupana*, is theobromine; no activity against paraxanthine was detected [49].

Theobromine synthase (EC2.1.1.159) is specific for the conversion of 7-methylxanthine to theobromine. In caffeine-accumulating species, in addition to caffeine synthase, theobromine synthase also appears to catalyze theobromine production. *CTS1* (AB034700), *CTS2* (AB054841), *CaMXMT1* (AB048794), and *CaMXMT2* (AB084126) from *C. arabica*, and *CcMXMT1* (JX978517) from *C. canephora* have been isolated as theobromine synthases [36, 42]. In theobromine-accumulating species, theobromine synthase plays a key role in purine alkaloid biosynthesis. *ICS1* (AB056108) from *Camellia irrawadiensis*, *PCS1* (AB207817) from

C. ptilophylla, and *BTS1* (AB096699) from *T. cacao* have also been isolated, and their recombinant protein activity was identified as that of a theobromine synthase. The accumulation of purine alkaloids is dependent on the substrate specificity of *N*-methyltransferases since aforementioned three species are theobromine-accumulating species.

SAM-dependent *N*-methyltransferases play important roles in the regulation of caffeine biosynthesis in *C. sinensis* [17]. The level of *TCS1* expression is higher in young leaves than in mature leaves [24, 30], which is consistent with the change in gene expression in coffee leaves [37, 38]. *TCS1* and three type of *N*-methyltransferase genes from coffee are also expressed in the flowers [24, 37, 38].

Recently, Huang et al. supposed the precursor of the caffeine was xanthine as distinct from xanthosine in *Theobroma cacao*, *Paullinia cupana* and *Citrus sinensis* [19]. Xanthine is the key substrate in the conventional purine catabolism pathway to CO_2 and NH_3 via uric acid, allantoin and allantonate [2]. This new hypothesis needs to be proved by physiological studies on the purine metabolism.

6.4 The Caffeine Synthase Gene Family in Plants

Genes with high sequences identity to caffeine synthase from five purine alkaloid-free *Camellia* species are listed in Table 6.1. Despite the lack of purine alkaloid biosynthetic activity, genes homologous to caffeine synthase are conserved in purine alkaloid-free species from the four subgenera, *Protocamellia*, *Camellia*, *Metacamellia*, and *Thea*, of the genus *Camellia*. The occurrence of transcripts of these genes suggests that they are functional genes and not pseudogenes. Recombinant enzymes derived from these genes have theobromine synthase activity [20]. These observations strongly suggest that caffeine synthase has evolved from theobromine synthase in *Camellia* plants. Yoneyama et al. [60] showed that amino acid H^{211} of theobromine synthase (PCS1) from *Camellia ptilophylla* plays a critical role in substrate discrimination. The corresponding amino acid residue of TCS1, which is a caffeine synthase, is R^{210}. The same critical amino acid is seen in the theobromine synthases from purine alkaloid-free *Camellia* species [20]. The theobromine synthase gene from purine alkaloid-free *Camellia* species shows higher expression levels in mature leaves than in young leaves, which is unlike the expression profile of caffeine synthase in caffeine-accumulating species [20]. However, the functions of some genes from purine alkaloid-accumulating species of *Camellia* homologous to caffeine synthase have yet to be determined. These genes, namely *TCS2* from *C. sinensis*, *ICS2* from *C. irrawadiensis*, and *PCS2* from *C. ptilophylla,* are placed in the same clade of a molecular phylogenetic tree [20, 41].

The caffeine synthases, theobromine synthases, and related proteins shown in Table 6.1 belong to the motif B′ methyltransferase family, members of which have a motif B′ and a YFFF region in the amino acid sequence, and are members of the SABATH family [12]. The three conserved motifs (A, B, and C) involved in binding the methyl donor SAM have been reported in most plant SAM-dependent methyltransferases [22]. Members of the motif B′ methyltransferase family share motif B′

(LNDL F/P XNDFN) instead of motif B between motif A and C, and have the characteristic YFFF region (AYXXQFXXDFXXFL), which is located downstream of motif C [27] (Fig. 6.3). Phylogenetic analysis of the major motif B′ methyltransferases, for which substrates have been identified, is shown in Fig. 6.4. The illustrated tree implies that the caffeine biosynthetic pathways in coffee, tea, guaraná, and cacao might have evolved in parallel with one another, consistent with the different catalytic properties of the enzyme involved. Studies on *C. canephora* genome demonstrated *N*-methyltransferases involved in caffeine biosynthesis expanded through sequential tandem duplications independently of genes from cacao and tea, suggesting that caffeine in eudicots is of polyphyletic origin [13]. It is noteworthy that nicotinic acid *N*-methyltransferase (CTgS1) from *Coffea arabica* has a high degree of sequence identity (82.3%, 80.8%, and 82.9%) with caffeine synthase (CCS1), theobromine synthase (CTS1), and 7-methylxanthosine synthase (CmXRS1), respectively [35]. CTgS1 is involved in the production of trigonelline in coffee. Baumann

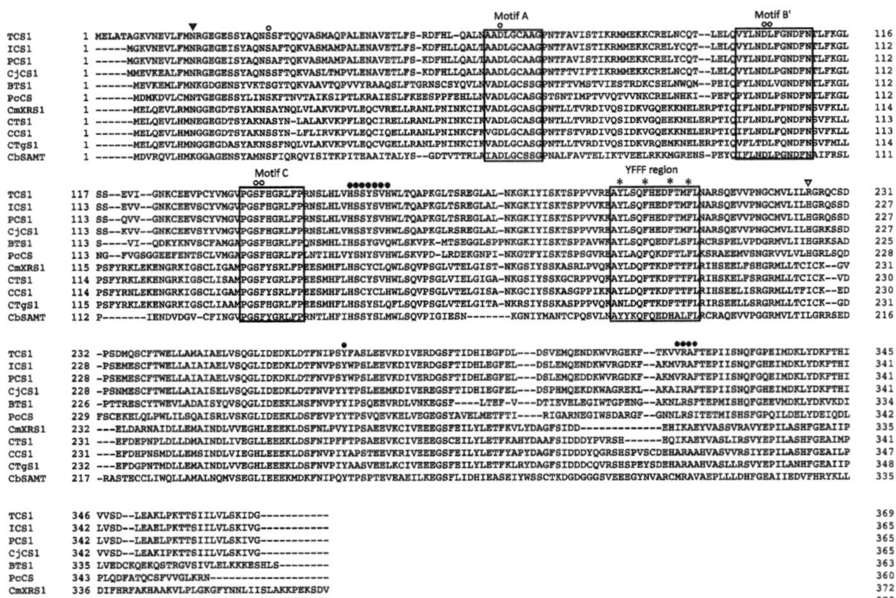

Fig. 6.3 Comparison of the amino acid sequences of caffeine synthases and related enzymes. The proposed SAM-binding motifs (A, B′ and C) and the conserved region nominated as "YFFF-region" are shown by open boxes. Asterisks indicate tyrosine (Y) or phenylalanine (F) residues in the region. The nominated amino acids in substrate binding are indicated by closed circles (methyl acceptor) and open circles (SAM), and additional active site residues are indicated by a closed arrowhead. The amino acid residue indicated by a open arrowhead plays a critical role in substrate discrimination in *Camellia* plants. The sources of sequences indicated by GenBank numbers are as follows: TCS1, AB031280; ICS1, AB056108; PCS1, AB207817; CjCS1, AB297451; BTS1, AB096699; PcCS, BK008796; CmXRS1, AB034699; CTS1, AB034700; CCS1, AB086414; CTgS1, AB054842; CbSAMT, AF133053

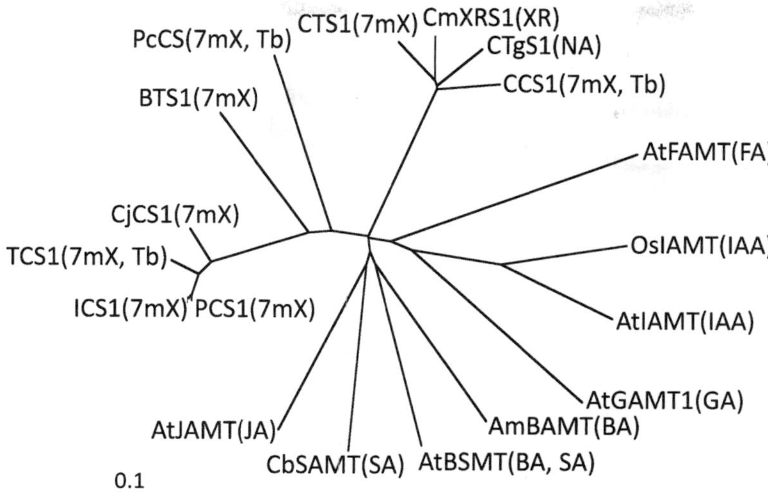

Fig. 6.4 Molecular phylogenetic tree of the motif B′ methyltransferase family. The unrooted tree was generated by the neighbor-joining method using Clustal W software [54]. The sources of sequences indicated by GenBank numbers are as follows: TCS1, AB031280; ICS1, AB056108; PCS1, AB207817; CjCS1, AB297451; CmXRS1, AB034699; CTS1, AB034700; CCS1, AB086414; CTgS1, AB054842; BTS1, AB096699; CbSAMT, AF133053; AmBAMT, AF198492; OsIAMT, EU375746; AtBSMT1, NM111981; AtJAMT, AY008434; AtIAMT1, NM124907; AtGAMT1, NM118775; AtFAMT, NM114355; PcCS, BK008796. Substrates of enzymes are shown in parentheses. Abbreviations of substrates are as follows: *XR* xanthosine, *7mX* 7-methylxantine, *Tb* theobromine, *NA* nicotinic acid, *FA* farnesoic acid, *IAA* indole 3-acetic acid, *GA*. gibberellic acid, *BA* benzoic acid, *SA* salicylic acid, *JA* jasmonic acid

proposes a speculative scheme that NAD is processed in parallel steps to both caffeine and trigonelline in coffee [9].

Well-characterized members of the motif B′ methyltransferase family are salicylic acid carboxyl methyltransferase (SAMT) [48], benzoic acid carboxyl methyltransferase (BAMT) [14], jasmonic acid carboxyl methyltransferase (JAMT) [51], indole-3-acetic acid methyltransferase (IAMT) [64], farnesoic acid methyltransferase (FAMT) [59], and gibberellic acid methyltransferase (GAMT) [56]. These members of the motif B′ methyltransferase family catalyze the formation of small molecule methyl esters by using SAM as a methyl group. Twenty-four and forty-one genes encoding motif B methyltransferase proteins have been identified in *Arabidopsis thaliana* and *Oryza sativa*, respectively [61]. However, the function of these genes has not been determined yet.

The three-dimensional crystal structure of several motif B′methyltransferases has been determined, including SAMT from *Clarkia breweri* [64], CcXMT1 and CcDXMT from *C. canephora* [33], and AtIAMT from *A. thaliana* [61]. The overall structure of the SAMT monomer consists of a globular domain containing an extended β-sheet and a unique α-helical cap that forms the top one-third of the active site cavity. SAMT exists as a homodimer in solution, and this dimeric arrangement is preserved in the crystal lattice [64]. This structure appears to be a common characteristic of motif B′ methyltransferase proteins. In addition, there are no residues located within the transmethylation pocket that act as a general acid/base for the methyl transfer reaction [33, 64].

A question yet to be addressed is how caffeine/theobromine synthases are differentiated from motif B′ methyltransferases. Further investigation into other motif B′ methyltransferases is needed to resolve this question.

6.5 Biotechnological Production of Caffeine by Genetic Engineering

The identification of caffeine synthase genes in purine alkaloid-producing plants led to produce transgenic caffeine-deficient coffee plants. Low-caffeine-containing coffee plants were generated by the RNA interference method [43, 44]. This method was effective in plantlets, although the suppression was not as complete as observed in embryogenic tissues of *C. arabica* [44].

The second approach is the engineering of microbial host for caffeine production. Caffeine production (0.38 mg/L) was reported by co-expression of *C. arabica* xanthosine methyltransferase (CaXMT) and *C. sinensis* caffeine synthase (TCS1) in *Saccharomyces cerevisiae* [21]. The authors also demonstrated the change in the sum yields of caffeine and theobromine by the expression of TCS 1 structure-guided mutants [21]. On the other hand, McKeague et al. improved by rational modifications to the native yeast central metabolic pathway the endogenous purine flux for the production of 7-methylxanthine, a key intermediate in caffeine biosynthesis (2016). They achieved titers of 0.27 mg/L, 0.061 mg/L, and 3.7 mg/L of caffeine, theophylline, and 3-methylxanthine, respectively, in 0.3-L bench-scale batch fermentation [34].

References

1. Ashihara H. Purine metabolism and the biosynthesis of caffeine in maté leaves. Phytochemistry. 1993;33:1427–30.
2. Ashihara H, Crozier A. Biosynthesis and metabolism of caffeine and related purine alkaloids in plants. Adv Bot Res. 1999;30:117–205.
3. Ashihara H, Gillies FM, Croizer A. Metabolism of caffeine and related purine alkaloids in leaves of tea (*Camellia sinensis* L.). Plant Cell Physiol. 1997;38:413–9.
4. Ashihara H, Kato M, Chuang-Xing Y. Biosynthesis and metabolism of purine alkaloids in leaves of cocoa tea (*Camellia ptilophylla*). J Plant Res. 1998;111:599–604.
5. Ashihara H, Kubota H. Biosynthesis of purine alkaloids in *Camellia* plants. Plant Cell Physiol. 1987;28:535–9.
6. Ashihara H, Kato M, Crozier A. Distribution, biosynthesis and catabolism of methylxanthines in plants. In: Fredholm BB, editor. Methylxanthines. Berlin/Heidelberg: Springer; 2011. p. 11–31.
7. Ashihara H, Sano H, Crizier A. Caffeine and related purine alkaloids: biosynthesis, catabolism, function and genetic engineering. Phytochemistry. 2008;69:841–56.

8. Ashihara H, Suzuki T. Distribution and biosynthesis of caffeine plants. Front Biosci. 2004;9:1864–76.
9. Baumann TW. Revisiting caffeine biosynthesis – speculations about the proximate source of its purine ring. Nat Prod Commun. 2015;10:793–7.
10. Baumann TW, Schulthess BH, Hanni K. Guarana (*Paullinia cupana*) rewards seed dispersers without intoxicating them by caffeine. Phytochemistry. 1995;39:1063–70.
11. Frischknecht PM, Ulmer-Dufex J, Baumann TW. Purine alkaloid formation in buds and developing leaflets of *Coffea arabica*: expression of an optimal defence strategy? Phytochemistry. 1986;25:613–6.
12. D'Auria JC, Chen F, Pichersky E. The SABATH family of MTs in *Arabidopsis thaliana* and other plant species. Recent Adv Phytochem. 2003;37:253–83.
13. Denoeud F, Carretero-Paulet L, Dereeper A, Droc G, Guyot R, Pietrella M, et al. The coffee genome provides insight into the convergent evolution of caffeine biosynthesis. Science. 2014;345:1181–4.
14. Dudareva N, Murfitt LM, Mann CJ, Gorenstein N, Kolosova N, Kish CM, et al. Developmental regulation of methyl benzoate biosynthesis and emission in snapdragon flowers. Plant Cell. 2000;12:949–61.
15. Fujimori N, Ashihara H. Adenine metabolism and the synthesis of purine alkaloids in flowers of *Camellia*. Phytochemistry. 1990;29:3513–6.
16. Fujimori N, Ashihara H. Biosynthesis of the theobromine and caffeine in developing leaves of *Coffea arabica*. Phytochemistry. 1994;36:1359–61.
17. Fujimori N, Suzuki T, Ashihara H. Seasonal variations in biosynthetic capacity for the synthesis of caffeine in tea leaves. Phytochemistry. 1991;30:2245–8.
18. Hammerstone JF, Romanczyk LJ, Aitken WM. Purine alkaloid distribution within *Herrania* and *Theobroma*. Phytochemistry. 1994;35:1237–40.
19. Huang R, O'Donnell AJ, Barboline JJ. Convergent evolution of caffeine in plants by co-option of exapted ancestral enzymes. Proc Natl Acad Sci U S A. 2016;113:10613–8.
20. Ishida M, Kitao N, Mizuno K, Tanikawa N, Kato M. Occurrence of theobromine synthase genes in purine alkaloid-free species of *Camellia* plants. Planta. 2009;229:559–68.
21. Jin L, Bhuiya MW, Li M, Liu X-Q, Han J, Deng W-W, et al. Metabolic engineering of *Saccharomyces cerevisiae* for caffeine and theobromine production. PLoS One. 2014;9:e105368.
22. Joshi CP, Chiang VL. Conserved sequence motifs in plant *S*-adenosyl-L-methionine-dependent methyltransferases. Plant Mol Biol. 1998;37:663–74.
23. Kato M, Kanehara T, Shimizu H, Suzuki T, Gillies FM, Crozier A, et al. Caffeine biosynthesis in young leaves of *Camellia sinensis*; *in vitro* studies on *N*-methyltransferase activity involved in the conversion of xanthosine to caffeine. Plant Physiol. 1996;98:629–36.
24. Kato M, Kitao N, Ishida M, Morimoto H, Irino F, Mizuno K. Expression for caffeine biosynthesis and related enzymes in *Camellia sinensis*. Z Naturforsch. 2010;65c:245–56.
25. Kato M, Mizuno K, Fujimura T, Iwama M, Irie M, Crozier A, et al. Purification and characterization of caffeine synthase from tea leaves. Plant Physiol. 1999;120:579–86.
26. Kato M, Mizuno K, Crozier A, Fujimura T, Ashihara H. Caffeine synthase gene from tea leaves. Nature. 2000;406:956–7.
27. Kato M, Mizuno K. Caffeine synthase and related methyltransferases in plants. Front Biosci. 2004;9:1833–42.
28. Koyama Y, Tomoda Y, Kato M, Ashihara H. Metabolism of purine bases, nucleosides and alkaloids in theobromine-forming *Theobroma cacao* leaves. Plant Physiol Biochem. 2003;41:977–84.
29. Kretschmar JA, Baumann TW. Caffeine in *Citrus* flowers. Phytochemistry. 1999;52:19–23.
30. Li Y, Ogita S, Keya CA, Ashihara H. Expression of caffeine biosynthesis gene in tea (*Camellia sinensis*). Z Naturforsch. 2008;63c:267–70.
31. Mazzafera P. Caffeine, theobromine and theophylline distribution in *Ilex paraguariensis*. Rev Bras Fisiol Veg. 1994;6:149–51.

32. Mazzafera P, Carvalho A. Breeding for low seed caffeine content of coffee (*Coffea* L.) by inter specific hybridization. Euphytica. 1992;59:55–60.
33. McCarthy AA, McCarthy JG. The structure of two *N*-methyltransferases from the caffeine biosynthetic pathway. Plant Physiol. 2007;144:879–89.
34. McKeague M, Wang Y-H, Cravens A, Win MN, Smolke CD. Engineering a microbial platform foe *de novo* biosynthesis of diverse methylxanthines. Metab Eng. 2016;38:191–203.
35. Mizuno K, Matsuzaki M, Kanazawa S, Tokiwano T, Yoshizawa Y, Kato M. Conversion of nicotinic acid to trigonelline is catalyzed by *N*-methyltransferase belonged to motif B' methyltransferase family in *Coffea arabica*. Biochem Biophys Res Commun. 2014;452:1060–6.
36. Mizuno K, Tanaka H, Kato M, Ashihara A, Fujimura T. cDNA cloning of caffeine (theobromine) synthases from coffee (*Coffea arabica* L.). In: International scientific colloquim on coffee. ASIC, Paris 2001;19:815–818.
37. Mizuno K, Okuda A, Kato M, Yoneyama N, Tanaka H, Ashihara A, et al. Isolation of a new dual-functional caffeine synthases gene encoding an enzyme for the conversion of 7-methylxanthosine to caffeine from coffee (*Coffea arabica* L.). FEBS Lett. 2003;534:75–81.
38. Mizuno K, Kato M, Irino F, Yoneyama N, Fujimura T, Ashihara H. The first committed step reaction of caffeine biosynthesis: 7-mehtlxanthosine synthase is closely homologous to caffeine synthases in coffee (*Coffea arabica* L.). FEBS Lett. 2003;547:56–60.
39. Nagata T, Sakai S. Differences in caffeine, flavanols and amino acids contents in leaves of cultivated species of *Camellia*. Jpn J Breed. 1984;34:459–67.
40. Nagata T, Sakai S. Purine base pattern of *Camellia irrawadiensis*. Phytochemistry. 1985;24:2271–2.
41. Nakayama F, Mizuno K, Kato M. Biosynthesis of caffeine underlying diversity of motif B' methyltransferase. Nat Prod Commun. 2015;10:799–801.
42. Ogawa M, Herai Y, Koizumi N, Kusano T, Sano H. 7-Methylxanthine methyltransferase of coffee plants. J Biol Chem. 2001;276:8213–8.
43. Ogita S, Uefuji H, Yamaguchi Y, Koizumi N, Sano H. Producing decaffeinated coffee plants. Nature. 2003;423:823.
44. Ogita S, Uefuji H, Morimoto M, Sano H. Application of RNAi to confirm theobromine as the major intermediate for caffeine biosynthesis in coffee plants with potential for construction of decaffeinated varieties. Plant Mol Biol. 2004;54:931–41.
45. Petermann JB, Baumann TW. Metabolic relations between methylxanthines and methyluric acids in *Coffea* L. Plant Physiol. 1983;73:961–4.
46. Petermann JB, Baumann TW, Wanner H. A new tetramethyluric acid from *Coffea liberica* and *C. dewevrei*. Phytochemistry. 1977;16:621–2.
47. Qian P, Guo H-B, Yue Y, Wang L, Yang X, Guo H. Understanding the catalytic mechanism of xanthosine methyltransferase in caffeine biosynthesis from QM/MM molecular dynamics and free energy simulations. J Chem Inf Model. 2016;56:1755–61.
48. Ross JR, Nam KH, D'Auria JC, Pichersky E. *S*-Adenosyl-L-methionine: salicylic acid carboxyl methyltransferase, an enzyme involved in floral scent production and plant defense, represents a new class of plant methyltransferases. Arch Biochem Biophys. 1999;367:9–16.
49. Schimpl FC, Kiyota E, Mayer JL, Gonçalves JF, da Silva JF, Mazzafera P. Molecular and biochemical characterization of caffeine synthase and purine alkaloid concentration in guarana fruit. Phytochemistry. 2014;105:25–36.
50. Senanayake UM, Wijesekera OB. Theobromine and caffeine content of the cocoa bean during its growth. J Sci Food Agric. 1971;22:262–3.
51. Seo HS, Song JT, Lee YH, Iwang I, Lee JS, Choi YD. Jasmonic acid carboxyl methyltransferase: a key enzyme for jasmonate-regulated plant responses. Proc Natl Acad Sci U S A. 2001;98:4788–93.
52. Schluthess BH, Morath P, Baumann TW. Caffeine biosynthesis starts with the metabolically channelled formation of 7-methyl-XMP – a new hypothesis. Phytochemistry. 1996;41:169–75.
53. Suzuki T. The participation of *S*-adenosylmethionine in the biosynthesis of caffeine in the tea plants. FEBS Lett. 1972;24:18–20.

54. Thompson JD, Higgins DG, Gibson TJ. CLUSTAL W: improving the sensitivity of progressive multiple sequence alignment through sequence weighting, position-specific gap penalties and weight matrix choice. Nucleic Acids Res. 1994;22:4673–80.
55. Uefuji H, Ogita S, Yamaguchi Y, Koizumi N, Sano H. Molecular cloning and functional characterization of three distinct *N*-methyltransferases involved in the caffeine biosynthetic pathway in coffee plants. Plant Physiol. 2003;132:372–80.
56. Varbanova M, Yamaguchi S, Yang Y, McKelvey K, Hanada A, Borochov R, et al. Methylation of gibberellins by *Arabidopsis* GAMT1 and GAMT2. Plant Cell. 2007;19:32–45.
57. Wanner H, Pešáková M, Baumann TW, Charubala R, Guggisberg A, Hesse M, et al. *O*(2),1,9-Trimethyluric acid and 1,3,7,9-tetramethyluric acid in leaves of different *Coffea* species. Phytochemistry. 1975;14:747–50.
58. Weckerle CS, Stuz MA, Baumann TW. Purine alkaloids in *Paullinia*. Phytochemistry. 2003;64:735–42.
59. Yang Y, Yuan JS, Ross J, Noel JP, Pichersky E, Chen F. An *Arabidopsis thaliana* methyltransferase capable of methylating farnesoic acid. Arch Biochem Biophys. 2006;448:123–32.
60. Yoneyama N, Morimoto H, Ye C, Ashihara H, Mizuno K, Kato M. Substrate specificity of *N*-methyltransferase involved in purine alkaloids synthesis is dependent upon one amino acid residue of the enzyme. Mol Gen Genomics. 2006;275:125–35.
61. Zhao N, Ferrer J-L, Ross J, Guan J, Yang Y, Pichersky E, Noel JP, Chen F. Structural, biochemical, and phylogenetic analyses suggest that indole-3-acetic acid methyltransferase is an evolutionarily ancient member of the SABATH family. Plant Physiol. 2008;146:455–67.
62. Zheng X-Q, Ashihara H. Distribution, biosynthesis and function of purine and pyridine alkaloids in *Coffea arabica* seedlings. Plant Sci. 2004;166:807–13.
63. Zheng X-Q, Ye C-X, Kato M, Crozier A, Ashihara H. Theacrine (1,3,7,9-tetramethyluric acid) synthesis in leaves of a Chinese tea, kucha (*Camellia assamica* var. kucha). Phytochemistry. 2002;60:129–34.
64. Zubieta C, Ross JR, Koscheski P, Yang Y, Pichersky E, Noel JP. Structural basis for substrate recognition in the salicylic acid carboxyl methyltransferase family. Plant Cell. 2003;15:1704–16.

Taxol® Biosynthesis and Production: From Forests to Fermenters

7

Christopher McElroy and Stefan Jennewein

Abbreviations

CPRs	Cytochrome P450 reductases
DMAPP	Dimethylallyl diphosphate
DXP	1-Deoxy-D-xylulose 5-phosphate pathway
GGPP	Geranylgeranyl diphosphate
HGT	Horizontal gene transfer
IPP	Isopentenyl diphosphate
MEP	2-C-methyl-D-erythritol 4-phosphate pathway
MVA	Mevalonate pathway
P450	Cytochrome P450 dependent mono-oxygenase

7.1 Introduction

Taxol® is a complex metabolite comprising a tetracyclic oxaheptadecane skeleton decorated with eight functional oxygen groups, two acyl groups and a benzyl group. This intricate molecule (Fig. 7.1) was first structurally characterized in 1971 [202]. It took 21 years before this highly desirable anti-cancer drug became available to the public following approval by the Food and Drug Administration (FDA). Despite the

C. McElroy
Institute for Molecular Biotechnology, RWTH Aachen University, Worringerweg 1, 52074 Aachen, Germany
e-mail: christopher.mcelroy@ime.fraunhofer.de

S. Jennewein (✉)
Department of Industrial Biotechnology, Fraunhofer Institute for Molecular Biology and Applied Ecology (IME), Forckenbeckstrasse 6, 52074 Aachen, Germany
e-mail: stefan.jennewein@ime.fraunhofer.de

W. Schwab et al. (eds.), *Biotechnology of Natural Products*,
https://doi.org/10.1007/978-3-319-67903-7_7

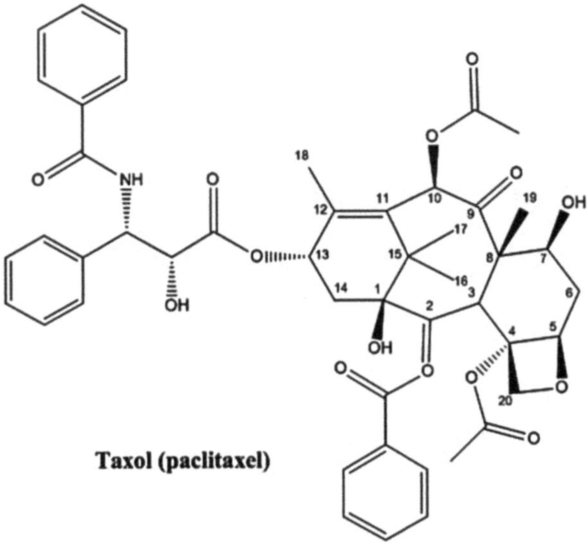

Fig. 7.1 The structure of taxol (paclitaxel)

difficult development process, taxol and related taxane analogues have become leading anti-cancer drugs with sales reported previously as exceeding five billion USD [34, 140]. The annual financial reports for 2016 from Sanofi and Celgene [24, 165], state that combined revenues from Taxotere®, Jevtana® and Abraxane® products summed in excess of one and a half billion USD. Although suggestive of a decrease, these figures do not include the sales of generics and other competitor flagship products that use paclitaxel or other toxic compounds linked to conjugate monoclonal antibodies such as Avastin® [160], which alone generated revenues greater than six billion USD in 2016. The lengthy development time largely reflected the lack of a good source, because the acquisition of 1 g of taxol was highly destructive. It required the bark of three adult Pacific yew (*Taxus brevifolia* Nutt) trees, which provided ~12 kg of bark material. Purification was laborious and complex due to the hydrophobic nature of the compound, which resulted in a yield of only 0.5 g, effectively 0.004% of the original dry weight [203]. The advancement of taxol also required delicate political manoeuvring due to conflicts with environmentalists, the relationship between privatized intellectual property and consumer drug costs, as well as the need to assess side effects [5, 60].

Taxol is a potent anti-neoplastic compound that binds to the β-tubulin subunits of microtubules, causing their polymerization even at low temperatures and in the presence of calcium concentrations that normally promote depolymerization [157]. Doses greater than 10 nM cause the accumulation of large, stable tubulin bundles, preventing cellular proliferation and thereby arresting cells in metaphase, whereas lower concentrations lead to apoptosis [166]. Studies *in vitro* with purified bovine brain tubulin and with HeLa cells confirmed these findings. In 1992, taxol was approved by the FDA for the treatment of refractory ovarian cancer. This was followed by its approval for the treatment of mammary cancer in 1994, for lung carcinoma in 1999 and adenocarcinoma in 2014 [203]. Indeed, since its approval, the drugs developed from taxol have been administered (independently or as part of a drug cocktail) to more than one million patients [203].

Taxol was discovered during a collaborative screening program in 1960 between the National Cancer Institute (NCI) and the US Department of Agriculture (USDA) led by Jonathan L. Hartwell. The purpose of the program was to identify cytotoxic natural products from plants. In the second year of the program, Arthur S. Barclay and three USDA assistants collected 650 plant samples, including stems, leaves, bark and fruit from the slow growing tree *Taxus brevifolia* [202, 203]. Crude extracts were then tested for cytotoxicity against 9 KB cell cultures isolated from nasopharynx tumours. Based on these initial findings, Hartwell was asked to send crude extracts to the medical chemists Monroe E. Wall and Mansukh C. Wani at the Research Triangle Institute (RTI). This ultimately led to this institute being given the assignment of extracting and purifying taxol from *T. brevifolia* samples, a laborious process which commenced in 1964 [203]. Purification was achieved by ethanol extraction, separation of the ethanol phase between chloroform and water, and many Craig countercurrent distribution treatments. The extraction steps were guided by *in vivo* inhibition assays using Walker WM solid tumours in order to prevent the loss of taxol [199]. Following purification, taxol was structurally characterized by chemical conversion followed by ^{1}H nuclear magnetic resonance (NMR) spectroscopy, X-ray diffraction and high-resolution mass spectrometry. Later, the ester located on the thirteenth carbon was found to be critical for antitumor activity and that its removal caused taxol to become inactive [202].

The current status of *T. brevifolia* is "near threatened". Its population has noticeably declined and continues to do so due to fires, logging and the exploitation of its bark for the acquisition of taxol [185]. Because these trees require ~300 years to reach maturity [158], it has long been appreciated that alternative sources or methods for taxol production must be developed to avoid the need to rely on destructive bark harvesting [159]. These include complete chemical synthesis, partial synthesis from taxol precursors such as baccatin III and 10-deacetylbaccatin III extracted from European yew needles, and the development of *Taxus* spp. cell tissue cultures for precursor synthesis. *Taxus* spp. cell cultures are currently used to produce taxol and related compounds from which taxol can be synthesized, but the metabolic engineering of microorganisms may replace this source in the future. *Taxus* spp. cell cultures are the current method of choice because they can be grown under controlled conditions, they are environmentally sustainable and also cost effective [132, 183]. However, as more potential uses for this compound are found, the demand will continue to increase, encouraging innovation and the development of advanced production methods with greater productivity. For the time being, such new methods will continue to be hampered by our lack of understanding of the taxol biosynthesis pathway. Once the entire pathway is well characterized, the entire pathway could be optimized and transferred into an industrially relevant microorganism which could be metabolically engineered for industrial-scale production. The following section summarizes our present knowledge of the taxol biosynthesis pathway.

7.2 The Biosynthesis of Taxol

The entire taxol biosynthesis pathway is not yet known, but it is widely agreed that it requires 19 enzymatic steps [88] from the precursor geranylgeranyl diphosphate (GGPP) to taxol itself [34]. Our present understanding of the genes and enzymes in this pathway mainly reflects the exceptional work carried out by Rodney Croteau, Robert M. Williams and their co-workers. Thus far, 14 enzymes have been well characterized (Table 7.1), although a candidate 15th enzyme was recently described in the form of a newly identified β-phenylalanine-CoA ligase identified by analysing the transcripts of *Taxus baccata* cells elicited with methyl jasmonate [155]. This is a known elicitor of taxol and baccatin III synthesis in cell cultures [108, 140, 219]. Furthermore, a list of potential candidate genes encoding the elusive four remaining enzymes was also provided, i.e. C1 and C9 hydroxylases, C4-C20 epoxidase, oxomutase and C9 oxidase [155].

Despite these exciting new revelations, these potential enzyme candidates require thorough testing using standard feed experiments with synthetic and isotopically-labelled substrates, followed by the structural determination of products and testing for the incorporation of these isolated intermediates into the taxol pathway *in vivo*.

The biosynthesis of taxol is a complex process involving eight oxidation steps, five acetyl/aroyl transferase steps, a C4β,C20-epoxidation reaction, a phenylalanine aminomutase step, *N*-benzoylation and two CoA esterifications [140] as shown in Fig. 7.2. Taxol is just one of ~400 taxoids (taxane diterpenoids) that share the same taxane skeleton produced by *Taxus* species [84], and individual *Taxus* species produce several taxoids at a time. Whereas the vast majority likely have clear biological roles such as discouraging predation due to their toxicity to mammals [139], insect anti-feeding activities [37] and fungal antibiotic activities [48], a small minority of these taxoids may be products of promiscuous acetyltransferases and oxygenases [214] which are possibly redirected into alternative taxoid pathway end products by as yet unknown enzymes. Due to the energy investment required to produce these diverse taxane diterpenoids, it has been suggested that increasing taxol yields in *Taxus* spp. cell cultures will need to "take into account these numerous and apparently diversionary taxoid biosynthetic side-routes and dead-ends" [34].

The source of the precursors, isopentenyl diphosphate (IPP) and dimethylallyl diphosphate (DMAPP) for the taxol biosynthesis pathway is the plastid 2-C-methyl-D-erythritol phosphate (MEP) pathway [46]. The conversion of three molecules of IPP and one of DMAPP into GGPP is catalysed by geranylgeranyl diphosphate synthase (GGPPS). Despite evidence suggesting that fosmidomycin and mevinolin inhibit the plastid and cytosolic IPP pathways of *T. baccata* cell cultures, respectively [36], both taxadiene synthase and GGPPS have native plastid targeting peptides suggesting a local IPP bias from the plastids [70, 210].

GGPP is then cyclized by taxadiene synthase (TS) into taxa-4(5),11(12)-diene (~95%) also known as taxadiene, taxa-4(20),11(12)-diene (~5%) and taxa-3(4),11(12)-diene (<1%) [65]. This terpene synthase is an interesting enzyme, which contains three α-helix domains, a class I terpenoid cyclase domain located at the C-terminus and a vestigial class II terpenoid cyclase formed via a tertiary domain and the N-terminus

Table 7.1 List of all known taxol biosynthesis enzymes and their original publications

Enzyme	Abbreviation	Size (kDa)	Probable localization	GenBank number	Reference
Geranylgeranyl diphosphate synthase	GGPPS	42	Plastids	AF081514	[70]
Taxadiene synthase	TS	98	Plastids	AY364469	[209]
Taxadiene-5α-hydroxylase	T5αOH	56	ER	AY289209	([87])
Taxadiene-13α-hydroxylase	T13αOH	54	ER	AY056019	[85]
Taxadiene-5α-ol-O-acetyl transferase	TAT	49	Cytosol	AF190130	[195]
Taxane-10β-hydroxylase	T10βOH	56	ER	AF318211	[167]
Taxane-10β-hydroxylase[a]	T10βOH	55	ER	AY563635	[88]
Taxane-9α-hydroxylase[b]	T9αOH	?	ER	–	–
Taxane-7β-hydroxylase	T7βOH	56	ER	AY307951	[28]
Taxadiene-2α-hydroxylase	T2αOH	55	ER	AY518383	[27]
Taxane-1β-hydroxylase	T1βOH	?	ER	–	–
Taxane-2α-O-benzoyl transferase	TBT	50	Cytosol	AF297618	[193]
C4β,C20-epoxidase	EPOX	?	Cytosol	–	–
Oxomutase	OXM	?	Cytosol	–	–
Taxane-9α-dehydrogenase	T9αDH	?	ER	–	–
10-deacetylbaccatin III-10-O-acetyl transferase	DBAT	49	Cytosol	AF193765	[194]
Phenylalanine aminomutase	PAM	76	Cytosol	AY582743	[198]
β-phenylalanoyl-CoA ligase[c]	PCL	59	Cytosol	KM593667	[155]
Baccatin III: 3-amino, 13-phenylpropanoyltransferase	BAPT	50	Cytosol	AY082804	[196]
Taxane-2'α-hydroxylase	T2'αOH	?	ER	–	–
N-benzoyl transferase	DBTNBT	49	Cytosol	AF466397	[197]
NADPH:cytochrome P450 reductase	NADPH: P450-Red	86	ER	AY571340	[89]

[a]A second taxane-10β-hydroxylase has also been found
[b]A presumed T9αOH has been identified but also requires further testing [34]
[c]The PCL protein listed here is a putative candidate isolated from *T. baccata* cultures, which is probably part of the pathway
ER endoplasmic reticulum

[100]. The formation of taxa-4(5),11(12)-diene by TS was rather unexpected, but enzyme assays using partially purified TS from *T. brevifolia* and different deuterium labelled substrates, [1-$^2H_2$, 20-$^2H_3$] and [20-$^2H_3$] geranylgeranyl diphosphate and putative olefin intermediates, conclusively showed the direct formation of the more stable endocyclic double bond isomer taxa-4(5),11(12)-diene [116]. Also experiments using heterologous expressed and purified enzyme confirmed taxa-4(5),11(12)-diene as the main conversion product of TS [210]. A comparison of the *in vitro* activity of TS and *Taxus canadensis* cell suspension cultures indicated that although the reaction is slow, it is not rate limiting within the pathway, nor does it cause a detectable accumulation of taxadiene [73]. After cyclization, the pathway diverges along a multitude of branches leading to the synthesis of other taxanes, via hydroxylations or acylations by other enzymes such as P450 taxane-14β-hydroxylase [86].

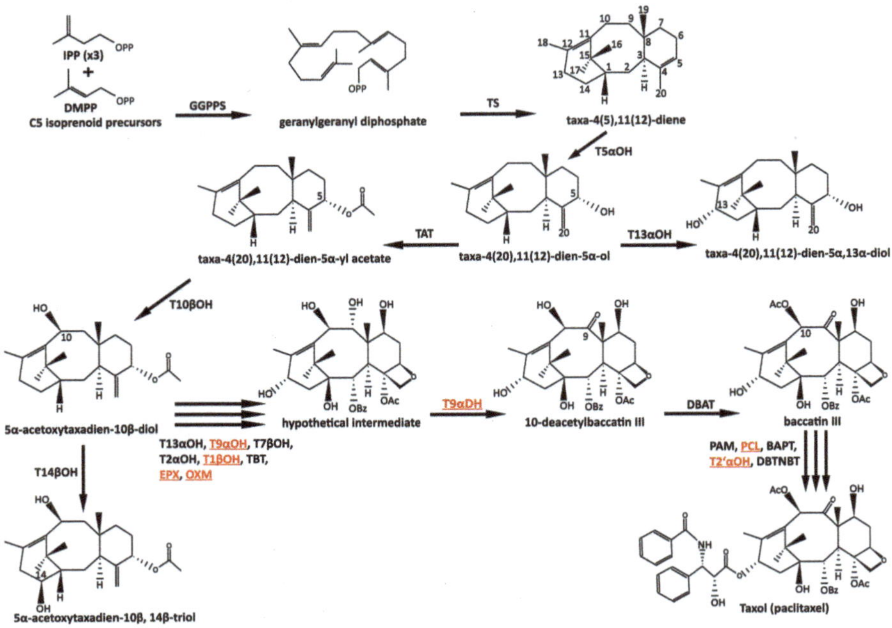

Fig. 7.2 Overview of taxol biosynthesis. Labelled steps (*underlined and marked by red colour*) indicate uncharacterized enzymatic steps. Multiple arrows represent interim sections of the pathway where several enzymes and the order of their reactions are unknown

Taxa-4(5),11(12)-diene is hydroxylated by a class II cytochrome P450 hydroxylase called taxadiene-5α-hydroxylase or CYP725A4 (T5αOH) bearing an N-terminal endoplasmic reticulum (ER) insertion sequence [87]. T5αOH converts taxa-4(5),11(12)-diene into taxa-4(20),11(12)-dien-5α-ol via hydroxylation at the fifth carbon and migration of the carbon double bond. The exact mechanism of this reaction is still unknown, and remains the topic of much debate. The original hypothesis was that T5αOH mediates hydrogen atom abstraction from the C20 methyl group of the 4(5)-olefin isomer, yielding an allylic radical which is subsequently oxygenated at the fifth carbon [65, 87]. Alternatively, taxadiene-4(5)-epoxide may facilitate the transition of taxa-4(5),11(12)-diene into taxa-4(20),11(12)-dien-5α-ol [10, 14, 45]. Investigations involving the heterologous expression of TS and T5αOH to increase taxa-4(20),11(12)-dien-5α-ol yields have focused this debate, providing insight into alternative taxol production platforms. The expression of the *T. brevifolia* TS and the *T. cuspidata* T5αOH in *Nicotiana sylvestris* surprisingly did not lead to the anticipated overproduction of taxadiene, but instead produced 5(12)-oxa-3(11)-cyclotaxane (OCT) [163]. OCT was the sole product from the conversion of taxadiene by T5αOH, instead of forming taxadien-5α-ol. This surprising result has also been reported in *Escherichia coli*, through the use of heterologous expression platforms and chimeric fusion enzymes made up of T5αOH and cytochrome P450 reductases (CPRs) [4, 214]. However, the product spectrum varied widely in these platforms, which is unsurprising considering the difficulties encountered when expressing plant P450s in *E. coli* [169], caused by the absence of an ER and of CPRs. Despite these limitations, the above approach has been successful in expressing a codon optimized synthetic T5αOH linked to a CPR, e.g. the chimeric At24T5αOH-tTCPR was able to convert 98% of the taxadiene into taxadien-5α-ol and OCT, in roughly equal amounts [4]. However, such chimeric enzymes can lead to abnormal product spectra such as the production of up to 16 other molecules within *E. coli*, of which taxadien-5α-ol is but a minor product drawing just 10% of the flux [214]. What is clear from these studies is that the chassis and enzymes used do not resemble the native *Taxus* system, because such promiscuity if attributable to the native T5αOH would have been detected and thus far no epoxide intermediates have been found in *Taxus* microsomes, although the adventitious taxa-4(20),11(12)-diene product has. Furthermore, attempts to find this hypothetical epoxide product in native physiological samples using [20-$^{2}H_3$] taxa-4(5),11(2)-diene were unable to identify the kinetic isotope effect after deprotonation [65]. Although recent evidence from *Taxus cuspidata* cell suspension cultures has shown the presence of an OCT epoxide intermediate, albeit in low amounts (0.5 mg L^{-1}) [45], these cultures were elicited with 40 μL L^{-1} methyl jasmonate to upregulate gene expression in the pathway, and when supplemented with as much as 260 mg L^{-1} of exogenous taxadiene the cells produced barely 3 mg L^{-1} OCT. This implies that if the supplemented taxadiene is fully incorporated into the host cells then taxadiene and (if present under non-elicited physiological conditions) OCT are quickly used within the native taxol pathway and its branches for the biosynthesis of other products.

The next part of the taxol pathway diverges, instigated by either a hydroxylation at the 13α position, or an acylation at the 5α position of the taxane skeleton of taxa-4(20),11(12)-dien-5α-ol [85, 195]. Taxadiene-5α-ol-*O*-acetyl transferase (TAT) has no organelle targeting sequences suggesting a cytosolic localization, and catalyses the acylation of the fifth carbon of taxa-4(20),11(12)-dien-5α-ol to form taxa-4(20),11(12)-dien-5α-yl-acetate. This acetylated substrate can then be further hydroxylated by a P450 of the CYP725A1 family, 10β-hydroxylase (T10βOH) [167], forming 5α-acetoxytaxadien-10β-ol, although with lower efficiency this P450 can also hydroxylate the 5α alcohol [34]. Concomitantly, the 13α-hydroxylase (T13αOH) P450 of the CYP725A family preferentially hydroxylates the thirteenth carbon of the taxa-4(20),11(12)-dien-5α-ol rather than that of the taxa-4(20),11(12)-dien-5α-yl-acetate substrate [205]. Interestingly, attempts to produce triols from these diols using the corresponding hydroxylases have proven unsuccessful [85]. A subsequent bifurcation introduced by the 14β-hydroxylase (T14βOH) leads to the formation of 5α-acetoxytaxadien-10β,14β-diol, which although not an intermediate of the taxol pathway does lead to the biosynthesis of other taxoids. The order of the subsequent catalytic steps in the pathway is not yet clear, with hydroxylations at positions C1, C2, C4, C7 and C9, a further oxidation at C9, and a C4β,C20 epoxidation. The 2α and 7β hydroxylases were identified using the surrogate substrate taxusin [27, 28], revealing that an intermediate in the middle of the pathway is first hydroxylated by the 7β-hydroxylase and then by the 2α-hydroxylase. Furthermore, these P450 hydroxylases are selective for acetylated and poly-oxygenated taxadiene substrates [34]. Contrasting this proposed sequence is the ratio of oxygenated taxoids found in *Taxus* spp. cell cultures, which suggests the sequence of hydroxylations is C5, C10, C13, C2, C9, C7 and C1 [192], although other sequences have also been proposed [78].

It is unclear whether the intermediates identified using *in vitro* tissue cultures are truly part of taxol biosynthesis or rather intermediates *en route* to other taxoids [140]. The oxygenation of the ninth carbon is believed to be an early event whereas the oxygenation of the first carbon occurs later [78], so the approximate hydroxylation order C5, C10, C13, C9, C7, C2 and C1 may be more accurate. However, this requires more investigation because the acylation of the poly-hydroxylated substrate is likely to occur during the formation of a hypothetical heptaol intermediate [34]. Additionally, it is unclear how the oxetane (trimethylene oxide) ring at C4 and C5 is formed [65], although it may involve the enzymatic epoxidation of the C4(20) double bond, subsequent migration of the C5 acetyl-oxy group to C4, and final oxetane group formation through oxirane (ethylene oxide) expansion. Neither the proposed C4β,C20-epoxidase (EPOX) nor the oxomutase (OXM) responsible for the latter oxirane conversion have been identified, but potential gene candidates have been discovered through the analysis of jasmonate-induced *T. baccata* cultures [155]. At the midpoint of the pathway, a hypothetical poly-hydroxylated and acylated taxadiene substrate undergoes oxidation at C9 to form a ketone, followed by the creation of the oxetane ring, the attachment of a hydroxyl group at the first carbon, and the benzoylation of the hydroxyl group at C2 [193]. The putative 9α-hydroxylase (T9αOH) is yet to be characterized, but a cDNA has been identified in feeding experiments using taxa-4(20),11(12)-diene-5a-ol as a substrate, although this work remains incomplete [196]. The timing of the hydroxylation at C1 is also unknown, and the anticipated 1β-hydroxylase (T1βOH) should be forthcoming once the appropriate surrogate substrates are acquired [34]. To reach 10-deacetylbaccatin III, a hypothetical poly-hydroxylated and acylated intermediate is benzoylated in a regiospecific manner by taxane-2α-O-benzoyl transferase (TBT), an enzyme which is also likely to be cytosolic [193].

Fortunately, the second half of the pathway is better understood, and the enzymes responsible for the acyl transfers and side chain attachments have been characterized (Fig. 7.3). The further trans-acetylation of 10-deacetylbaccatin III at the C10 position by 10-deacetylbaccatin III-10-*O*-acetyl transferase (DBAT) [194] produces baccatin III, an important intermediate in the taxol pathway. Baccatin III and 10-deacetylbaccatin III are major substrates for the chemical semi-synthesis of taxol. They can be isolated from the needles of yew trees [15, 135, 136] and can be modified with synthesized side chains to produce taxol and its analogues [77, 208]. The β-phenylalanoyl-CoA side chain is synthesized naturally in a two-step process: first β-phenylalanine is produced from α-phenylalanine by phenylalanine aminomutase (PAM) [78, 198] and then a CoA ligase is predicted to activate the side chain into β-phenylalanoyl-CoA. This β-phenylalanoyl-CoA ligase (PCL) could be encoded by the *T. baccata* TB768 gene, which encodes a cytoplasmic PCL that can convert β-phenylalanine into β-phenylalanoyl-CoA [155].

Subsequently the esterification of the β-phenylalanoyl-CoA side chain to the C13 hydroxyl group of baccatin III is mediated by baccatin III: 3-amino, 13-phenylpropanoyltransferase (BAPT), producing 3′-*N*-debenzoyl-2′-deoxy-taxol

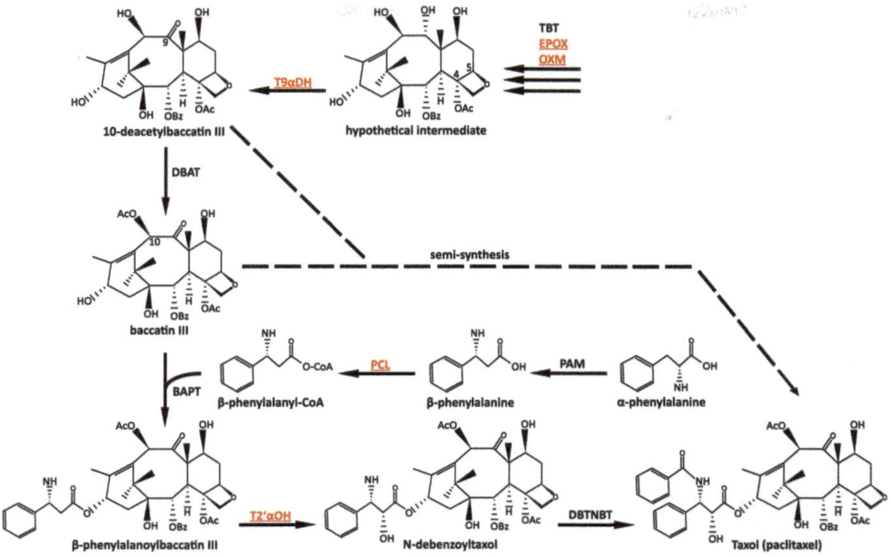

Fig. 7.3 Terminal steps in the taxol biosynthesis pathway. The dashed arrows represent semi-synthetic "short-cuts" through the pathway. Steps (*underlined and marked by red colour*) indicate uncharacterized enzymatic reactions

[196]. The last two steps are proposed to require the hydroxylation of the second carbon of the β-phenylalanine side chain and terminal *N*-benzoylation. The precise timing of the hydroxylation step is not known but it is assumed to occur prior to the benzoylation step because the *N*-benzoyl transferase (DBTNBT) [119, 120, 197] has a substrate preference for hydroxylated *N*-debenzoyl-taxol rather than debenzoyl-2′-deoxy-taxol based upon tests with both substrates. The hydroxylase taxane-2′α-hydroxylase (T2α'OH) is likely to be identified soon but its position within the pathway is unknown because the hydroxylation of free β-phenylalanyl-CoA has also not yet been attempted, and further investigation will require analysis with suitable substrates [155]. The pathway appears to be divided among different compartments in *Taxus* spp. cells, with GGPP synthesis localized to the plastids using available DMPP with local and translocated cytosolic IPP precursors for core taxane synthesis via TS. Following this, the multitude of hydroxylation and acylation reactions would suggest either free movement or trafficking between the ER and the cytosolic acetyltransferases. It is possible that some of the acylation steps may help to control flux through the pathway, theoretically limiting the accumulation of intermediates for taxol biosynthesis, thereby providing precursors for other taxoid species likely also possessing distinct biological roles. This is supported by the many acyltransferase genes responsible for numerous structural side chain modifications found among taxoid variants [34].

Extraction of advanced taxoids from plant material and semi-synthesis has served as a practical solution for the commercial supply of taxol and Taxotere (a second-generation clinical taxoid) for clinical applications [64, 106, 215]. Baccatin III and 10-deacetylbaccatin III can be extracted from the needles of *T. baccata* or

Taxus spp. cell cultures [6]. Thus, semi-synthesis provided a renewable alternative, supplanting the need for intensive deforestation of the pacific yew.

Due to the complexity of taxol possessing a unique tricyclic core, several chiral centres and a high degree of oxygenation, total synthesis is unlikely to be an efficient alternative to semi-synthesis or biotechnological solutions. However, the total synthesis of taxol presented organic chemists with a formidable challenge with which to demonstrate their synthetic skills and ingenuity. In 1994, two groups, one headed by Nicolaou, the other by Holten described the total chemical synthesis of taxol [79, 80, 136]. Both groups achieved the total chemical synthesis of taxol in 1994 simultaneously, albeit with extremely low yields of 2.7% and 0.07% respectively [77, 136]. The Nicolaou group developed a convergent synthesis method facilitated by Shapiro and McMurry coupling, requiring at least 40 steps, whereas the Holten group converted (−)-borneol into an unsaturated ketone for further modification. Many other synthesis methods have since been developed, with a recent method involving 37 steps from an easily available substrate [75]. For wider reading about the chemical synthesis of taxol, and how these developments have been translated into potential medical applications, readers are referred to [99].

Despite the development of chemical synthesis techniques (semi-synthesis and total synthesis), the inherent complexity and costs will likely discourage their adoption for high-volume production. In order to meet the predicted demands, metabolic engineering of *Taxus* spp. cells using gene editing techniques will help to avoid the loss of flux through branches and alternative end products. Eventually, industrial biotechnology will provide a practical and sustainable alternative but for this to become a reality the last few gaps in the pathway must be identified before it can be engineered in a suitable host organism that allows controlled, reliable and cost-effective production.

7.3 Taxol Production by Endophytes

Endophytes are considered asymptomatic bacterial and fungal microorganisms that for at least a part of their life cycle inhabit the intercellular spaces in plant tissues [68, 170]. However, the original definition referred to any species living together, such as microorganisms found within a plant [9]. Like many biological terms, this one is constantly evolving as our knowledge of plant microbiomes and host interactions increases [211]. Recent evidence for bacterial endophytes within the cytoplasm and the periplasm suggests that the specification of intercellular spaces should be updated, at least for bacterial endophytes [149, 186, 206]. It has been predicted that at least one species is likely to be present within each of the 350,699 [150] different plant species on Earth [181]. Endophytic fungi cause symptomless infections but are closely related to biotrophic and necrotrophic pathogens. Indeed, some species appear to have swapped roles multiple times during their evolutionary history [40]. Interestingly, infected plants acquire selectable advantages resulting from endophyte colonization, including improved growth and increased resilience to abiotic and biotic stress [33, 40, 161].

Over the last few decades, several genera of endophytic fungi have been shown to synthesize a plethora of valuable natural products [133]. The pool of secondary metabolites produced by endophytic fungi is vast and includes phenolic acids such as tyrosol and p-coumaric acids [102], quinones [217], hundreds of terpenoids [174], plant hormones such as gibberellins and indoleacetic acid [94, 204], as well as antimicrobial compounds such as Hsp90 inhibitors [126]. Endophytes have also been proposed to synthesize the potent cytotoxic compound taxol [177]. The finding that *Taxomyces adreanae* isolated from the phloem of *T. brevifolia* was capable of taxol production caused an explosion of subsequent publications reporting similar capabilities in other endophytic microorganisms [177]. Approximately 200 different fungi representing diverse orders are thought to produce taxol [54, 69]. Reported productivity has ranged from as little as 0.001 ng mL^{-1} [111] to ~800 ng mL^{-1} [118] in various isolates, but so far these results have not been independently reproduced [54]. Therefore, it is still unclear whether some of these species can actually produce taxanes, or whether the detected compounds were more likely artefacts of laboratory culture methods, or misidentified by the analytical and immunological methods deployed in the original studies [54, 71, 176]. The most intriguing aspect is how so many fungi could acquire the independent ability to synthesize this complex diterpenoid when at least 19 enzymatic steps are required, and which other compounds they can produce [34].

In many of the early publications claiming taxol production from isolated endophytes, the compound was detected using a competitive inhibition enzyme immunoassay. Without the necessary positive and negative controls, this method lacks stringency when only ng levels of a target compound are present [71]. Indeed, approximately 10 times the concentration of taxanes was detected in the tobacco (*Nicotiana tabacum*) plants used as a negative control than endophytes, which did not produce taxol but had carryover from their plant hosts. The authors proposed that the signal detected in the negative control was likely to reflect the cross-reactivity of the polyclonal antibody, because tobacco is not known to produce taxol or any other taxoids. In contrast, the positive control extract from *T. baccata* needles contained nearly 50,000-fold more taxanes. The authors also analysed isolates of *T. andreanae* (CBS 279.92) [180], UPH-12 (NRRL 30405) [76] and H10BA2 (NRRL 21209) [178] along with candidate taxol-producing endophytes by LC-MS/MS. Contrary to the earlier reports, no taxol was detected in *T. andreanae*, UPH-12 or H10BA2 using either of the analytical methods [72]. Furthermore, after sequencing the genomes of *T. andreanae* and EF0021, no evidence for a taxol biosynthesis pathway was found, providing a logical basis for the absence of taxol biosynthesis [71]. For *T. andreanae* not even a diterpene cyclase could be identify in the genome sequence, thus also excluding the possibility of taxoid biosynthesis via an alternative independently developed biosynthetic pathway. The relatively weak analytical evidence for taxol synthesis in some reported isolates [66, 147, 224] suggests that an overreliance on HPLC (and even HPLC-MS) and immunological methods has led to the misidentification of taxol [54]. Furthermore, LC-MS/MS methods for the accurate detection of taxol were described several years ago, but this method has been curiously neglected. The misidentification of taxol is supported by studies that have scrutinized the products of fungi isolated from *Taxus* species, e.g. one group found three novel harziane tetracyclic diterpenes, structurally similar to taxanes, in *Trichoderma atroviridae* [3].

The loss of taxol biosynthesis in endophytes was also proposed to reflect laboratory cultivation or storage in biorepositories, but this would not explain how the data for all these reports was acquired with such unstable isolates. The proteins responsible for taxol production should at least possess key structural features and thus should retain some similarity at the amino acid and nucleotide levels [71]. Predicted conserved sequences in genes encoding TS, 10-deacetylbaccatin III-10-O-acetyltransferase and C-13 phenylpropanoid side chain-CoA acyltransferase were therefore used for Southern blot analysis. The results indicated the presence of introns within the amplicons, suggesting that *Taxus* genomic sequences had been amplified rather than endophyte cDNAs. Furthermore, some reports include 'endophyte' taxol genes with up to 99% identity with the corresponding *Taxus* genes [222] suggesting host DNA contamination is the most likely source of these products. Additionally the sheer commonality of taxol biosynthesis within large clades of distant endophytic fungal and even bacterial species leads some researchers to question the likelihood of such an extreme case of convergent evolution. Isolates from increasingly surprising sources are being published, such as from the giant panda *Ailuropoda melanoleuca*, capable of producing 1.5 mg L^{-1} 'taxol' [63]. Something else that has also not been explained is the lack of intermediates or side products, which one would normally associate with any biosynthesis pathway. The taxoid biosynthesis pathway is neither simple nor fully understood in *Taxus* and yet many intermediates, branches and alternative end products have been identified, but these have not been identified in endophytes (Fig. 7.4).

Despite these discrepancies, which at the very least should prompt inter-laboratory comparisons for independent verification although this has not yet been implemented [174], few published reports have countered these findings or established concrete evidence for an endophytic pathway through the use of knockouts and protein characterization with *in vitro* substrate kinetic analysis. If the volume of publications is to be used itself as evidence then the pathway should be investigated in more detail to enrich our understanding of endophytic terpenoid synthesis. Two main hypotheses have been proposed to explain the evolution of endophytic taxol production, one based on the transfer of genetic information and the other based on convergent evolution. The first hypothesis proposes that horizontal gene transfer (HGT) events could be responsible for the transfer of metabolic capability from the host plant to the endophyte, but the dispersal of the corresponding genes makes this an unlikely explanation if one assumes that complete pathway transfer allows immediate taxol production [104]. Furthermore, recently published genomics and transcriptomics data suggest that biosynthesis genes in the 'taxol' producing microbe *Penicillium aurantiogriseum* NRRL 62431 evolved independently from its host, hazel [216]. This was based on the identification of putative transcripts that shared according to the authors "some similarity" with taxol biosynthesis genes. However, without a stringent knockout study including these sequences and any other putative fungal taxol biosynthesis genes, the ability of endophytes to produce

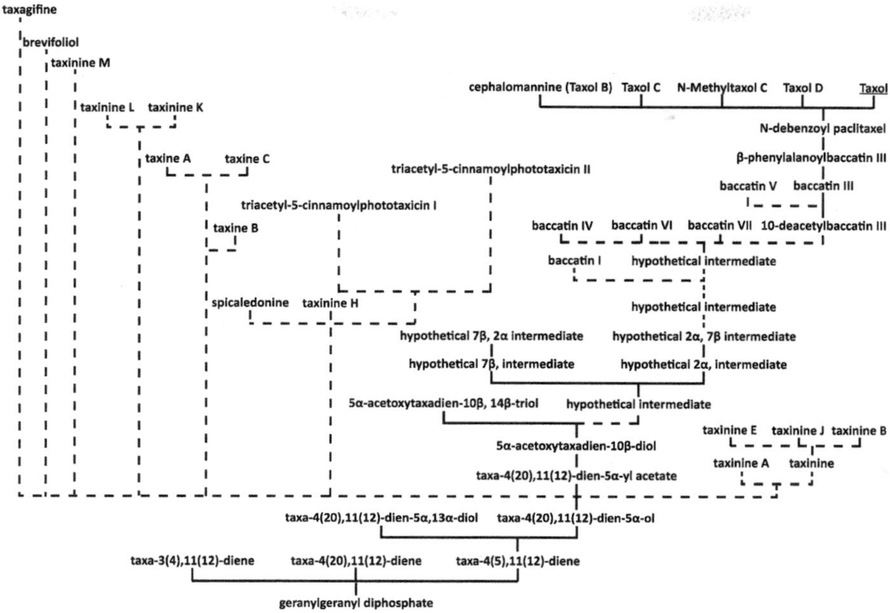

Fig. 7.4 A simplified taxoid biosynthesis pathway. The dashed lines represent a small number of alternative taxoid products. The filled line is the known route to taxol and a small selection of its analogues

taxol will remain in doubt. If HGT does explain current observations it must have been an early event that radiated into many different fungal clades, or it must have happened on many occasions during endophyte evolution. Perhaps an ancestor of the *Pinales* family *Taxaceae* was the first host to an ancient predecessor of the first taxol-synthesizing endophytes, but if so then such an organism would have proven extremely successful. Presumably this organism evolved the taxol pathway in response to pathogen colonization, immediately conferring an advantage over its competing flora and allowing it to colonize the northern hemisphere. This adaptation would have elicited a strong selection pressure for widespread tissue taxol concentrations, resulting in the commonly observed resistance phenotype within endophyte populations [172].

Another prediction from the HGT hypothesis is that geographically disparate endophytes producing the same secondary metabolite should display genetic divergence equal to their hosts, in relation to sequence conservation in the taxol biosynthesis pathway, in contrast to other endophytes in different biomes and different hosts. Some taxol-producing endophytes and their hosts have been analysed by comparing genomic sequences representing known pathway genes in different organisms, and as stated above such comparisons failed to identify any comparable genes or gene clusters proven to be involved in taxol biosynthesis. This work could be extended further by comparing genomic and proteomic data representing the TS, T5αOH, T13αOH, TAT, T10βOH, TBT, DBAT, PAM, BAPT and DBTNBT sequences. This would provide a definitive basis for investigating lineages in different regional populations to determine the origin of taxol-producing endophytes [216]. Furthermore, at least one endophyte genome and transcriptome has been analysed showing at best seven hits with homology to archetypal *Taxus* pathway genes, but their variation disproves at least in this fungus the likelihood of HGT [216], suggesting divergent evolution generated these similar sequences if they are involved in taxol biosynthesis. As these sequences have been identified, they should be easily and rapidly investigated, e.g. by mutation or heterologous expression in another host as described for *Taxus* TS. Endophytic taxol biosynthesis may also reflect convergent evolution, although one cannot help but ask how an organism would benefit by evolving the ability to synthesize a toxin already present in its environment, when another would surely be more favourable. Alternatively, for hosts that cannot produce taxol themselves, a taxol-producing endophyte and its host would benefit from maintaining symbiosis, but the vast majority of 'taxol-producing' endophytes have been isolated from *Taxus*. Furthermore, it is unclear why taxol-producing endophytes isolated from taxol-producing hosts would lose most of their productivity after several passages in culture. This would suggest there are environmental pressures eliciting and/or maintaining production which are not present under laboratory conditions resulting in the silencing of the pathway or genomic rearrangements. These issues need to be addressed for endophytic production to become a realistic opportunity.

Alternatively, the ability of endophytes to produce taxol could be a remarkable case of convergent evolution, with the fungus independently evolving its own biosynthesis pathway as proposed for fungal gibberellic acid biosynthesis [16]. Interactions between the plant host and fungus may also influence taxol production [152] and may depend on specific host and environmental factors such as host age, tissue, season and local environment [171]. Furthermore one study argues that these symbiotic fungi might act as part of a plant defence system, offering a curious explanation as to why some taxol-producing fungi are found within hosts that produce taxol already [173]. Perhaps in the absence of stimuli from the host (or a pathogen), the endophyte product spectrum may be attenuated, thus explaining the loss of productivity in culture [152]. Should this hypothesis prove correct, industrial production using endophytes as production hosts will no doubt prove challenging until the relationship is fully understood. It is deeply troubling that so little third-party confirmatory work has been carried out to probe what is clearly a conflicted subject, and so long as this continues our knowledge will stagnate. Given the commercial and ethical interests involved with high-profile pharmaceutical compounds like taxol, the investigation of this phenomenon would prove a wise investment.

7.4 Taxol Production by Plant Cell Cultures

Most taxol is currently produced by semi-synthesis, which is achieved by modifying the late taxol precursors baccatin III and 10-deacetylbaccatin III isolated from *Taxus* cell cultures (**see also Chap. 8 of this book**). This has replaced semi-synthesis methods using precursors extracted from *Taxus* needles, which in turn replaced bark extraction, given that taxol comprises as little as 0.02% of the total dry weight of the bark and 2000–3000 adult *T. brevifolia* trees are therefore required for the commercial production of 1 kg of taxol [190]. The extraction of precursors from needles nevertheless requires a constant supply of *T. brevifolia* trees, which is dependent on external factors. The first semi-synthesis method for the production of taxol from 10-deacetylbaccatin III was reported in 1988 [41].

An alternative means of production is the use of *Taxus* spp. cell suspension cultures [59, 92, 175]. This has many advantages, e.g. productivity is consistent because optimized culture conditions can be maintained, large-scale production platforms allow the production of more biomass, and the whole platform is ecologically sustainable [58]. The addition of 200 μM methyl jasmonate to *Taxus* cell suspension can increase the total taxane yield to 23.4 mg L d^{-1}, with taxol representing up to 20% of the total [92]. Finally, production strains can be manipulated and genetically modified to increase yields, e.g. by overexpressing rate-limiting enzymes or blocking/removing competing pathways [52, 220]. Better techniques for the engineering of production strains have emerged over the years because transformation methods are becoming more efficient through the refinement of gene transfer and cell culture techniques [7]. The most widely used transformation methods involve *Agrobacterium tumefaciens* or particle bombardment, chiefly due to technological developments in vector design that have enabled these technologies to produce marker-free transgenic plants using processes such as transposition and site specific recombination. Furthermore, the development of binary bacterial artificial chromosomes and superbinary vectors has enabled the transformation of previously recalcitrant species [7]. For metabolic engineering, precision genome editing tools are now available, such as zinc finger nucleases (ZFNs), transcription activator-like effector nucleases (TALENs) and clustered regularly interspaced short palindromic repeats (CRISPRs) with CRISPR-associated (Cas) proteins [50, 53]. However successful production platforms not only require a tailored biosynthesis pathway and optimized regulatory networks, because important intermediates or the end-products themselves may not pass unhindered across cell membranes, highlighting the need for membrane trafficking optimization achieved by engineering the translocation machinery [138, 218].

Although cell and tissue culture provides a potentially reliable and sustainable production platform, it is difficult to maintain high rates of secondary metabolite biosynthesis [42, 67, 131, 148, 168]. The yield of secondary metabolites in cell suspension cultures must typically increase by several orders of magnitude before it reaches commercially acceptable levels, and this often requires the simultaneous use of many yield improvement strategies such as the manipulation of biotic and abiotic stimuli, scale-up techniques and the optimization of downstream processing. The following sections will describe the most recent findings concerning the utilization of plant cell cultures for the production of taxol and the discovery of further pathway proteins (**see also Chap. 8 of this book**).

The development of *Taxus* cell cultures was first described in an abstract and a patent application [29, 30]. The earliest yields of 1–3 mg L^{-1} increased to levels between 77.46 and 153.3 mg L^{-1} [19, 30, 140, 219]. Many advances were achieved by the optimization of culture conditions, such as the use of elicitors to increase secondary metabolite synthesis, the modification of gas composition, and varying the osmotic pressure, medium composition and cultivation method. Many other factors can also influence productivity including temperature, pH, light and the presence of specific metals. An *in vitro Taxus* culture is established by inducing callus formation, i.e. undifferentiated masses of plant cells on solid media which act as the inoculum for cell suspension cultures. Callus induction in *Taxus* spp. was first reported by [162], but another 22 years would pass before these techniques were used for the production of taxol, when *Taxus chinensis* cells achieved a yield of 153.3 mg L^{-1} after 42 days [19]. The repetitive elicitation of *T. chinensis* suspension cultures was later shown to achieve taxoid yields of 612 mg L^{-1} confirming that higher yields could be achieved by optimizing the elicitation process [201].

The use of biotic and abiotic elicitors to stimulate plant stress response pathways can induce the production of desirable compounds [156]. These desirable compounds are produced in native plants to allow their adaptation to changing environmental conditions, such as predation or drought. Elicitation can be abiotic or biotic, with many subcategories. For example, abiotic elicitors include physical stimuli such as thermal stress, drought or UV radiation, or exogenous hormones, or chemical stimuli such as the presence of heavy metals or salts [134]. Alternatively, biotic elicitors can include any substances of biological origin, the best-characterized of which are polysaccharides from bacteria, fungi and yeast [57]. Elicitors have been thoroughly reviewed recently so we will only discuss key findings related to taxol production, but the interested reader can consult the following reviews [2, 121, 134, 156].

Jasmonates are a family of oxylipins produced by plants and some fungi, and the best known are jasmonic acid, methyl jasmonate and jasmonyl-isoleucine. Physiological analysis has shown that (+)-7-iso-jasmonoyl-L-isoleucine is involved in Arabidopsis leaf jasmonate signalling, and jasmonyl-L-tryptophan is an inhibitor of Arabidopsis root auxin signalling [2]. Several elicitors induce taxane biosynthesis in *Taxus* spp. [117, 128, 129, 219] including methyl jasmonate, jasmonic acid, coronatine, vanadium sulfate, silver nitrate, salicylic acid, ammonium citrate, cobalt chloride, cyclodextrin and even fungal extracts in conjunction with methyl jasmonate and salicylic acid [35, 95, 129, 141, 154, 219]. The structures of methyl jasmonate and coronatine are shown in Fig. 7.5. Some of these elicitors will be described in detail in this section. The production rate and yield of taxol can be dramatically influenced by elicitation, e.g. *Taxus media* cultures elicited with 100 μM methyl jasmonate have been reported to produce 110 mg L^{-1} taxol within 14 days [219]. The mechanism of jasmonate-mediated elicitation of secondary metabolites has been reviewed [38].

Specialized production methods such as cell immobilization, two-phase culture techniques, *in situ* taxane removal and large-scale bioreactors with volumes up to 75,000 L have all been developed to increase the production of valuable taxanes [212]. Cell immobilization methods can circumvent the issues of variable productivity related to aggregation and product-related feedback inhibition, shear

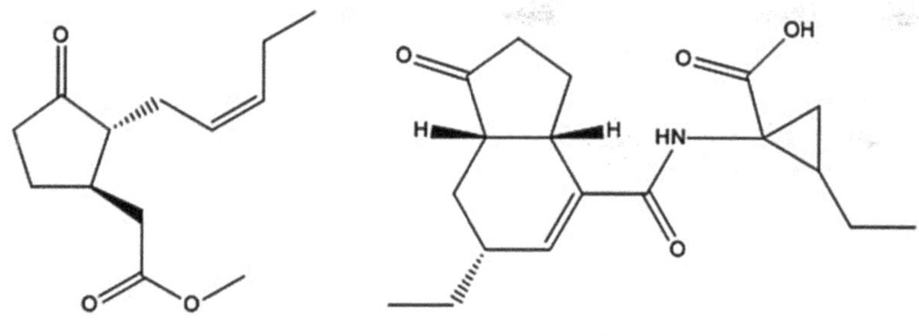

Fig. 7.5 Structures of the elicitors methyl jasmonate and coronatine

force susceptibility and slow growth [20, 21, 44]. Cells have been encapsulated within a range of matrices, such as agar, gelatin, polyacrylamide, hollow fibre membranes and polyurethane foam. Production rates of 2.71 mg L^{-1} d^{-1} were achieved using *T. baccata* cells immobilized in calcium alginate beads in a stirred bioreactor, peaking at 43.43 mg L^{-1} after 16 days of culture [11] with jasmonate elicitation increasing the yields further [18]. More recently, the capacity for taxane production in free and immobilized *T. globosa* cells in two media containing different plant growth regulators elicited with methyl jasmonate resulted in a two-fold increase in taxol production compared to free cells, yielding 130 μg L^{-1} d^{-1} when treated with picloram and kinetin [143]. The *in situ* removal of taxol, baccatin III and other taxanes from cultivation media increases yields, because *in vitro* culture conditions can inhibit growth probably due to intracellular and/or extracellular taxane accumulation [51, 96]. The implementation of *in situ* removal has achieved some promising results. For example, the development of two-phase (aqueous-organic) cultivation techniques for *T. chinensis* was reported to increase taxol yields by a factor of six when the two-phase cultivation system was supported with sucrose feeding [200]. Different carbon sources can also affect the synthesis of secondary metabolite, with sucrose and fructose being beneficial to yields [47, 98]. Furthermore, environmental factors such as temperature and gas composition within the culture can also effect the taxane content and production rates, e.g. high carbon dioxide levels prevented taxol biosynthesis, probably due to the inhibition of P450 catalytic activity, whereas low oxygen levels promoted the early onset of secondary metabolite production [127].

Finally, plant cells in culture can aggregate, forming disparate and heterogeneous populations with varied morphology and taxol productivity. Cell aggregates of 400 μm have been shown to produce 6 mg L^{-1} taxol, 20-fold more than aggregates roughly twice their size, after elicitation with methyl jasmonate [101]. Large-scale bioreactors containing *T. chinensis* cells have been used successfully for commercial taxol production for several years [132]. However, although culture methods are important, specialized production cell lines with higher titres can be engineered, reducing the need for larger culture systems, limiting costs and also production times.

Rational approaches for the further improvement of taxol biosynthesis include the generation of cell lines with desirable characteristics, the identification and manipulation of flux regulators, and the use of metabolic engineering methods to modify natural producers or integrate the taxol biosynthesis pathway into alternative hosts. The variable secondary metabolite yields reported in cell suspension cultures argues for the selection and/or generation of stable, fast-growing cell lines capable of taxol biosynthesis [17, 105, 121]. Indeed, *T. baccata* cell cultures produce different taxol yields even when cultivated under the same conditions, if initiated with explants taken from different parts of the same mother plant [22].

Genetic modification is likely to be faster than selection for natural mutations when it comes to the isolation of productive cell lines. For example, stable transgenic *T. cuspidata* cell lines produced by *Agrobacterium*-mediated transformation were viable even after 20 months of continuous culture [93]. Furthermore, particle bombardment has been used to achieve transient expression in *T. cuspidata* cell lines [191], and the upregulation of abscisic acid in *T. chinensis* cells led to a 2.7-fold increase in taxol yields [114]. Taxol production has been increased by the overexpression of genes such as *DBAT* [223] and the redirection of flux away from competitive branches by antisense silencing of the *T14βOH* gene in *Taxus media* [110]. The overexpression of *TS* in *T. media* cell lines achieved a 265% increase in taxane production compared to an untransformed control line, resulting in yields of up to 70 mg L^{-1} taxol [52, 140]. Despite these achievements, previous investigations have largely focused on the modification of key genes rather than the entire pathway, and this serves only to reveal bottlenecks in the remainder of the pathway, including the unknown steps. Additionally the genetic engineering of *Taxus* cells has previously been achieved only with very low transformation efficiencies i.e. 1% at best [93]. Fortunately, more recent work by [123] has demonstrated a greatly improved methodology conveying a 75% transformation efficiency by exchanging kanamycin for paromomycin as the selectable marker. Unfortunately the effect upon the taxol biosynthesis from these *Taxus* transformants was not disclosed in this paper. A potential reason for such a delay between reported improvements is likely due to fears of patent conflicts which restrict the generation of transgenic *Taxus* cell lines with modified taxol biosynthesis pathways [124]. This patent has recently been assigned to Diana U.S. INC., Oregon, of which Diana Plant Sciences, INC., Oregon is a subsidiary. Unfortunately, little has been achieved in this field with regards to strain development, substantial yield improvement and reduction of side product accumulation despite the presence of technology that could likely achieve this. For example the use of CRISPR/Cas9 for channelling flux into the pathway branches of interest could lead to increased taxol titres. Crucially, before this method can be attempted the enzyme sequences responsible for these side branches must be elucidated. Thus, until this research is conducted cell line improvement strategies will be restricted to those that do not require knowledge of the entire biosynthesis pathway, such strategies include increasing known gene copy number, GGPP pool expansion, elicitation, upstream bioprocess optimization and genetic modification of regulatory elements. Therefore a more holistic approach is required to achieve substantial gains, based on the global modulation of host cell metabolism.

Several recent investigations have focussed on the global taxol pathway by analysing the effects of elicitors on the transcriptomes of *T. media* [141], *T. baccata* [155] and *T. globosa* cell cultures [154]. Such experiments can facilitate gene discovery, even in the absence of genomic sequencing data [32, 61]. The effect of elicitation with coronatine (1 μM) or methyl jasmonate (100 μM) on the expression of *TS, T13αOH, T7βOH, T2αOH, DBAT, PAM, BAPT* and *DBTNBT* was determined by quantitative RT-PCR, revealing the differential transcription of these genes in a transgenic *T. media* cell line expressing *T. baccata TS* [52] up to 4 days after elicitation [141]. *TS* was induced 5.2-fold by methyl jasmonate after 4 days, whereas *TS* transcript levels peaked 24 h after elicitation with coronatine, with a 4.8 fold increase compared to the control [141]. The three P450 transcripts peaked 2 and 4 days after elicitation with coronatine and methyl jasmonate, respectively, and the peak response of each elicitor was approximately twice as strong as the other elicitor at the same time point. The *PAM, BAPT* and *DBTNBT* transcripts behaved in a similar manner, but the *DBAT* transcript (encoding the enzyme that converts 10-deacetylbaccatin III into baccatin III) was strongly induced just 12 h after elicitation with methyl jasmonate and maintained the peak until 24 h after elicitation, at double the level achieved by coronatine and triple the level of the control culture [141]. This suggests that flux may be less inhibited by side chain assembly and attachment to baccatin III when cultures are elicited with coronatine, because *BAPT, PAM* and *DBTNBT* are strongly expressed at earlier time points compared to the control culture and elicitation with methyl jasmonate. Furthermore, the unknown T2'αOH may also be induced when comparing the expression profiles of the three P450s. A follow-up study using a similar methodology has also included a likely *PAM* candidate, which is also induced by at least three-fold in the presence of elicitors, but this gene could still represent a rate-limiting step in the pathway [154]. This follow-up study explored various elicitation strategies in two-stage cultures of a transgenic *T. baccata* cell line [52] and *T. globosa* cell lines. The elicitors coronatine (1 μM) and cyclodextrin (50 mM) presented together promoted the highest accumulation of taxanes, with yields of 35 mg L^{-1} in transgenic *T. globosa* cultures and 70 mg L^{-1} in *T. baccata* [154].

The elicitor cyclodextrin forms inclusion complexes around hydrophobic compounds such as taxol to facilitate excretion [23]. This should alleviate feedback inhibition caused by the intracellular accumulation of taxol and also prevent degradation [213]. The extraction of taxol is expensive and labour-intensive, but cyclodextrin inclusion complexes protect taxol from modification and degradation, thus prolonging cultures, simplifying downstream processing and reducing costs [164]. When used in conjunction with methyl jasmonate, cyclodextrin has been shown to induce a 60-fold increase in taxol production [164]. Interestingly, a large proportion of the taxanes produced when *T. globosa* cells were elicited with cyclodextrin or cyclodextrin plus coronatine was represented by cephalomannine, which is similar to taxol but has a tigloylation group instead of a benzoylation at the C3 position of the side chain. Although cephalomannine can be used for the semi-synthesis of taxol analogues, the enzyme responsible for its creation could be identified and knocked out, allowing the *T. globosa* cell line in this study to produce more taxol.

The *T. globosa* cell line has a faster growth rate and capacity for product excretion than the transgenic *T. baccata* cell line. Alternatively, a desirable hybrid could be selected from protoplast fusion experiments. Once identified, other candidate sequences such as the T2'αOH could also be tested to alleviate other bottlenecks within the pathway. In contrast to the elicited lines, the control culture expression profile indicated that only *TS* and *DBAT* are strongly expressed, with peaks one day after the test cultures were elicited. Overall this investigation demonstrated that coronatine elicits a stronger and faster transcriptional response than methyl jasmonate, albeit at the cost of reducing the proportion of taxol in the total taxane content [141]. If the aim were to accumulate as much taxol as possible at the expense of other valuable precursors, multiple overlapping cultures with coronatine elicitation running for just 4 days after elicitation would produce at least 30 mg L^{-1} taxol each, in contrast to a single suspension cell culture running for 16 days after elicitation yielding 77 mg L^{-1}. Alternatively, a two-stage cell culture could be used to produce high levels of biomass under optimal growth conditions, followed by transfer into an optimized production medium containing elicitors [154]. However, the *TS* and *DBAT* expression profiles suggest that coronatine elicitation will not circumvent bottlenecks in the pathway because the flux is determined by the slowest step. Therefore, until the entire pathway is characterized, the rate for each step determined, and the regulatory networks understood, any manipulations will be hampered by our lack of knowledge.

Pathway regulation has been explored for taxol biosynthesis at the transcriptional and post-translational levels. It is unsurprising that such a complex pathway must be tightly regulated in order to conserve precious resources. As previously stated, jasmonates are elicitors that affect many physiological processes in plants, including the synthesis of defence compounds. Jasmonates are phytohormones derived from oxygenated tri-unsaturated fatty acids, with jasmonic acid, methyl jasmonate and jasmonoyl-isoleucine playing key regulatory roles [2, 38]. The jasmonate elicitation mechanism is well conserved across the plant kingdom, suggesting early evolution within a distant common ancestor. Species radiation then led to diverse species-dependent terpenoid, alkaloid and phenylpropanoid pathways that are directly or indirectly (via induced precursor biosynthesis) regulated by jasmonates [38]. In *Taxus*, some of the regulatory components controlling taxol biosynthesis have been identified, including transcriptional activators, repressors and evidence of a 46-amino-acid, cysteine-rich signalling peptide [108, 113, 142, 221]. The first transcriptional regulator of the taxol pathway to be characterized was a WRKY transcription factor (TcWRKY1) that activates the *T. chinensis DBAT* gene [113]. This was identified by the deletion analysis of *cis*-acting elements in the *DBAT* promoter, and these elements were subsequently used as bait for yeast one-hybrid screening of a *T. chinensis* cDNA library. TcWRKY1 binds to two W-box *cis*-acting elements in the upstream *DBAT* promoter. Furthermore, the elicitation of *T. chinensis* with methyl jasmonate induces this transcription factor, which in turn upregulates the transcription of the *DBAT* gene [113]. Similar methods were used to investigate the *T. chinensis TS* gene [221]. Deletion analysis revealed the jasmonate-response region using a β-glucuronidase (GUS) reporter gene expressed in *T.*

chinensis cells. Several deletion constructs attached to the reporter were tested in the presence or absence of 100 μM methyl jasmonate, showing that deleting the promoter sequence to position −131 abolished jasmonate induction. Further analysis revealed that induction was abolished by the deletion of an ethylene-response element known as a GCC-box between positions −150 and −131 [56] or by the loss of the sequences containing an E-box upstream of position –219. The GCC-box sequence was used as bait in a yeast one-hybrid screen as above, resulting in the identification of two candidates TcERF12 and TcERF15. TcERF12 was similar to the previously identified ethylene-responsive repressor AtERF3 [56] whereas TcERF15 was similar to the known activator ORA59 [151]. The binding of these proteins was confirmed *in vivo* based upon the expression of the *HIS2* selectable marker under the control of the *TS* promoter. Confirmation of the roles of TcERF12 and TcERF15 as a repressor and an activator, respectively, was achieved by the elicitation of *T. chinensis* cell suspension cultures with methyl jasmonate followed by RT-PCR analysis, by overexpressing the transcription factors in *Taxus* cells and subsequent analysis of *TS* gene expression, and by the transient expression of GUS reporter constructs in cells co-transformed with plasmids carrying the *TcERF12* or *TcERF15* genes [221]. Elicitation with 100 μM methyl jasmonate induced the expression of *TcERF12* by 47-fold within 1 h, followed by a gradual decline over the next 12 h, whereas *TcERF15* levels increased by 4.5-fold after 30 min and remained at high levels for the next 3–6 h before declining. The effect of the two transcriptional regulators on *TS* gene expression was determined by northern blot and quantitative RT-PCR. The overexpression of *TcERF15* induced *TS* expression by 2.5-fold, whereas *TcERF12* overexpression caused a 2-fold reduction in *TS* gene expression compared with controls. These results were supported by reporter gene activity assays [221]. Three further methyl jasmonate-inducible transcriptional regulators were later identified in *Taxus cuspidate*, and although they were similar to the Arabidopsis transcriptional activator AtMYC2, all three were shown to act as transcriptional repressors [1, 108]. GUS assays were carried out using the promoter regions of the *TS*, *T5αOH*, *TBT*, *DBAT*, *PAM*, *BAPT* and *DBTNBT* genes. The constructs were introduced into *T. cuspidata* along with a control construct containing the firefly luciferase (*LUC*) gene driven by the *Cauliflower mosaic virus* (CaMV) 35S promoter, which does not respond to jasmonates and therefore provides a control for transformation efficiency. Methyl jasmonate elicitation was shown to induce GUS activity driven by all seven taxol pathway promoters by 1.5-fold, whereas there was no effect on the construct driven by the CaMV 35S promoter [108]. The promoter sequences were then analysed *in silico* using PLACE (http://www.dna.affrc.go.jp/PLACE/) [74] which identified multiple E-boxes (CANNTG), suggesting the presence of jasmonate-responsive MYC regulatory proteins. Degenerate primers were designed according to the reference sequences of the known regulators MYC2 and JAMYC10, yielding a 172-bp cDNA fragment that was used to obtain a full length cDNA for sequence comparison with known methyl jasmonate-sensitive bHLH proteins. After confirming its similarity with other MYC proteins, the new *T. cuspidata* gene was named *TcJAMYC1*. The induction of *TcJAMYC1* following elicitation with methyl jasmonate was confirmed by semi-quantitative

RT-PCR, and two other transcripts (*TcJAMYC2* and *TcJAMYC4*) were found to be upregulated in the same manner. The binding of the putative transcription factor TcJAMYC1 to the E-boxes in the taxol gene promoters was confirmed using a competitive electrophoretic mobility shift assay (EMSA) with a mutated sequence lacking an E-box as a control, further revealing that the preferential E-box sequence was CACGTG [108]. Transcriptional regulation by TcJAMYC1, TcJAMYC2 and TcJAMYC4 was tested by co-expressing the corresponding expression vectors with the GUS reporter genes discussed above in the presence and absence of 100 μM methyl jasmonate, revealing that TcJAMYC1, TcJAMYC2 and TcJAMYC4 expressed individually were able to suppress reporter gene expression [108]. Interestingly, mock elicitation with ethanol also allowed TcJAMYC1 to suppress by 3-fold the *TBT*, *BAPT* and *DBTNBT* reporter constructs, but none of the others. Conversely, TcJAMYC2 induced the *T5αOH* reporter by 1.5-fold and the *PAM* promoter by 2.5-fold, whereas the *BAPT* construct was marginally repressed. TcJAMYC4 suppressed the *TBT*, *DBAT*, *PAM*, *BAPT* and *DBTNBT* promoters but the *TS* promoter was slightly induced. The authors stated that reporter activity was measured 48 h after bombardment, equivalent to 54 h post-elicitation, when the endogenous regulatory machinery was likely to be active. This suggests that the positive regulation they detected was probably less potent than would be the case at an earlier time point, and that the overexpression of these repressors coincided with the expression of native repressors, potentially repressing the taxol gene promoters more strongly than would normally be observed. This agrees with the reported induction of taxol gene expression within 24 h of elicitation [137], but the authors suggested that downregulation occurs after this first 24-h period, returning the cultures to their basal state by 48 h post-elicitation [108]. In contrast to this hypothesis, increased taxol gene expression has been observed 4 days after methyl jasmonate elicitation [141], but this could reflect differences in the cell line and cultivation conditions even though the same final concentration of methyl jasmonate was used. TcJAMYC1, TcJAMYC2 and TcJAMYC4 were proposed either to repress the taxol gene promoters by direct binding or to activate the genes encoding other transcriptional repressors, similar to the regulation of WRKY26 and ERF11 by MYC2 in Arabidopsis. It is also likely that the regulation of the taxol pathway involves a complex regulatory mechanism with multiple transcription factors, as reported for other jasmonate-regulated pathways [38]. This is supported by the presence of putative WRKY binding sites (TGAC) in the *TS*, *T5αOH*, *TBT*, *DBAT*, *PAM*, *BAPT* and *DBTNBT* promoters [108].

In addition to jasmonate, defence pathways can also be regulated by small secretory or non-secretory peptides that range in size from 5 to 75 amino acids. These are perceived by receptors in signalling cascades that may overlap with those elicited by plant hormones, or the peptides themselves may be expressed in response to stimuli such as insect oral secretions, which indicate an attack by herbivores [82]. The mature signalling peptides can in turn elicit the production of jasmonates or ethylene, which ultimate cause the transcriptional activation of plant defence genes [82]. Taxane production by *Taxus* cultures can be elicited synergistically with active phytosulfokine-α and methyl jasmonate, prompting the investigation of other *Taxus* signalling peptides [97]. In relation to taxol, the signalling peptide Taximin (*TB595*) was identified in *T. baccata* cells, and was found to be highly conserved among higher plants [142]. The 73-amino-acid pre-protein sequence included a 27-amino-acid N-terminal signal peptide, which was predicted to pass through the secretory system and localize the peptide extracellularly or in the vacuoles and plasma membrane. The native localization of the peptide was investigated by fusing the C-terminus of the peptide to the Venus yellow fluorescent protein, followed by transiently expression in *Nicotiana benthamiana.* Attempts to determine the effect of exogenous Taximin on *T. baccata* cells required the synthetic surrogate HyproTaximin (in which proline residues were replaced with hyroxyproline) because the original could not be synthesized, and HyproTaximin was designed to mimic post-transitional proline hydroxylation which is believed to occur often with native signalling peptides [125]. Taxol and baccatin III levels were determined over a 21-day period in relation to the addition of ethanol (mock elicitation), HyproTaximin, methyl jasmonate or a combination of methyl jasmonate and HyproTaximin. Taxane production peaked 7 days after treatment. The addition of HyproTaximin alone increased baccatin III production by 1.4-fold and taxol by 2-fold compared to ethanol controls, and in agreement with earlier reports [36], methyl jasmonate alone increased baccatin III production by 4-fold and taxol by 4.7-fold. The combined treatment with methyl jasmonate and HyproTaximin was the most potent, increasing baccatin III production by 4.6-fold and taxol by 6.6-fold compared to ethanol controls. However, by day 21 there was no significant difference between the elicited and control cultures, suggesting that Taximin has a transient positive influence on taxane production that is additive to the effect of methyl jasmonate in *T. baccata* cell cultures [142]. They also noted that Taximin induced the synthesis of alkaloids in tobacco but did not affect the transcript levels of alkaloid biosynthesis genes, suggesting the peptide does not act as a transcriptional activator. The role of Taximin should therefore be determined before it is used as a target for the manipulation of *Taxus* cell lines to increase taxol production. Furthermore the roles of plant defence hormones, and their impact upon taxol biosynthesis within *Taxus* cell cultures could present novel targets leading to higher titres. There is an antagonistic relationship between the plant hormones jasmonic acid and salicylic acid in the context of jasmonate-responsive gene expression [188]. The presence of a GCC-box in jasmonate-responsive promoters is sufficient for salicylic acid mediated suppression of jasmonic acid induced gene expression. This possibly provides further targets for the regulatory fine tuning of taxol producing cell lines, because such defence responses can hinder the production of useful compounds in controlled aseptic environments [188]. Current knowledge concerning the regulation of taxol biosynthesis is summarized in Fig. 7.6. Coronatine insensitive 1 receptors, Skp-Cullin-F-box-type E3 ubiquitin ligase and JAZ protein homologues are likely to be involved but their role in the regulation of taxol production is unclear.

The identification of positive and negative regulators and their binding sites provides potential targets for the metabolic engineering of production strains by gene knockout or RNA silencing. Furthermore, these investigations illustrate the

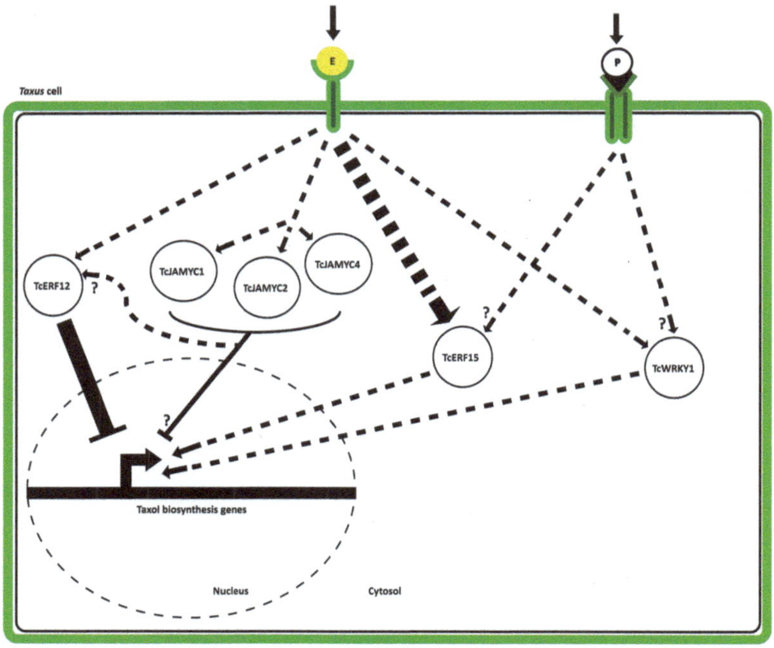

Fig. 7.6 Major players in the taxol biosynthesis regulatory network. Arrows (dashed) indicate positive transcriptional regulation and blunt lines (filled) indicate negative transcriptional regulation. Lines (heavy) indicate a strong regulatory response. E = elicitor such as methyl jasmonate, P = peptide such as Taximin,? = undetermined mechanism

promising use of transcriptomics in the context of *in vitro* cultures with and without elicitation, not only for the identification of unknown genes that are co-expressed with genes known to be involved in taxol biosynthesis, but also for mapping regulatory networks at the transcriptional level [153].

Metabolic engineering has also been used for the heterologous expression of taxol biosynthesis genes in Arabidopsis [12], tomato [103] and *Nicotiana sylvestris* [163] plants with varying success. The heterologous expression of *TS* in Arabidopsis using constitutive and inducible promoters resulted in the accumulation of taxadiene to levels of 20 and 600 ng g^{-1} dry weight, respectively [12]. The host plants suffered stunting and decreased pigmentation, which the authors suggested may reflect the suppression of other GGPP-dependent pathways by direct precursor competition. The *TS* gene has also been expressed in the tomato *yellow flesh* mutant, which lacks fruit phytoene synthase activity [55, 103]. Phytoene synthase normally uses GGDP for the synthesis of carotenoids, which can account for up to 2% of tomato fruit dry weight [103]. This mutant therefore provides an ideal background for GGPP-demanding pathways because the excess GGPP can be channelled into taxadiene synthesis via TS and the absence of carotenoids facilitates product extraction. Accordingly, a taxadiene yield of 160 mg/kg freeze dried tomato fruit was achieved with a purity of >95% after extraction [103]. Interestingly, attempts to express *TS* and *T5αOH* within tobacco (*Nicotiana sylvestris*) a year later showed that the transfer of P450s into another host does not guarantee the native activity [163]. Although preliminary analysis indicated that the leaves contained 20 μg g^{-1} taxadiene, the next product was expected to be taxa-4(20),11(12)-diene-5a-ol but the novel taxane OCT was detected instead [163]. Another study investigated the heterologous expression of *TS* in *Artemisia annua*, a fast-growing herb that naturally synthesizes the important anti-malarial compound artemisinin [112]. *Agrobacterium*-mediated delivery of a construct containing the *TS* gene and a hygromycin phosphotransferase selectable marker (*hptII*) resulted in taxadiene yields of up to 130 μg g^{-1} dry weight, with a concomitant decrease in the levels of artemisinin [112].

The challenges associated with the heterologous expression of plant P450s must be overcome before efficient taxol production can be achieved in a surrogate metabolic chassis. Perhaps a combination of genome editing and metabolic engineering will provide the means to streamline a *Taxus* cell line, with competitive pathways silenced e.g. by disrupting the *T14βOH* gene, and regulatory mechanisms removed by mutating E-boxes in the promoters of taxol biosynthesis genes or integrating similar genes from other *Taxus* species. Coupling this designer *Taxus* cell line with optimized cultivation methods, elicitation strategies and the use of cyclodextrin or the improvement of taxol export machinery could provide the best conditions for taxol production.

7.5 Microbial Biotechnology for the Production of Taxol

Despite the impressive yields achieved using plant tissue culture methods and cell suspension cultures, limitations such as slow growth, high costs and contamination risks are difficult to overcome. Some researchers have therefore investigated the practicality of transferring this highly complex pathway into a microbial chassis. The growth rate of bacteria and yeast is faster, doubling at least once per hour rather than the once per day typical for plant cells, and the productivity of microbes is also superior. Progress has been made in two major candidates for heterologous production, the bacterium *E. coli* and the yeast *Saccharomyces cerevisiae*. In both cases, metabolic engineering can be achieved by the stable integration of heterologous genes and each species benefits from a range of genetic tools that facilitate their manipulation. Both species are also easy to handle, grow rapidly on inexpensive carbon sources and are metabolically well characterized [212]. Accordingly, both species successfully express *TS* and *T5αOH* (with appropriate modification, such as the removal of plastid-targeting peptides), and at least eight *T. canadensis* genes involved in taxol biosynthesis have been expressed episomally in *S. cerevisiae*, namely *GGPPS*, *TS*, *T5αOH*, *T10βOH*, *T13αOH*, *TAT*, *TBT* and *DBAT* [4, 39]. The first successful investigation of *in vivo* heterologous taxadiene synthesis involved a non-optimized *E. coli* strain overexpressing IPP isomerase, *GGPPS* and *TS*, resulting in a yield of 1.3 mg L^{-1} taxadiene [81]. The highest heterologous taxadiene yields achieved thus far are ~1 g L^{-1} in *E. coli* [4] and 72.8 mg L^{-1} in *S. cerevisiae* [43]. Recently, the highest yield for oxygenated taxane production was achieved in *E. coli* (~570 mg L^{-1}) of which ~16% (~91 mg L^{-1}) was taxadiene-5α-ol, ~62% was iso-OCT, ~13% was OCT and ~8% was an unidentified oxygenated taxane [13]. The expression of eukaryotic P450s in bacteria was achieved by the optimization of a P450 pathway module, opening a new chapter in the development of microbial taxol, as well as other oxygenation-dependent chemistries in *E. coli*. However, the low proportional production of taxadiene-5α-ol suggests that too much flux is lost: at least seven further oxygenases are required to produce baccatin III. Clearly the present art is still very far away from the development of microbially produced taxol or even just baccatin III. This section will discuss the major milestones and strategies used to overcome the problems posed by non-native chassis utilization.

Choosing an appropriate host for a particular process requires a solid understanding of the pathway biochemistry and any predictable incompatibilities which might arise from it, so that potential solutions can be considered. In the case of microbial taxol production, this process has been facilitated by chimeric P450 redox partner proteins, co-culture fermentations, host metabolism optimization, protein engineering, novel reductase partner interactions, computer assisted modelling and the use of multivariate operons for pathway expression. Until recently, the cloning and functional expression of plant CYP450s in *E. coli* has been daunting because of physiological inconsistencies, such as the absence of CPRs for electron transfer, the localization signals usually attached to plant CYP450s, and the absence of intracellular compartments usually required for CYP450s in plants [146, 169]. However, these issues have been addressed by the co-expression of P450s and CPRs, or the expression of truncated P450s with N-terminal modifications [4, 25, 26, 109]. Recent statistical modelling suggests that the 1-deoxy-D-xylulose 5-phosphate (DXP) pathway has the theoretical potential to supply nearly 50% more IPP than the MVA pathway in yeast [62]. The DXP pathway also requires less oxygen, is redox balanced, and can theoretically produce more IPP from glucose than the MVA pathway [207]. Conversely, the MVA pathway has produced superior terpene yields, with highly-engineered systems producing impressively high titres [122, 144, 207]. A *S. cerevisiae* strain capable of producing 25 g L^{-1} artemisinic acid was successfully constructed for an artemisinin semi-synthesis pipeline by extensive pathway optimization and metabolic engineering of the mevalonate pathway for the overproduction of FPP [145]. The MVA pathway can also be improved, with a carbon efficiency equivalent to or exceeding that of the MEP pathway as well as lower oxygen requirements [107].

Combined strategies have recently been used for the successful functional expression of plant CYP450s in *E. coli* [13]. This was achieved by using codon-optimized truncated proteins, N-terminal sequence modification with a membrane-associated octapeptide [8, 184], genomic integration of MEP pathway genes (*dxs, idi, ispDF*) and cyclase modules (*TS* and *GGPPS*) regulated by the T7 promoter, and finally by balancing CPR and T5αOH expression. In the latter case, optimal performance was accomplished by uncoupling T5αOH and its redox partner and expressing them using a five-copy plasmid under the control of a weak *Trc* promoter, recreating the unbalanced (1:15) host CPR to CYP450 ratios hypothetically due to inefficiencies with cellular resources and NADPH use [13, 90]. Furthermore, the hydrophilicity conferred by N-terminal signal modification reduced monooxygenase activity, probably due to the interruption of F-G loop-mediated substrate acceptance [31], effectively reducing the oxygenated taxane yield by at least 50% in small-scale cultures [13].

The use of yeast for terpenoid synthesis has several advantages over bacteria because the subcellular compartments required for CYP450 activity are already present, as are the native reductase partners, although these alone are not sufficient for high levels of taxol oxygenase catalysis [39]. Yeast cells are capable of high-density growth, the post-translational modifications are similar to higher eukaryotes and products are efficiently secreted, although it may be possible to transfer some eukaryotic post-translational capacity into *E. coli* [187]. The earliest attempt to reconstruct the first half of the taxol pathway in *S. cerevisiae* involved the expression of five enzymes, from GGPPS to DBAT, yielding 1 mg L^{-1} taxadiene and ~25 μg L^{-1} taxadiene-5α-ol in selective medium [39]. Flux was restricted at the T5αOH step, potentially due to weak expression from the corresponding promoter and the absence of its native NADPH:cytochrome P450 reductase, which was previously shown to increase hydroxylase activity [89]. The next great advance in the creation of a yeast production platform required multifaceted flux rechannelling, which was achieved by overcoming aerobic sterol exclusion by mutating the transcription factor UPC2 (which is normally responsible for preventing sterol uptake under aerobic conditions). Site-directed mutagenesis led to the creation of UPC2.1, which reverses this process, reducing the amount of metabolic flux diverted to steroid biosynthesis. Additionally, the creation of a truncated 3-hydroxy-3-methylglutaryl-CoA reductase (tHMG1) by removing the N-terminal regulatory domain abolished feedback inhibition of the mevalonate pathway, resulting in higher levels of the precursor IPP. Finally, to reduce precursor competition between taxadiene and steroid biosynthesis, the GGPPS from *Sulfolobus acidocaldarius* was expressed along with *upc2.1*, *thmgr1* and codon-optimized *TS*. The taxadiene yield was 8.7 mg L^{-1} plus ~33 mg L^{-1} GGPP, suggesting there is potential for even greater taxadiene production [49]. Computer assisted design and modelling has been used to increase this taxadiene concentration by more than 7-fold [43]. This was achieved by screening the FPP binding affinity and catalytic efficiency of six *GGPPS* genes with diverse origins (*Ginkgo biloba*, *Rana catesbeiana*, *Erwinia herbicola*, *Chlamydomonas reinhardtii*, *T. baccata* x *T. cuspidate* and *S. cerevisiae*) facilitated by *in silico* proofing and *in vivo* confirmation, eventually leading to the selection and expression of the *GGPPS* from a *T. baccata* x *T. cuspidate* hybrid in a *S. cerevisiae* chassis capable of increased FPP synthesis via the integrated overexpression of *erg20* and *thmgr*. The reported yield of ~73 mg L^{-1} taxadiene was achieved in this strain by expressing *TS* and *GGPPS* using a high-copy-number plasmid [43]. Co-culture fermentation systems for terpenoid production have also been attempted, by combining an 'upstream' *E. coli* strain producing taxadiene with a 'downstream' *S. cerevisiae* strain expressing the P450s TAT, T5αOH and T10βOH, resulting in 33 mg L^{-1} of oxygenated taxanes [225]. Although this is an innovative concept for taxol biosynthesis, the practicality of such a method for industrial-scale production is unclear [107].

The combination of powerful new technologies such as synthetic biology, CRISPR/Cas9 editing [115, 179], the development and refinement of new artificial chassis for metabolic engineering [83], and the applications of *in silico* design [189], should make it possible to achieve the production of microbial taxol within a few years, at least via the semi-synthetic route. However for this goal to be realized the rest of the missing steps of the pathway need to be elucidated, their required proteins characterized and then successfully expressed within the desired host. Alternatively, perhaps rather than transferring an unknown pathway into another organism it might instead prove preferential to merely screen for productive *Taxus sp.* mutants following N-methyl-N′-nitro-N-nitrosoguanidine (NTG) or ethyl methanesulfonate (EMS) mutagenesis [91, 182]. The mutagenized cells could then be cultivated individually within 96-well plates and the resulting cultures screened using a simple method easily modified for 96 well plate format such as a taxol dependent fluorescence assay [130]. After initial determination, candidate productive mutants could be further characterized by stringent LC-MS/MS quantification of taxane product profiles. Multiple rounds of mutagenesis and selection could eventually produce production strains capable of significantly superior taxol titres.

7.6 Concluding Remarks

The discovery, production and applications of taxol have emerged from decades of research, and the story will continue for as long as taxol continues to benefit society. This chapter has highlighted research milestones that have enabled the harnessing of this powerful secondary metabolite for modern medicine, although the story is not without its share of controversy. Twenty years after the first taxol-producing fungus was reported, no industrial processes have been realized even though such fungi are abundant in different genera. Furthermore, the discovery of taxol-producing plants that are not members of the *Taxaceae* is also a surprising, yet potentially useful, revelation. Total security of supply cannot be achieved using natural sources, and the scientific community must therefore strive to develop new methods to increase the productivity of both the native pathway in *Taxus* cells and the imported pathway in heterologous production hosts. The remaining steps in the pathway remain to be characterized and the heterologous expression of P450s has produced some unexpected results, so the transfer of the pathway to non-native genetic backgrounds is likely to be challenging. Metabolic engineering and synthetic biology offer a potential route to overcome these issues and provide an ideal natural or synthetic chassis for the synthesis of this valuable secondary metabolite.

Acknowledgments The authors would like to thank Dr Richard M Twyman and Dr Birgit Orthen for assistance with the manuscript.

References

1. Abe H, Urao T, Ito T, Seki M, Shinozaki K, Yamaguchi-Shinozaki K. Arabidopsis AtMYC2 (bHLH) and AtMYB2 (MYB) function as transcriptional activators in abscisic acid signaling. Plant Cell. 2003;15(1):63–78. https://doi.org/10.1105/tpc.006130.
2. Acosta IF, Farmer EE. Jasmonates. The Arabidopsis Book. 2009;8:e0129. https://doi.org/10.1199/tab.0129.
3. Adelin E, Servy C, Martin MT, Arcile G, Iorga BI, Retailleau P, Bonfill M, Ouazzani J. Bicyclic and tetracyclic diterpenes from a Trichoderma symbiont of *Taxus baccata*. Phytochemistry. 2014;97:55–61. https://doi.org/10.1016/j.phytochem.2013.10.016.
4. Ajikumar PK, Xiao WH, Tyo KE, Wang Y, Simeon F, Leonard E, Mucha O, Phon TH, Pfeifer B, Stephanopoulos G. Isoprenoid pathway optimization for taxol precursor overproduction in *Escherichia coli*. Science. 2010;330(6000):70–4. https://doi.org/10.1126/science.1191652.
5. Arbuck SG, Blaylock BA. Taxol: clinical results and current issues in development. In: Suffness M, editor. Taxol: science and applications. Boca Raton: CRC Press; 1995. p. 379.
6. Baloglu E, Kingston DG. A new semisynthesis of paclitaxel from baccatin III. J Nat Prod. 1999;62(7):1068–71. https://doi.org/10.1021/np990040k.
7. Barampuram S, Zhang ZJ. Recent advances in plant transformation. In: Plant chromosome engineering: methods and protocols. Methods Mol Biol. 2011;701:1–35. https://doi.org/10.1007/978-1-61737-957-4_1.
8. Barnes HJ, Arlotto MP, Waterman MR. Expression and enzymatic-activity of recombinant cytochrome-P450 17-alpha-hydroxylase in *Escherichia coli*. P Natl Acad Sci USA. 1991;88(13):5597–601. https://doi.org/10.1073/pnas.88.13.5597.
9. Barry DA. Morphologie und Physiologie der Pilze, Flechten, und Myxomyceten. Vol. 2. In: Hofmeister's Handbook of Physiological Botany. Engelmann W, Leipzig; 1866.
10. Barton NA, Marsh BJ, Lewis W, Narraidoo N, Seymour GB, Fray R, Hayes CJ. Accessing low-oxidation state taxanes: is taxadiene-4 (5)-epoxide on the taxol biosynthetic pathway? Chem Sci. 2016;7(5):3102–7. https://doi.org/10.1039/C5SC03463A.
11. Bentebibel S, Moyano E, Palazon J, Cusido RM, Bonfill M, Eibl R, Pinol MT. Effects of immobilization by entrapment in alginate and scale-up on paclitaxel and baccatin III production in cell suspension cultures of *Taxus baccata*. Biotechnol Bioeng. 2005;89(6):647–55. https://doi.org/10.1002/bit.20321.
12. Besumbes O, Sauret-Gueto S, Phillips MA, Imperial S, Rodriguez-Concepcion M, Boronat A. Metabolic engineering of isoprenoid biosynthesis in Arabidopsis for the production of taxadiene, the first committed precursor of taxol. Biotechnol Bioeng. 2004;88(2):168–75. https://doi.org/10.1002/bit.20237.
13. Biggs BW, Lim CG, Sagliani K, Shankar S, Stephanopoulos G, De Mey M, Ajikumar PK. Overcoming heterologous protein interdependency to optimize P450-mediated taxol precursor synthesis in *Escherichia coli*. P Natl Acad Sci USA. 2016;113(12):3209–14. https://doi.org/10.1073/pnas.1515826113.
14. Biggs BW, Rouck JE, Kambalyal A, Arnold W, Lim CG, De Mey M, O'Neil-Johnson M, Starks CM, Das A, Ajikumar PK. Orthogonal assays clarify the oxidative biochemistry of taxol P450 CYP725A4. ACS Chem Biol. 2016. https://doi.org/10.1021/acschembio.5b00968.
15. Boa A, Jenkins P, Lawrence N. Recent progress in the synthesis of taxanes. Contemp Org Synth. 1994;1(1):47–75. https://doi.org/10.1039/CO9940100047.
16. Bomke C, Tudzynski B. Diversity, regulation, and evolution of the gibberellin biosynthetic pathway in fungi compared to plants and bacteria. Phytochemistry. 2009;70(15–16):1876–93. https://doi.org/10.1016/j.phytochem.2009.05.020.
17. Bonfill M, Exposito O, Moyano E, Cusido RM, Palazon J, Pinol MT. Manipulation by culture mixing and elicitation of paclitaxel and baccatin III production in *Taxus baccata* suspension cultures. Vitro Cell Dev-Pl. 2006;42(5):422–6. https://doi.org/10.1079/Ivp2006761.

18. Bonfill M, Bentebibel S, Moyano E, Palazon J, Cusido RM, Eibl R, Pinol MT. Paclitaxel and baccatin III production induced by methyl jasmonate in free and immobilized cells of *Taxus baccata*. Biol Plant. 2007;51(4):647–52. https://doi.org/10.1007/s10535-007-0137-2.
19. Bringi V, Kadkade PG, Prince CL, Schubmehl BF, Kane EJ, Roach B. Enhanced production of taxol and taxanes by cell cultures of Taxus species. U.S. Patent No. 5,407,816. 1995.
20. Brodelius P. The potential role of immobilization in plant-cell biotechnology. Trends Biotechnol. 1985;3(11):280–5. https://doi.org/10.1016/0167-7799(85)90003-4.
21. Brodelius P. Permeabilization of plant-cells for release of intracellularly stored products – viability studies. Appl Microbiol Biotechnol. 1988;27(5–6):561–6. https://doi.org/10.1007/BF00451632.
22. Bruňáková K, Babincova Z, Čellárová E. Selection of callus cultures of *Taxus baccata* L. as a potential source of paclitaxel production. Eng Life Sci. 2004;4(5):465–9. https://doi.org/10.1002/elsc.200420050.
23. Cai Z, Kastell A, Knorr D, Smetanska I. Exudation: an expanding technique for continuous production and release of secondary metabolites from plant cell suspension and hairy root cultures. Plant Cell Rep. 2012;31(3):461–77. https://doi.org/10.1007/s00299-011-1165-0.
24. Celgene Corporation. Celgene 2016 Annual Report. 2016. Retrieved from http://files.shareholder.com/downloads/AMDA-262QUJ/4420895141x0x939757/E2F8A415-6D24-4387-9CEA-D0620358E8E1/Celgene_2016_Annual_Report.pdf
25. Chang MC, Keasling JD. Production of isoprenoid pharmaceuticals by engineered microbes. Nat Chem Biol. 2006;2(12):674–81. https://doi.org/10.1038/nchembio836.
26. Chang MCY, Eachus RA, Trieu W, Ro DK, Keasling JD. Engineering *Escherichia coli* for production of functionalized terpenoids using plant P450s. Nat Chem Biol. 2007;3(5):274–7. https://doi.org/10.1038/nchembio875.
27. Chau M, Croteau R. Molecular cloning and characterization of a cytochrome P450 taxoid 2 alpha-hydroxylase involved in taxol biosynthesis. Arch Biochem Biophys. 2004;427(1):48–57. https://doi.org/10.1016/j.abb.2004.04.016.
28. Chau M, Jennewein S, Walker K, Croteau R. Taxol biosynthesis: molecular cloning and of a cytochrome p450 characterization taxoid 7 beta-hydroxylase. Chem Biol. 2004;11(5):663–72. https://doi.org/10.1016/j.chembiol.2004.02.025.
29. Christen A, Bland J, Gibson D Cell cultures as a means to produce taxol. In: Proceedings of the American Association for Cancer Research; 1989. p 566.
30. Christen AA, Gibson DM, Bland J. Production of taxol or taxol-like compounds in cell culture. 1991. U.S. Patent No. 5,019,504.
31. Cojocaru V, Winn PJ, Wade RC. The ins and outs of cytochrome P450s. Biochim Biophys Acta-Gen Subj. 2007;1770(3):390–401. https://doi.org/10.1016/j.bbagen.2006.07.005.
32. Colling J, Pollier J, Makunga NP, Goossens A cDNA-AFLP-based transcript profiling for genome-wide expression analysis of jasmonate-treated plants and plant cultures. Jasmonate Signal Methods Protocols. 2013; 287–303. doi:https://doi.org/10.1007/978-1-62703-414-2_23.
33. Compant S, van der Heijden MGA, Sessitsch A. Climate change effects on beneficial plant-microorganism interactions. FEMS Microbiol Ecol. 2010;73(2):197–214. https://doi.org/10.1111/j.1574-6941.2010.00900.
34. Croteau R, Ketchum RE, Long RM, Kaspera R, Wildung MR. Taxol biosynthesis and molecular genetics. Phytochem Rev. 2006;5(1):75–97. https://doi.org/10.1007/s11101-005-3748-2.
35. Cusido RM, Palazon J, Navia-Osorio A, Mallol A, Bonfill M, Morales C, Pinol MT. Production of taxol((R)) and baccatin III by a selected *Taxus baccata* callus line and its derived cell suspension culture. Plant Sci. 1999;146(2):101–7. https://doi.org/10.1016/S0168-9452(99)00093-X.
36. Cusido RM, Palazon J, Bonfill M, Exposito O, Moyano E, Pinol MT. Source of isopentenyl diphosphate for taxol and baccatin III biosynthesis in cell cultures of *Taxus baccata*. Biochem Eng J. 2007;33(2):159–67. https://doi.org/10.1016/j.bej.2006.10.016.
37. Daniewski WM, Gumulka M, Anczewski W, Masnyk M, Bloszyk E, Gupta KK. Why the yew tree (*Taxus baccata*) is not attacked by insects. Phytochemistry. 1998;49(5):1279–82. https://doi.org/10.1016/S0031-9422(98)00102-2.

38. De Geyter N, Gholami A, Goormachtig S, Goossens A. Transcriptional machineries in jasmonate-elicited plant secondary metabolism. Trends Plant Sci. 2012;17(6):349–59. https://doi.org/10.1016/j.tplants.2012.03.001.
39. DeJong JM, Liu YL, Bollon AP, Long RM, Jennewein S, Williams D, Croteau RB. Genetic engineering of taxol biosynthetic genes in *Saccharomyces cerevisiae*. Biotechnol Bioeng. 2006;93(2):212–24. https://doi.org/10.1002/bit.20694.
40. Delaye L, García-Guzmán G, Heil M. Endophytes versus biotrophic and necrotrophic pathogens – are fungal lifestyles evolutionarily stable traits? Fungal Divers. 2013;60(1):125–35. https://doi.org/10.1007/s13225-013-0240-y.
41. Denis JN, Greene AE, Guenard D, Gueritte-Voegelein F, Mangatal L, Potier P. Highly efficient, practical approach to natural taxol. J Am Chem Soc. 1988;110(17):5917–9. https://doi.org/10.1021/ja00225a063.
42. Deusneumann B, Zenk MH. Instability of indole alkaloid production in *Catharanthus roseus* cell-suspension cultures. Planta Med. 1984;50(5):427–31. https://doi.org/10.1055/s-2007-969755.
43. Ding MZ, Yan HF, Li LF, Zhai F, Shang LQ, Yin Z, Yuan YJ. Biosynthesis of taxadiene in *Saccharomyces cerevisiae* : selection of geranylgeranyl diphosphate synthase directed by a computer-aided docking strategy. PLoS One. 2014;9(10). https://doi.org/10.1371/journal.pone.0109348.
44. Dornenburg H, Knorr D. Strategies for the improvement of secondary metabolite production in plant cell cultures. Enzym Microb Technol. 1995;17(8):674–84. https://doi.org/10.1016/0141-0229(94)00108-4.
45. Edgar S, Zhou K, Qao KJ, King JR, Simpson JH, Stephanopoulos G. Mechanistic insights into taxadiene epoxidation by taxadiene-5 alpha-hydroxylase. ACS Chem Biol. 2016;11(2):460–9. https://doi.org/10.1021/acschembio.5b00767.
46. Eisenreich W, Menhard B, Hylands PJ, Zenk MH, Bacher A. Studies on the biosynthesis of taxol: the taxane carbon skeleton is not of mevalonoid origin. P Natl Acad Sci USA. 1996;93(13):6431–6. https://doi.org/10.1073/pnas.93.13.6431.
47. Ellis DD, Zeldin EL, Brodhagen M, Russin WA, McCown BH. Taxol production in nodule cultures of Taxus. J Nat Prod. 1996;59(3):246–50. https://doi.org/10.1021/np960104g.
48. Elmer W, Mattina M, MacEachern G. Sensitivity of plant pathogenic fungi to taxane extracts from ornamental yews. Phytopathology. 1994;84(10):1179–85. https://doi.org/10.1094/Phyto-84-1179.
49. Engels B, Dahm P, Jennewein S. Metabolic engineering of taxadiene biosynthesis in yeast as a first step towards taxol (Paclitaxel) production. Metab Eng. 2008;10(3–4):201–6. https://doi.org/10.1016/j.ymben.2008.03.001.
50. Esvelt KM, Wang HH. Genome-scale engineering for systems and synthetic biology. Mol Syst Biol. 2013;9(1):641. https://doi.org/10.1038/msb.2012.66.
51. Exposito O, Bonfill M, Onrubia M, Jane A, Moyano E, Cusido RM, Palazon J, Pinol MT. Effect of taxol feeding on taxol and related taxane production in *Taxus baccata* suspension cultures. New Biotechnol. 2009;25(4):252–9. https://doi.org/10.1016/j.nbt.2008.11.001.
52. Exposito O, Syklowska-Baranek K, Moyano E, Onrubia M, Bonfill M, Palazon J, Cusido RM. Metabolic responses of *Taxus media* transformed cell cultures to the addition of methyl Jasmonate. Biotechnol Prog. 2010;26(4):1145–53. https://doi.org/10.1002/btpr.424.
53. Fichtner F, Castellanos RU, Ülker B. Precision genetic modifications: a new era in molecular biology and crop improvement. Planta. 2014;239(4):921–39. https://doi.org/10.1007/s00425-014-2029-y.
54. Flores-Bustamante ZR, Rivera-Orduna FN, Martinez-Cardenas A, Flores-Cotera LB. Microbial paclitaxel: advances and perspectives. J Antibiot. 2010;63(8):460–7. https://doi.org/10.1038/ja.2010.83.
55. Fray RG, Grierson D. Identification and genetic analysis of normal and mutant phytoene synthase genes of tomato by sequencing, complementation and co-suppression. Plant Mol Biol. 1993;22(4):589–602. https://doi.org/10.1007/BF00047400.
56. Fujimoto SY, Ohta M, Usui A, Shinshi H, Ohme-Takagi M. Arabidopsis ethylene-responsive element binding factors act as transcriptional activators or repressors of GCC box–mediated gene expression. Plant Cell. 2000;12(3):393–404. https://doi.org/10.1105/tpc.12.3.393.

57. Gadzovska Simic S, Tusevski O, Maury S, Delaunay A, Joseph C, Hagège D. Effects of polysaccharide elicitors on secondary metabolite production and antioxidant response in *Hypericum perforatum* L. shoot cultures. ScientificWorldJ. 2014. https://doi.org/10.1155/2014/609649.
58. Georgiev MI, Weber J, Maciuk A. Bioprocessing of plant cell cultures for mass production of targeted compounds. Appl Microbiol Biotechnol. 2009;83(5):809–23. https://doi.org/10.1007/s00253-009-2049-x.
59. Gibson D, Ketchum R, Vance N, Christen A. Initiation and growth of cell lines of *Taxus brevifolia* (Pacific yew). Plant Cell Rep. 1993;12(9):479–82. https://doi.org/10.1007/BF00236091.
60. Goodman J, Walsh V. The story of taxol : nature and politics in the pursuit of an anti-cancer drug. New York: Cambridge University Press; 2001.
61. Goossens A. It is easy to get huge candidate gene lists for plant metabolism now, but how to get beyond? Mol Plant. 2015;8(1):2–5. https://doi.org/10.1016/j.molp.2014.08.001.
62. Gruchattka E, Hadicke O, Klamt S, Schutz V, Kayser O. In silico profiling of *Escherichia coli* and *Saccharomyces cerevisiae* as terpenoid factories. Microb Cell Factories. 2013;12. https://doi.org/10.1186/1475-2859-12-84.
63. Gu Y, Wang YL, Ma XP, Wang CD, Yue GZ, Zhang YT, Zhang YY, Li SS, Ling SS, Liu XM, Wen XT, Cao SJ, Huang XB, Deng JL, Zuo ZC, Yu SM, Shen LH, Wu R. Greater taxol yield of fungus *Pestalotiopsis hainanensis* from dermatitic scurf of the giant panda (*Ailuropoda melanoleuca*). Appl Biochem Biotechnol. 2015;175(1):155–65. https://doi.org/10.1007/s12010-014-1254-y.
64. Gueritte-Voegelein F, Guenard D, Dubois J, Wahl A, Potier P. Chemical and biological studies on taxol (Paclitaxel) and Taxotere (Docetaxel), new antineoplastic agents. J Pharm Belg. 1993;49(3):193–205.
65. Guerra-Bubb J, Croteau R, Williams RM. The early stages of taxol biosynthesis: an interim report on the synthesis and identification of early pathway metabolites. Nat Prod Rep. 2012;29(6):683–96. https://doi.org/10.1039/c2np20021j.
66. Guo BH, Wang YC, Zhou XW, Hu K, Tan F, Miao ZQ, Tang KX. An endophytic taxol-producing fungus BT2 isolated from *Taxus chinensis var. mairei*. Afr J Biotechnol. 2006;5(10):875–7.
67. Hall RD, Yeoman MM. Intercellular and intercultural heterogeneity in secondary metabolite accumulation in cultures of *Catharanthus roseus* following cell-line selection. J Exp Bot. 1987;38(193):1391–8. https://doi.org/10.1093/jxb/38.8.1391.
68. Hallmann J. Plant interactions with endophytic bacteria. New York: CABI Publishing; 2001.
69. Hao X, Pan J, Zhu X. Taxol producing fungi. In: Natural products: Springer; 2013. p. 2797–812.
70. Hefner J, Ketchum REB, Croteau R. Cloning and functional expression of a cDNA encoding geranylgeranyl diphosphate synthase from *Taxus canadensis* and assessment of the role of this prenyltransferase in cells induced for taxol production. Arch Biochem Biophys. 1998;360(1):62–74. https://doi.org/10.1006/abbi.1998.0926.
71. Heinig U, Scholz S, Jennewein S. Getting to the bottom of taxol biosynthesis by fungi. Fungal Divers. 2013;60(1):161–70. https://doi.org/10.1007/s13225-013-0228-7.
72. Heinig UH. Studies on the evolution of complex natural products biosynthetic pathways on basis of taxol biosynthesis in plants and endophytic fungi: RWTH Aachen University; 2012.
73. Hezari M, Ketchum REB, Gibson DM, Croteau R. Taxol production and taxadiene synthase activity in *Taxus canadensis* cell suspension cultures. Arch Biochem Biophys. 1997;337(2):185–90. https://doi.org/10.1006/abbi.1996.9772.
74. Higo K, Ugawa Y, Iwamoto M, Korenaga T. Plant cis-acting regulatory DNA elements (PLACE) database: 1999. Nucleic Acids Res. 1999;27(1):297–300.
75. Hirai S, Utsugi M, Iwamoto M, Nakada M. Formal total synthesis of (−)-taxol through Pd-catalyzed eight-membered carbocyclic ring formation. Chem Eur J. 2015;21(1):355–9. https://doi.org/10.1002/chem.201404295.
76. Hoffman A Methods for obtaining taxanes. 2003. U.S. Patent No. 6,638,742.
77. Holten R, Samoza C, Kim H, Liang F, Biediger R, Boatman P, Shindo M, Smith C, Kim S. First total synthesis of taxol. J Am Chem Soc. 1994;116:1587–600. https://doi.org/10.1021/ja00083a066.

78. Holton R, Biediger RJ, Boatman PD. Semisynthesis of taxol and taxotere. In: Suffness M, editor. Taxol: science and applications, vol. 22. Boca Raton: CRC Press; 1995. p. 97–121.
79. Holton RA, Kim HB, Somoza C, Liang F, Biediger RJ, Boatman PD, Shindo M, Smith CC, Kim SC, Nadizadeh H, Suzuki Y, Tao CL, Vu P, Tang SH, Zhang PS, Murthi KK, Gentile LN, Liu JH. First total synthesis of taxol. 2. Completion of the C-ring and D-ring. J Am Chem Soc. 1994;116(4):1599–600. https://doi.org/10.1021/ja00083a067.
80. Holton RA, Somoza C, Kim HB, Liang F, Biediger RJ, Boatman PD, Shindo M, Smith CC, Kim SC, Nadizadeh H, Suzuki Y, Tao CL, Vu P, Tang SH, Zhang PS, Murthi KK, Gentile LN, Liu JH. First total synthesis of taxol. 1. Functionalization of the B-ring. J Am Chem Soc. 1994;116(4):1597–8. https://doi.org/10.1021/ja00083a066.
81. Huang QL, Roessner CA, Croteau R, Scott AI. Engineering *Escherichia coli* for the synthesis of taxadiene, a key intermediate in the biosynthesis of taxol. Bioorg Med Chem. 2001;9(9):2237–42. https://doi.org/10.1016/S0968-0896(01)00072-4.
82. Huffaker A, Pearce G, Veyrat N, Erb M, Turlings TCJ, Sartor R, Shen ZX, Briggs SP, Vaughan MM, Alborn HT, Teal PEA, Schmelz EA. Plant elicitor peptides are conserved signals regulating direct and indirect antiherbivore defense. P Natl Acad Sci USA. 2013;110(14):5707–12. https://doi.org/10.1073/pnas.1214668110.
83. Hutchison CA, Chuang R-Y, Noskov VN, Assad-Garcia N, Deerinck TJ, Ellisman MH, Gill J, Kannan K, Karas BJ, Ma L. Design and synthesis of a minimal bacterial genome. Science. 2016;351(6280):aad6253. https://doi.org/10.1126/science.aad6253.
84. Itokawa H, Lee K-H. Taxus: the genus taxus. In: Medicinal and aromatic plants-industrial profiles, vol. 32. London/New York: Taylor & Francis; 2003.
85. Jennewein S, Rithner CD, Williams RM, Croteau RB. Taxol biosynthesis: Taxane 13 alpha-hydroxylase is a cytochrome P450-dependent monooxygenase. P Natl Acad Sci USA. 2001;98(24):13595–600. https://doi.org/10.1073/pnas.251539398.
86. Jennewein S, Rithner CD, Williams RM, Croteau R. Taxoid metabolism: Taxoid 14 beta-hydroxylase is a cytochrome P450-dependent monooxygenase. Arch Biochem Biophys. 2003;413(2):262–70. https://doi.org/10.1016/S0003-9861(03)00090-0.
87. Jennewein S, Long RM, Williams RM, Croteau R. Cytochrome P450 taxadiene 5 alpha-hydroxylase, a mechanistically unusual monooxygenase catalyzing the first oxygenation step of taxol biosynthesis. Chem Biol. 2004;11(3):379–87. https://doi.org/10.1016/j.chembiol.2004.02.022.
88. Jennewein S, Wildung MR, Chau M, Walker K, Croteau R. Random sequencing of an induced Taxus cell cDNA library for identification of clones involved in taxol biosynthesis. P Natl Acad Sci USA. 2004;101(24):9149–54. https://doi.org/10.1073/pnas.0403009101.
89. Jennewein S, Park H, DeJong JM, Long RM, Bollon AP, Croteau RB. Coexpression in yeast of Taxus cytochrome p450 reductase with cytochrorne P450 oxygenases involved in taxol biosynthesis. Biotechnol Bioeng. 2005;89(5):588–98. https://doi.org/10.1002/bit.20390.
90. Jensen K, Møller BL. Plant NADPH-cytochrome P450 oxidoreductases. Phytochemistry. 2010;71(2):132–41. https://doi.org/10.1016/j.phytochem.2009.10.017.
91. Kaul BL. The effect of some treatment conditions on the radiomimetic activity of 1-methyl-3-nitro-1-nitrosoguanidine in plants. Mutat Res. 1969;7(1):43–9.
92. Ketchum RE, Gibson DM, Croteau RB, Shuler ML. The kinetics of taxoid accumulation in cell suspension cultures of Taxus following elicitation with methyl jasmonate. Biotechnol Bioeng. 1999;62(1):97–105. https://doi.org/10.1002/(SICI)1097-0290(19990105)62:1<97::AID-BIT11>3.0.CO;2-C.
93. Ketchum REB, Wherland L, Croteau RB. Stable transformation and long-term maintenance of transgenic Taxus cell suspension cultures. Plant Cell Rep. 2007;26(7):1025–33. https://doi.org/10.1007/s00299-007-0323-x.
94. Khan AL, Hamayun M, Kang SM, Kim YH, Jung HY, Lee JH, Lee IJ. Endophytic fungal association via gibberellins and indole acetic acid can improve plant growth under abiotic stress: an example of *Paecilomyces formosus* LHL10. BMC Microbiol. 2012;12. https://doi.org/10.1186/1471-2180-12-3.

95. Khosroushahi AY, Valizadeh M, Ghasempour A, Khosrowshahli M, Naghdibadi H, Dadpour MR, Omidi Y. Improved taxol production by combination of inducing factors in suspension cell culture of *Taxus baccata*. Cell Biol Int. 2006;30(3):262–9. https://doi.org/10.1016/j.cellbi.2005.11.004.
96. Kim BJ, Gibson DM, Shuler ML. Effect of subculture and elicitation on instability of taxol production in Taxus sp suspension cultures. Biotechnol Prog. 2004;20(6):1666–73. https://doi.org/10.1021/bp034274c.
97. Kim BJ, Gibson DM, Shuler ML. Effect of the plant peptide regulator, phytosulfokine-alpha on the growth and taxol production from Taxus sp suspension cultures. Biotechnol Bioeng. 2006;95(1):8–14. https://doi.org/10.1002/bit.20934.
98. Kim JH, Yun JH, Hwang YS, Byun SY, Kim DI. Production of taxol and related Taxanes in *Taxus brevifolia* cell cultures: effect of sugar. Biotechnol Lett. 1995;17(1):101–6. https://doi.org/10.1007/Bf00134204.
99. Kingston DG. Taxol and its analogs. In: Cragg GM, Kingston DG, Newman DJ, editors. Anticancer agents from natural products. Boca Raton: CRC press; 2011. p. 123–76.
100. Koksal M, Jin YH, Coates RM, Croteau R, Christianson DW. Taxadiene synthase structure and evolution of modular architecture in terpene biosynthesis. Nature. 2011;469(7328):116–U138. https://doi.org/10.1038/nature09628.
101. Kolewe ME, Henson MA, Roberts SC. Analysis of aggregate size as a process variable affecting paclitaxel accumulation in taxus suspension cultures. Biotechnol Prog. 2011;27(5):1365–72. https://doi.org/10.1002/btpr.655.
102. Koshino H, Terada S-I, Yoshihara T, Sakamura S, Shimanuki T, Sato T, Tajimi A. Three phenolic acid derivatives from stromata of *Epichloe typhina* on *Phleum pratense*. Phytochemistry. 1988;27(5):1333–8. https://doi.org/10.1016/0031-9422(88)80188-2.
103. Kovacs K, Zhang L, Linforth RST, Whittaker B, Hayes CJ, Fray RG. Redirection of carotenoid metabolism for the efficient production of taxadiene [taxa-4(5),11(12)-diene] in transgenic tomato fruit. Transgenic Res. 2007;16(1):121–6. https://doi.org/10.1007/s11248-006-9039-x.
104. Kusari S, Singh S, Jayabaskaran C. Rethinking production of taxol (R) (paclitaxel) using endophyte biotechnology. Trends Biotechnol. 2014;32(6):304–11. https://doi.org/10.1016/j.tibtech.2014.03.011.
105. Larkin PJ, Scowcroft WR. Somaclonal variationa – novel source of variability from cell cultures for plant improvement. Theor Appl Genet. 1981;60(4):197–214. https://doi.org/10.1007/BF02342540.
106. Lavelle F, Gueritte-Voegelein F, Guenard D. Taxotere: from yew's needles to clinical practice. Bull Cancer. 1993;80(4):326–38.
107. Leavell MD, McPhee DJ, Paddon CJ. Developing fermentative terpenoid production for commercial usage. Curr Opin Biotechnol. 2016;37:114–9. https://doi.org/10.1016/j.copbio.2015.10.007.
108. Lenka SK, Nims NE, Vongpaseuth K, Boshar RA, Roberts SC, Walker EL. Jasmonate-responsive expression of paclitaxel biosynthesis genes in *Taxus cuspidata* cultured cells is negatively regulated by the bHLH transcription factors TcJAMYC1, TcJAMYC2, and TcJAMYC4. Front Plant Sci. 2015;6. https://doi.org/10.3389/fpls.2015.00115.
109. Leonard E, Koffas MA. Engineering of artificial plant cytochrome P450 enzymes for synthesis of isoflavones by *Escherichia coli*. Appl Environ Microbiol. 2007;73(22):7246–51. https://doi.org/10.1128/AEM.01411-07.
110. Li F-L, Ma X-J, X-L H, Hoffman A, Dai J-G, D-Y Q. Antisense-induced suppression of taxoid 14β-hydroxylase gene expression in transgenic *Taxus*× *media* cells. Afr J Biotechnol. 2013;10(44):8720–8. https://doi.org/10.5897/AJB11.319.
111. Li JY, Strobel G, Sidhu R, Hess WM, Ford EJ. Endophytic taxol-producing fungi from bald cypress, *Taxodium distichum*. Microbiol-UK. 1996;142:2223–6. https://doi.org/10.1099/13500872-142-8-2223.
112. Li MY, Jiang FS, Yu XL, Miao ZQ. Engineering isoprenoid biosynthesis in *Artemisia annua* L. for the production of Taxadiene: a key intermediate of taxol. Biomed Res Int. 2015. https://doi.org/10.1155/2015/504932.

113. Li S, Zhang P, Zhang M, Fu C, Yu L. Functional analysis of a WRKY transcription factor involved in transcriptional activation of the DBAT gene in *Taxus chinensis*. Plant Biol. 2013;15(1):19–26. https://doi.org/10.1111/j.1438-8677.2012.00611.
114. Li ST, CH F, Zhang M, Zhang Y, Xie S, LJ Y. Enhancing taxol biosynthesis by overexpressing a 9-cis-epoxycarotenoid dioxygenase gene in transgenic cell lines of *Taxus chinensis*. Plant Mol Biol Report. 2012;30(5):1125–30. https://doi.org/10.1007/s11105-012-0436-4.
115. Li YF, Lin ZQ, Huang C, Zhang Y, Wang ZW, Tang YJ, Chen T, Zhao XM. Metabolic engineering of *Escherichia coli* using CRISPR-Cas9 meditated genome editing. Metab Eng. 2015;31:13–21. https://doi.org/10.1016/j.ymben.2015.06.006.
116. Lin XY, Hezari M, Koepp AE, Floss HG, Croteau R. Mechanism of taxadiene synthase, a diterpene cyclase that catalyzes the first step of taxol biosynthesis in Pacific yew. Biochemistry-US. 1996;35(9):2968–77. https://doi.org/10.1021/bi9526239.
117. Linden JC, Phisalaphong M. Oligosaccharides potentiate methyl jasmonate-induced production of paclitaxel in *Taxus canadensis*. Plant Sci. 2000;158(1–2):41–51. https://doi.org/10.1016/S0168-9452(00)00306-X.
118. Liu KH, Ding XW, Deng BW, Chen WQ. Isolation and characterization of endophytic taxol-producing fungi from *Taxus chinensis*. J Ind Microbiol Biotechnol. 2009;36(9):1171–7. https://doi.org/10.1007/s10295-009-0598-8.
119. Long RM, Croteau R. Preliminary assessment of the C13-side chain 2′-hydroxylase involved in taxol biosynthesis. Biochem Bioph Res Commun. 2005;338(1):410–7. https://doi.org/10.1016/j.bbrc.2005.08.119.
120. Long RM, Lagisetti C, Coates RM, Croteau RB. Specificity of the N-benzoyl transferase responsible for the last step of taxol biosynthesis. Arch Biochem Biophys. 2008;477(2):384–9. https://doi.org/10.1016/j.abb.2008.06.021.
121. Malik S, Cusido RM, Mirjalili MH, Moyano E, Palazon J, Bonfill M. Production of the anticancer drug taxol in *Taxus baccata* suspension cultures: a review. Process Biochem. 2011;46(1):23–34. https://doi.org/10.1016/j.procbio.2010.09.004.
122. Martin VJJ, Pitera DJ, Withers ST, Newman JD, Keasling JD. Engineering a mevalonate pathway in *Escherichia coli* for production of terpenoids. Nat Biotechnol. 2003;21(7):796–802. https://doi.org/10.1038/nbt833.
123. Martínez-Márquez A, Morante-Carriel J, Ramírez-Estrada K, Cusido R, Sellés-Marchart S, Palazon J, Pedreño MA, Bru-Martínez R. A reliable protocol for the stable transformation of non-embryogenic cells cultures of grapevine (Vitis vinifera L.) and Taxus x media. J Biol Methods. 2015;2(2):e21.
124. Mathews HV, Reddy VS, Lammers AH. Taxus transformation transformed cells, and related compositions and methods. 2011. U.S. Patent No. 8,053,637.
125. Matsubayashi Y. Recent progress in research on small post-translationally modified peptide signals in plants. Genes Cells. 2012;17(1):1–10. https://doi.org/10.1111/j.1365-2443.2011.01569.x.
126. McLellan CA, Turbyville TJ, Wijeratne EMK, Kerschen A, Vierling E, Queitsch C, Whitesell L, Gunatilaka AAL. A rhizosphere fungus enhances Arabidopsis thermotolerance through production of an HSP90 inhibitor. Plant Physiol. 2007;145(1):174–82. https://doi.org/10.1104/pp.107.101808.
127. Mirjalili N, Linden JC. Gas phase composition effects on suspension cultures of *Taxus cuspidata*. Biotechnol Bioeng. 1995;48(2):123–32. https://doi.org/10.1002/bit.260480206.
128. Mirjalili N, Linden JC. Methyl jasmonate induced production of taxol in suspension cultures of *Taxus cuspidata*: ethylene interaction and induction models. Biotechnol Prog. 1996;12(1):110–8. https://doi.org/10.1021/bp9500831.
129. Moon WJ, Yoo BS, Kim DI, Byun SY. Elicitation kinetics of taxane production in suspension cultures of *Taxus baccata* Pendula. Biotechnol Tech. 1998;12(1):79–81. https://doi.org/10.1023/A:1008867828951.
130. Morais S, Pandey PC, Chen W, Mulchandani A. A novel bioassay for screening and quantification of taxanes. Chem Commun. 2003;10:1188–9.
131. Morris P, Rudge K, Cresswell R, Fowler M. Regulation of product synthesis in cell cultures of *Catharanthus roseus*. V. Long-term maintenance of cells on a production medium. Plant Cell Tissue Organ Cult. 1989;17(2–3):79–90. https://doi.org/10.1007/BF00046853.

132. Mountford PG. The taxol® story–development ofa green synthesis via plant cell fermentation. Green Chem Pharmaceut Ind. 2010. https://doi.org/10.1002/9783527629688.ch7.
133. Mousa WK, Raizada MN. The diversity of ant microbial secondary metabolites produced by fungal endophytes: an interdisciplinary perspective. Front Microbiol. 2013;4. https://doi.org/10.3389/fmicb.2013.00065.
134. Naik PM, Al-Khayri JM. Abiotic and biotic elicitors–role in secondary metabolites production through in vitro culture of medicinal plants. In: Shanker AK and Shankar C (eds). Abiotic and biotic stress in plants - recent advances and future perspectives. InTech, Croatia. 2016: 247–77. https://doi.org/10.5772/61442.
135. Nicolaou K, Yang Z, Liu J, Ueno H, Nantermet P, Guy R, Claiborne C, Renaud J, Couladouros E, Paulvannan K. Total synthesis of taxol. Nature. 1994;367(6464):630–4. https://doi.org/10.1038/367630a0.
136. Nicolaou KC, Dai WM, Guy RK. Chemistry and biology of taxol. Angew Chem Int Ed Engl. 1994;33(1):15–44. https://doi.org/10.1002/anie.199400151.
137. Nims E, Dubois CP, Roberts SC, Walker EL. Expression profiling of genes involved in paclitaxel biosynthesis for targeted metabolic engineering. Metab Eng. 2006;8(5):385–94. https://doi.org/10.1016/j.ymben.2006.04.001.
138. Nour-Eldin HH, Halkier BA. The emerging field of transport engineering of plant specialized metabolites. Curr Opin Biotechnol. 2013;24(2):263–70. https://doi.org/10.1016/j.copbio.2012.09.006.
139. Ogden L. Taxus (yews) – a highly toxic plant. Vet Hum Toxicol. 1988;30(6):563–4.
140. Onrubia M, Cusido RM, Ramirez K, Hernandez-Vazquez L, Moyano E, Bonfill M, Palazon J. Bioprocessing of plant in vitro Systems for the Mass Production of pharmaceutically important metabolites: paclitaxel and its derivatives. Curr Med Chem. 2013;20(7):880–91. https://doi.org/10.2174/0929867113805219064.
141. Onrubia M, Moyano E, Bonfill M, Cusido RM, Goossens A, Palazon J. Coronatine, a more powerful elicitor for inducing taxane biosynthesis in *Taxus media* cell cultures than methyl jasmonate. J Plant Physiol. 2013;170(2):211–9. https://doi.org/10.1016/j.jplph.2012.09.004.
142. Onrubia M, Pollier J, Vanden Bossche R, Goethals M, Gevaert K, Moyano E, Vidal-Limon H, Cusido RM, Palazon J, Goossens A. Taximin, a conserved plant-specific peptide is involved in the modulation of plant-specialized metabolism. Plant Biotechnol J. 2014;12(7):971–83. https://doi.org/10.1111/pbi.12205.
143. Osuna L, Tapia N, Cusido R, Palazon J, Bonfill M, Zamilpa A, Lopez-Upton J, Cruz-Sosa F. Taxane production induced by methyl jasmonate in free and immobilized cell cultures of Mexican yew (*Taxus globosa* Schltdl). Acta Physiol Plant. 2015;37(10). https://doi.org/10.1007/s11738-015-1947-z.
144. Paddon CJ, Westfall PJ, Pitera D, Benjamin K, Fisher K, McPhee D, Leavell M, Tai A, Main A, Eng D. High-level semi-synthetic production of the potent antimalarial artemisinin. Nature. 2013;496(7446):528–32. https://doi.org/10.1038/nature12051.
145. Paddon CJ, Keasling JD. Semi-synthetic artemisinin: a model for the use of synthetic biology in pharmaceutical development. Nat Rev Microbiol. 2014;12(5):355–67. https://doi.org/10.1038/nrmicro3240.
146. Pandey RP, Parajuli P, Koffas MA, Sohng JK. Microbial production of natural and non-natural flavonoids: pathway engineering, directed evolution and systems/synthetic biology. Biotechnol Adv. 2016. https://doi.org/10.1016/j.biotechadv.2016.02.012.
147. Pandi M, Kumaran RS, Choi YK, Kim HJ, Muthumary J. Isolation and detection of taxol, an anticancer drug produced from *Lasiodiplodia theobromae*, an endophytic fungus of the medicinal plant *Morinda citrifolia*. Afr J Biotechnol. 2011;10(8):1428–35. https://doi.org/10.5897/AJB10.950.
148. Parr AJ, Payne J, Eagles J, Chapman BT, Robins RJ, Rhodes MJC. Variation in tropane alkaloid accumulation within the Solanaceae and strategies for its exploitation. Phytochemistry. 1990;29(8):2545–50. https://doi.org/10.1016/0031-9422(90)85185-I.

149. Paungfoo-Lonhienne C, Rentsch D, Robatzek S, Webb RI, Sagulenko E, Nasholm T, Schmidt S, Lonhienne TGA. Turning the table: plants consume microbes as a source of nutrients. PLoS One. 2010;5(7). https://doi.org/10.1371/journal.pone.0011915.
150. The Plant List Version 1.1. Published on the Internet. http://www.theplantlist.org/. 2013. Accessed 9 Mar 2016.
151. Pre M, Atallah M, Champion A, De Vos M, Pieterse CMJ, Memelink J. The AP2/ERF domain transcription factor ORA59 integrates jasmonic acid and ethylene signals in plant defense. Plant Physiol. 2008;147(3):1347–57. https://doi.org/10.1104/pp.108.117523.
152. Priti V, Ramesha BT, Singh S, Ravikanth G, Ganeshaiah KN, Suryanarayanan TS, Shaanker RU. How promising are endophytic fungi as alternative sources of plant secondary metabolites? CURRENT SCIENCE. 2009;7(4):477.
153. Prosser GA, Larrouy-Maumus G, de Carvalho LPS. Metabolomic strategies for the identification of new enzyme functions and metabolic pathways. EMBO Rep. 2014;15(6):657–69. 10.15252/embr.201338283.
154. Ramirez-Estrada K, Osuna L, Moyano E, Bonfill M, Tapia N, Cusido RM, Palazon J. Changes in gene transcription and taxane production in elicited cell cultures of *Taxus* x *media* and *Taxus globosa*. Phytochemistry. 2015;117:174–84. https://doi.org/10.1016/j.phytochem.2015.06.013.
155. Ramirez-Estrada K, Altabella T, Onrubia M, Moyano E, Notredame C, Osuna L, Bossche RV, Goossens A, Cusido RM, Palazon J. Transcript profiling of jasmonate-elicited Taxus cells reveals a beta-phenylalanine-CoA ligase. Plant Biotechnol J. 2016;14(1):85–96. https://doi.org/10.1111/pbi.12359.
156. Ramirez-Estrada K, Vidal-Limon H, Hidalgo D, Moyano E, Golenioswki M, Cusido RM, Palazon J. Elicitation, an effective strategy for the biotechnological production of bioactive high-added value compounds in plant cell factories. Molecules. 2016;21(2). https://doi.org/10.3390/molecules21020182.
157. Rao S, He L, Chakravarty S, Ojima I, Orr GA, Horwitz SB. Characterization of the taxol binding site on the microtubule identification of Arg282 in β-tubulin as the site of photoincorporation of a 7-benzophenone analogue of taxol. J Biol Chem. 1999;274(53):37990–4. https://doi.org/10.1074/jbc.274.53.37990.
158. Raunkiær C, Gilvert-Carter H, Fausbøll A, Tansley AG. The life forms of plants and statistical plant geography. Oxford: The Clarendon press; 1934.
159. Rikhari HC, Palni LMS, Sharma S, Nandi SK. Himalayan yew: stand structure, canopy damage, regeneration and conservation strategy. Environ Conserv. 1998;25(4):334–41. https://doi.org/10.1017/S0376892998000411.
160. Roche Holdings Ltd. Roche Finance Report 2016. 2016. Retrieved from http://www.roche.com/dam/jcr:6ddcec16-c658-48b2-82b5-4ed426c14ac8/en/fb16e.pdf
161. Rodriguez R, White J Jr, Arnold A, Redman R. Fungal endophytes: diversity and functional roles. New Phytol. 2009;182(2):314–30. https://doi.org/10.1111/j.1469-8137.2009.02773.
162. Rohr R. Production de cals par les gamétophytes mâles de Taxus baccata L. cultivés sur un milieu artificiel. Etude en microscopie photonique et électronique. Caryologia. 1973;25(sup1):177–89. https://doi.org/10.1080/00087114.1973.10797122.
163. Rontein D, Onillon S, Herbette G, Lesot A, Werck-Reichhart D, Sallaud C, Tissier A. CYP725A4 from yew catalyzes complex structural rearrangement of taxa-4(5),11(12)-diene into the cyclic ether 5(12)-oxa-3(11)-cyclotaxane. J Biol Chem. 2008;283(10):6067–75. https://doi.org/10.1074/jbc.M708950200.
164. Sabater-Jara AB, Onrubia M, Moyano E, Bonfill M, Palazon J, Pedreno MA, Cusido RM. Synergistic effect of cyclodextrins and methyl jasmonate on taxane production in *Taxus* x *media* cell cultures. Plant Biotechnol J. 2014;12(8):1075–84. https://doi.org/10.1111/pbi.12214.
165. Sanofi. Sanofi annual report on form 20-F 2016. 2016. Retrieved from http://en.sanofi.com/Images/49288_20-F_2016.pdf
166. Schiff PB, Horwitz SB. Taxol stabilizes microtubules in mouse fibroblast cells. P Natl Acad Sci-Biol. 1980;77(3):1561–5. https://doi.org/10.1073/pnas.77.3.1561.

167. Schoendorf A, Rithner CD, Williams RM, Croteau RB. Molecular cloning of a cytochrome P450 taxane 10 beta-hydroxylase cDNA from Taxus and functional expression in yeast. P Natl Acad Sci USA. 2001;98(4):1501–6. https://doi.org/10.1073/pnas.98.4.1501.
168. Schripsema J, Verpoorte R. Search for factors related to the indole alkaloid production in cell suspension cultures of *Tabernaemontana divaricata*. Planta Med. 1992;58(3):245–9. https://doi.org/10.1055/s-2006-961445.
169. Schuler MA, Werck-Reichhart D. Functional genomics of P450s. Annu Rev Plant Biol. 2003;54:629–67. https://doi.org/10.1146/annurev.arplant.54.031902.134840.
170. Schulz B. Mutualistic interactions with fungal root endophytes. In: Schulz BJE, Boyle CJC, Sieber TN, editors. Microbial root endophytes. Berlin/Heidelberg: Springer; 2006. p. 261–79. https://doi.org/10.1007/3-540-33526-9_15.
171. Shwab EK, Keller NP. Regulation of secondary metabolite production in filamentous ascomycetes. Mycol Res. 2008;112(2):225–30.
172. Soliman SS, Trobacher CP, Tsao R, Greenwood JS, Raizada MN. A fungal endophyte induces transcription of genes encoding a redundant fungicide pathway in its host plant. BMC Plant Biol. 2013;13(1):1. https://doi.org/10.1186/1471-2229-13-93.
173. Soliman SSM, Greenwood JS, Bombarely A, Mueller LA, Tsao R, Mosser DD, Raizada MN. An Endophyte constructs fungicide-containing extracellular barriers for its host plant. Curr Biol. 2015;25(19):2570–6. https://doi.org/10.1016/j.cub.2015.08.027.
174. Souza JJ, Vieira IJ, Rodrigues-Filho E, Braz-Filho R. Terpenoids from endophytic fungi. Molecules. 2011;16(12):10604–18. https://doi.org/10.3390/molecules161210604.
175. Srinivasan V, Pestchanker L, Moser S, Hirasuna TJ, Taticek RA, Shuler ML. Taxol production in bioreactors: kinetics of biomass accumulation, nutrient uptake, and taxol production by cell suspensions of *Taxus baccata*. Biotechnol Bioeng. 1995;47(6):666–76. https://doi.org/10.1002/bit.260470607.
176. Staniek A, Woerdenbag HJ, Kayser O. Taxomyces andreanae: a presumed paclitaxel producer demystified? Planta Med. 2009;75(15):1561–6. https://doi.org/10.1055/s-0029-1186181.
177. Stierle A, Strobel G, Stierle D. Taxol and taxane production by *Taxomyces andreanae*, an endophytic fungus of Pacific yew. Science. 1993;260(5105):214–6. https://doi.org/10.1126/science.8097061.
178. Stierle A, Stierle D. Bioactive compounds from four endophytic *Penicillium* sp. of a northwest Pacific yew tree. Stud Nat Prod Chem. 2000;24:933–77. https://doi.org/10.1016/S1572-5995(00)80058-7.
179. Stovicek V, Borodina I, Forster J. CRISPR–Cas system enables fast and simple genome editing of industrial *Saccharomyces cerevisiae* strains. Metabol Eng Commun. 2015;2:13–22. https://doi.org/10.1016/j.meteno.2015.03.001.
180. Strobel G, Stierle A, Hess WM. The stimulation of taxol production in *Taxus brevifolia* by various growth-retardants. Plant Sci. 1994;101(2):115–24. https://doi.org/10.1016/0168-9452(94)90247-X.
181. Strobel GA. Endophytes as sources of bioactive products. Microbes Infect. 2003;5(6):535–44. https://doi.org/10.1016/S1286-4579(03)00073-X.
182. Sung ZR. Mutagenesis of cultured plant cells. Genetics. 1976;84(1):51–7.
183. Tabata, H. (2004). Paclitaxel production by plant-cell-culture technology. In Biomanufacturing (pp. 1-23). Springer Berlin Heidelberg.
184. Tang ZM, Salamanca-Pinzon SG, ZL W, Xiao Y, Guengerich FP. Human cytochrome P450 4F11: heterologous expression in bacteria, purification, and characterization of catalytic function. Arch Biochem Biophys. 2010;494(1):86–93. https://doi.org/10.1016/j.abb.2009.11.017.
185. Thomas, P. *Taxus brevifolia*. The IUCN red list of threatened species 2013. 2013. https://doi.org/10.2305/IUCN.UK.2013-1.RLTS.T34041A2841142.en. Accessed 29 Apr 2016.
186. Thomas P, Sekhar AC. Live cell imaging reveals extensive intracellular cytoplasmic colonization of banana by normally non-cultivable endophytic bacteria. Aob Plants. 2014;6. https://doi.org/10.1093/aobpla/plu002.

187. Valderrama-Rincon JD, Fisher AC, Merritt JH, Fan YY, Reading CA, Chhiba K, Heiss C, Azadi P, Aebi M, DeLisa MP. An engineered eukaryotic protein glycosylation pathway in *Escherichia coli*. Nat Chem Biol. 2012;8(5):434–6. https://doi.org/10.1038/Nchembio.921.
188. Van der Does D, Leon-Reyes A, Koornneef A, Van Verk MC, Rodenburg N, Pauwels L, Goossens A, Korbes AP, Memelink J, Ritsema T, Van Wees SCM, Pieterse CMJ. Salicylic acid suppresses Jasmonic acid signaling downstream of SCFCOI1-JAZ by targeting GCC promoter motifs via transcription factor ORA59. Plant Cell. 2013;25(2):744–61. https://doi.org/10.1105/tpc.112.108548.
189. Van Dien S. From the first drop to the first truckload: commercialization of microbial processes for renewable chemicals. Curr Opin Biotechnol. 2013;24(6):1061–8. https://doi.org/10.1016/j.copbio.2013.03.002.
190. Vidensek N, Lim P, Campbell A, Carlson C. Taxol content in bark, wood, root, leaf, twig, and seedling from several Taxus species. J Nat Prod. 1990;53(6):1609–10. https://doi.org/10.1021/np50072a039.
191. Vongpaseuth K, Nims E, Amand MS, Walker EL, Roberts SC. Development of a particle bombardment-mediated transient transformation system for Taxus spp. cells in culture. Biotechnol Prog. 2007;23(5):1180–5. https://doi.org/10.1021/bp0700307.
192. Vongpaseuth K, Roberts SC. Advancements in the understanding of paclitaxel metabolism in tissue culture. Curr Pharm Biotechnol. 2007;8(4):219–36. https://doi.org/10.2174/138920107781387393.
193. Walker K, Croteau R. Taxol biosynthesis: molecular cloning of a benzoyl-CoA : taxane 2 alpha-O-benzoyltransferase cDNA from Taxus and functional expression in *Escherichia coli*. P Natl Acad Sci USA. 2000a;97(25):13591–6. https://doi.org/10.1073/pnas.250491997.
194. Walker K, Croteau R. Molecular cloning of a 10-deacetylbaccatin III-10-O-acetyl transferase cDNA from Taxus and functional expression in *Escherichia coli*. P Natl Acad Sci USA. 2000b;97(2):583–7. https://doi.org/10.1073/pnas.97.2.583.
195. Walker K, Schoendorf A, Croteau R. Molecular cloning of a taxa-4(20),11(12)-dien-5 alpha-ol-O-acetyl transferase cDNA from Taxus and functional expression in *Escherichia coli*. Arch Biochem Biophys. 2000;374(2):371–80. https://doi.org/10.1006/abbi.1999.1609.
196. Walker K, Fujisaki S, Long R, Croteau R. Molecular cloning and heterologous expression of the C-13 phenylpropanoid side chain-CoA acyltransferase that functions in taxol biosynthesis. P Natl Acad Sci USA. 2002;99(20):12715–20. https://doi.org/10.1073/pnas.192463699.
197. Walker K, Long R, Croteau R. The final acylation step in taxol biosynthesis: cloning of the taxoid C13-side-chain N-benzoyltransferase from Taxus. P Natl Acad Sci USA. 2002;99(14):9166–71. https://doi.org/10.1073/pnas.082115799.
198. Walker KD, Klettke K, Akiyama T, Croteau R. Cloning, heterologous expression, and characterization of a phenylalanine aminomutase involved in taxol biosynthesis. J Biol Chem. 2004;279(52):53947–54. https://doi.org/10.1074/jbc.M411215200.
199. Wall ME, Wani MC. Camptothecin and taxol: discovery to clinic – thirteenth Bruce F. Cain Memorial Award Lecture. Cancer Res. 1995;55(4):753–60.
200. Wang CG, JY W, Mei XG. Enhanced taxol production and release in *Taxus chinensis* cell suspension cultures with selected organic solvents and sucrose feeding. Biotechnol Prog. 2001;17(1):89–94. https://doi.org/10.1021/bp0001359.
201. Wang ZY, Zhong JJ. Repeated elicitation enhances taxane production in suspension cultures of *Taxus chinensis* in bioreactors. Biotechnol Lett. 2002;24(6):445–8. https://doi.org/10.1023/A:1014549008516.
202. Wani MC, Taylor HL, Wall ME, Coggon P, McPhail AT. Plant antitumor agents. VI. The isolation and structure of taxol, a novel antileukemic and antitumor agent from *Taxus brevifolia*. J Am Chem Soc. 1971;93(9):2325–7. https://doi.org/10.1021/ja00738a045.
203. Wani MC, Horwitz SB. Nature as a remarkable chemist: a personal story of the discovery and development of taxol. Anti-Cancer Drug. 2014;25(5):482–7. https://doi.org/10.1097/Cad.0000000000000063.

204. Waqas M, Khan AL, Kamran M, Hamayun M, Kang SM, Kim YH, Lee IJ. Endophytic fungi produce gibberellins and indoleacetic acid and promotes host-plant growth during stress. Molecules. 2012;17(9):10754–73. https://doi.org/10.3390/molecules170910754.
205. Wheeler AL, Long RM, Ketchum REB, Rithner CD, Williams RM, Croteau R. Taxol biosynthesis: differential transformations of taxadien-5 alpha-ol and its acetate ester by cytochrome P450 hydroxylases from Taxus suspension cells. Arch Biochem Biophys. 2001;390(2):265–78. https://doi.org/10.1006/abbi.2001.2377.
206. White JF, Torres MS, Somu MP, Johnson H, Irizarry I, Chen Q, Zhang N, Walsh E, Tadych M, Bergen M. Hydrogen peroxide staining to visualize intracellular bacterial infections of seedling root cells. Microsc Res Tech. 2014;77(8):566–73. https://doi.org/10.1002/jemt.22375.
207. Whited GM, Feher FJ, Benko DA, Cervin MA, Chotani GK, McAuliffe JC, LaDuca RJ, Ben-Shoshan EA, Sanford KJ. Technology update: development of a gas-phase bioprocess for isoprene-monomer production using metabolic pathway engineering. Ind Biotechnol. 2010;6(3):152–63. https://doi.org/10.1089/ind.2010.6.152.
208. Wilding B, Vesela AB, Perry JJB, Black GW, Zhang M, Martinkova L, Klempier N. An investigation of nitrile transforming enzymes in the chemo-enzymatic synthesis of the taxol sidechain. Org Biomol Chem. 2015;13(28):7803–12. https://doi.org/10.1039/c5ob01191d.
209. Wildung MR, Croteau R. A cDNA clone for taxadiene synthase, the diterpene cyclase that catalyzes the committed step of taxol biosynthesis. J Biol Chem. 1996;271(16):9201–4. https://doi.org/10.1074/jbc.271.16.9201.
210. Williams DC, Wildung MR, Jin AQ, Dalal D, Oliver JS, Coates RM, Croteau R. Heterologous expression and characterization of a "Pseudomature" form of taxadiene synthase involved in paclitaxel (taxol) biosynthesis and evaluation of a potential intermediate and inhibitors of the multistep diterpene cyclization reaction. Arch Biochem Biophys. 2000;379(1):137–46. https://doi.org/10.1006/abbi.2000.1865.
211. Wilson D. Endophyte: the evolution of a term, and clarification of its use and definition. Oikos. 1995:274–6. https://doi.org/10.2307/3545919.
212. Wilson SA, Roberts SC. Recent advances towards development and commercialization of plant cell culture processes for the synthesis of biomolecules. Plant Biotechnol J. 2012;10(3):249–68. https://doi.org/10.1111/j.1467-7652.2011.00664.x.
213. ZL W, Yuan YJ, Liu JX, Xuan HY, ZD H, Sun AC, CX H. Study on enhanced production of taxol from *Taxus chinensis* var. mairei in biphasic-liquid culture. Acta Bot Sin. 1999;41(10):1108–13.
214. Yadav VG. Unraveling the multispecificity and catalytic promiscuity of taxadiene monooxygenase. J Mol Catal B-Enzym. 2014;110:154–64. https://doi.org/10.1016/j.molcatb.2014.10,004.
215. Yanagi M, Ninomiya R, Ueda Y, Furuta T, Yamada T, Sunazuka T, Kawabata T. Organocatalytic site-selective acylation of 10-deacetylbaccatin III. Chem Pharm Bull. 2016. https://doi.org/10.1248/cpb.c16-00037.
216. Yang Y, Zhao H, Barrero RA, Zhang B, Sun G, Wilson IW, Xie F, Walker KD, Parks JW, Bruce R. Genome sequencing and analysis of the paclitaxel-producing endophytic fungus *Penicillium aurantiogriseum* NRRL 62431. BMC Genomics. 2014;15(1):1. https://doi.org/10.1186/1471-2164-15-69.
217. Yarwood C. *Ampelomyces quisqualis* on clover mildew. Phytopathology. 1932;22:31.
218. Yuan L, Grotewold E. Metabolic engineering to enhance the value of plants as green factories. Metab Eng. 2015;27:83–91. https://doi.org/10.1016/j.ymben.2014.11.005.
219. Yukimune Y, Tabata H, Higashi Y, Hara Y. Methyl jasmonate-induced overproduction of paclitaxel and baccatin III in Taxus cell suspension cultures. Nat Biotechnol. 1996;14(9):1129–32. https://doi.org/10.1038/nbt0996-1129.
220. Zhang JF, Gong S, Guo ZG. Effects of different elicitors on 10-deacetylbaccatin III-10-O-acetyltransferase activity and cytochrome P450 monooxygenase content in suspension cultures of *Taxus cuspidata* cells. Cell Biol Int Rep. 2011;18(1):7–13. https://doi.org/10.1042/CBR20110001.

221. Zhang M, Li ST, Nie L, Chen QP, XP X, LJ Y, CH F. Two jasmonate-responsive factors, TcERF12 and TcERF15, respectively act as repressor and activator of tasy gene of taxol biosynthesis in *Taxus chinensis*. Plant Mol Biol. 2015;89(4–5):463–73. https://doi.org/10.1007/s11103-015-0382-2.
222. Zhang P, Zhou P-P, L-J Y. An endophytic taxol-producing fungus from *Taxus* x *media*, *Aspergillus candidus* MD3. FEMS Microbiol Lett. 2009;293(2):155–9. https://doi.org/10.1111/j.1574-6968.2009.01481.x.
223. Zhang P, Li S-T, Liu T-T, C-H F, Zhou P-P, Zhao C-F, L-J Y. Overexpression of a 10-deacetylbaccatin III-10 β-O-acetyltransferase gene leads to increased taxol yield in cells of *Taxus chinensis*. Plant Cell Tissue Organ Cult (PCTOC). 2011;106(1):63–70. https://doi.org/10.1007/s11240-010-9894-2.
224. Zhao K, Ping W, Li Q, Hao S, Zhao L, Gao T, Zhou D. *Aspergillus niger* var. taxi, a new species variant of taxol-producing fungus isolated from *Taxus cuspidata* in China. J Appl Microbiol. 2009;107(4):1202–7. https://doi.org/10.1111/j.1365-2672.2009.04305.x.
225. Zhou K, Qiao KJ, Edgar S, Stephanopoulos G. Distributing a metabolic pathway among a microbial consortium enhances production of natural products. Nat Biotechnol. 2015;33(4):377–U157. https://doi.org/10.1038/nbt.3095.

Part II

Technologies for Metabolic, Enzyme and Process Engineering

Commercial-Scale Tissue Culture for the Production of Plant Natural Products: Successes, Failures and Outlook

8

Bernd Markus Lange

Abbreviation

PNP Plant natural product

8.1 Introduction

Plant tissue culture can be broadly defined as the *in vitro* aseptic maintenance of cells, tissues or organs under defined conditions. The pioneering developments for sustaining isolated plant cells date back to the early 1900s [1]. Haberlandt already introduced the concept of totipotency, which refers to the unique genetic potential to regenerate a whole plant from a single somatic plant cell (validated in the 1960s [2]). During the decades following the initial discoveries, tissue cultures were established from seed embryos, cambial tissue, roots, and many additional plant parts (reviewed in [3]). Biotechnological applications of plant tissue culture for the production of medicinally relevant plant natural products (PNPs) emerged in the 1960s and 1970s [4]. Over the last forty years, several companies have had research and development programs aimed at optimizing plant tissue culture [5]. However, the large-scale market introduction of plant tissue culture products has mostly been confined to (i) early successes with isolated metabolites for the cosmetic (shikonin) [6] and pharmaceutical (paclitaxel) [7] (**see also Chap. 7 of this book**) industries and (ii) more recent uses of whole cell extracts as drinks, dietary supplements and food additives [5] (Table 8.1). The production of therapeutically relevant proteins in plant tissue cultures, which is also an area of very active research and development efforts, has been reviewed recently [8] and will not be covered here.

B.M. Lange (✉)
Institute of Biological Chemistry and M.J. Murdock Metabolomics Laboratory, Washington State University, Pullman, WA 99164-6340, USA
e-mail: lange-m@wsu.edu

W. Schwab et al. (eds.), *Biotechnology of Natural Products*,
https://doi.org/10.1007/978-3-319-67903-7_8

Table 8.1 Commercial-scale production of PNPs by tissue culture technologies

Product	Production system	Manufacturer, trademark, and scale of production	Uses	Source
Berberine (alkaloid)	*Coptis japonica*	Mitsui Chemicals Inc. (Japan) (not commercialized) (750 l fermenter)	Pharmaceutical	[6]
Bilberry cells (polyphenol extract)	*Vaccinium* sp., cell suspension cultures	Diana Plant Sciences (USA), ActivBerry® (to be launched soon)	Dietary supplement	
Cocoa cells (polyphenolic extract)	*Theobroma cacao* L., cell suspension cells	Diana Plant Sciences (USA), Cocovanol® (launched in 2013)	Dietary supplement	www.diana-group.com/
Echinacea cells (polysaccharide extract)	*Echinacea* sp.	Diversa (Germany) (not commercialized) (75,000 l fermenters)	Pharmaceutical (immunostimulant)	[5]
Ginseng cells (ginsenoside extract)	*Panax ginseng* L., cell suspension cells	Nitto Denko Corp. (Japan), (since 1989; 25,000 l fermenters)	Food additive	www.nitto.com [15]
	Panax ginseng L., stem cell cultures	Unhwa Corp. (Korea), Ddobyul®, LifeAde (since 2011)	Energy drink, Cosmetics (creams)	www.unhwa.com [16]

(continued)

Paclitaxel (diterpene alkaloid)	*Taxus chinensis* L., cell suspension cultures	Phyton Biotech Inc. (USA/Germany) and Bristol-Myers Squibb (USA)	Pharmaceutical (several types of cancer)	www.phytonbiotech.com/ [7]
		Taxol® (since 2002)		
		(75,000 l fermenters at Phyton)		
		Samyang Biopharm (Korea)		www.samyangbiopharm.com
		Genexol® (since 2007)		[11]
Shikonin (naphthoquinone)	*Lithospermum erythrorhizon* Siebold & Zucc., cell suspension cultures	Mitsui Chemicals Inc. (Japan) and Kenebo Cosmetics Inc. (Japan)	Cosmetics	www.kanebo.com [6]
		Lady 80 BIO lipstick (started in 1983; 750 l fermenter)		

8.2 Commercial Products from Plant Tissue Culture

Shikonin The naphthoquinone pigment shikonin (Fig. 8.1) was produced by Mitsui Petrochemical Industries Ltd. (now Mitsui Chemicals Inc.) in *Lithospermum erythrorhizon* Siebold & Zucc. cell suspension cultures, formulated into the Lady 80 BIO lipstick by Kanebo Cosmetics Inc., and introduced to the market in 1983 (www.kanebo.com/aboutus/history/). High shikonin yields (up to 23% (w/w) of dry biomass) were achieved in 750 L bioreactors using batch processing [6]. It is difficult to assess how long Mitsui pursued the biotechnological production of shikonin, but it appears that a combination of unfavorable factors - delays in the regulatory approval of the product, a limited market size (150 kg per year), a decrease in product price ($ 4000 per kg in 1988), and high operating costs of fermenters - led to a termination of the project [9].

Paclitaxel Phyton Catalytic (now Phyton Biotech; owned by DFB Pharmaceuticals) began the scale-up of cell suspension cultures *Taxus chinensis* (Pilg.) Rehder in the early 1990s with the goal of a large-scale production of paclitaxel (taxol®) (see also Chap. 7 of this book) (Fig. 8.1), which had been in high demand for the treatment of

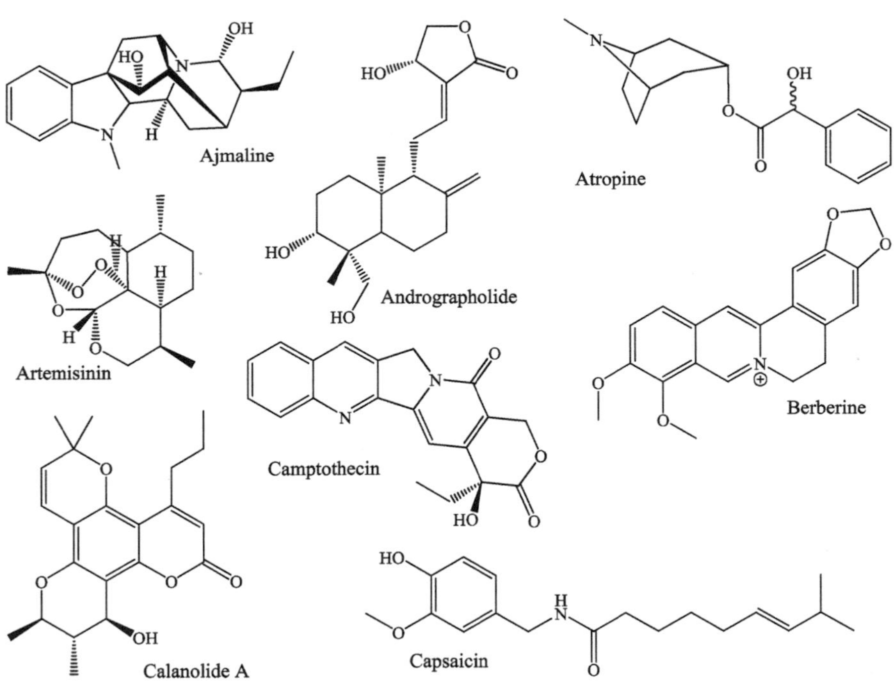

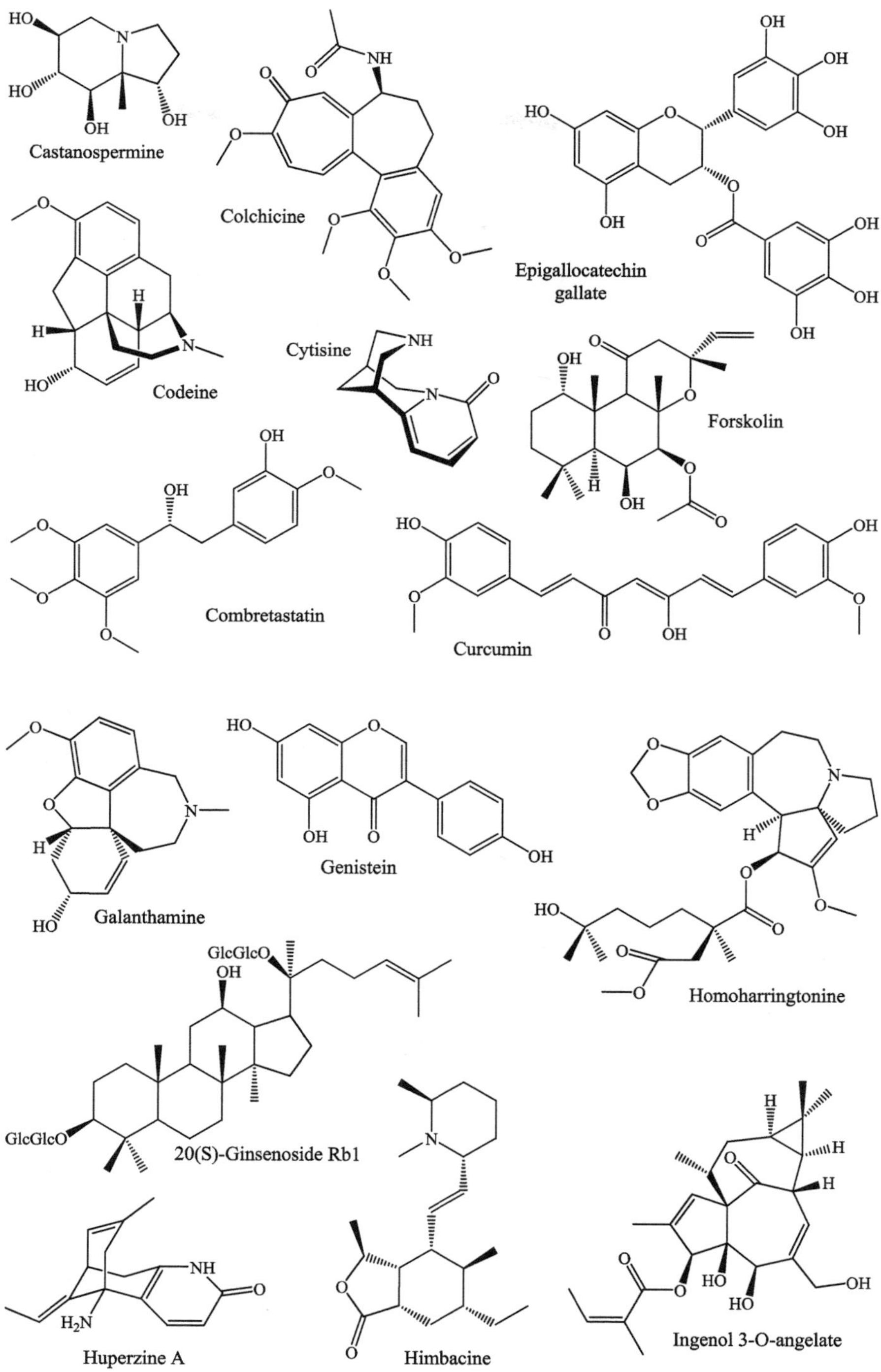
Castanospermine
Colchicine
Epigallocatechin gallate
Codeine
Cytisine
Forskolin
Combretastatin
Curcumin
Galanthamine
Genistein
Homoharringtonine
20(S)-Ginsenoside Rb1
Huperzine A
Himbacine
Ingenol 3-O-angelate

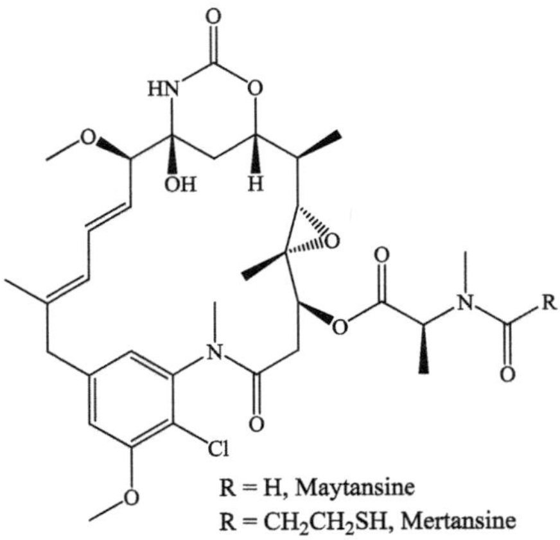

R = H, Maytansine
R = CH_2CH_2SH, Mertansine

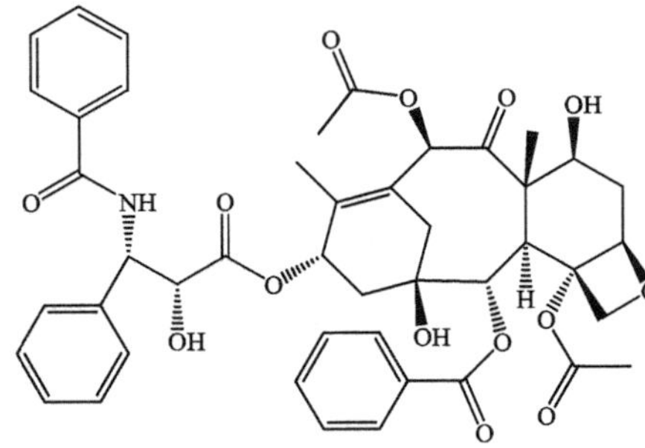

Paclitaxel (taxol™)

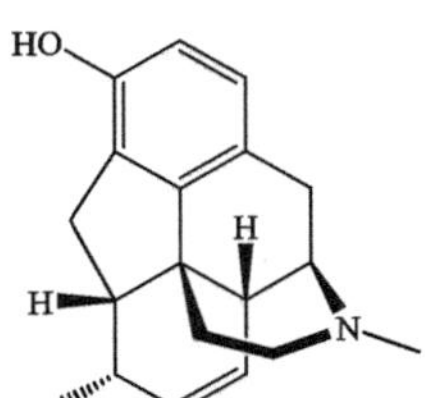

Morphine

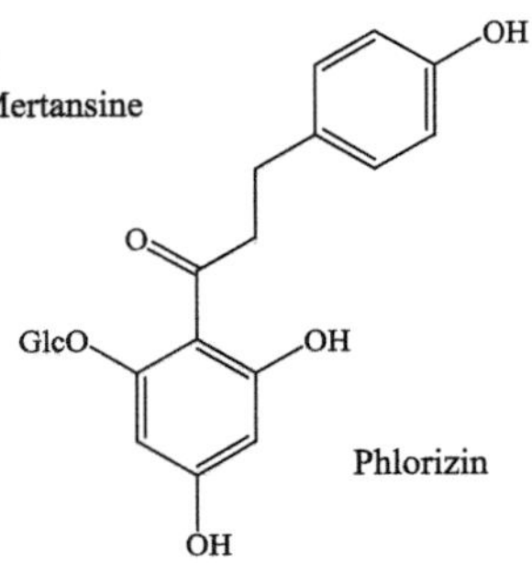

Phlorizin

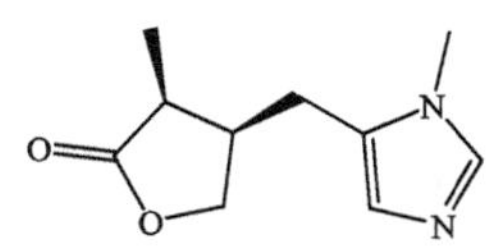

Pilocarpine

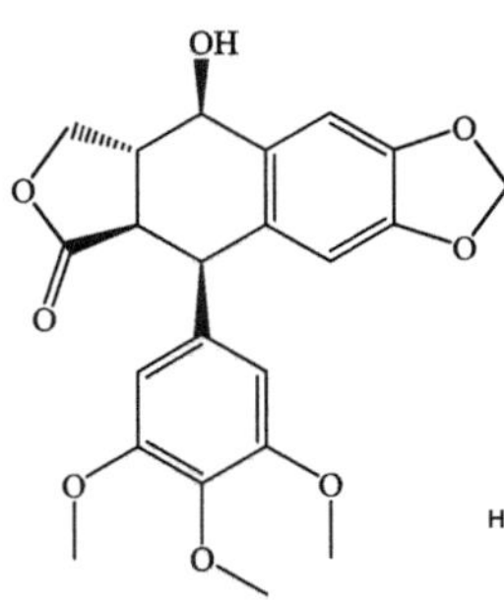

Podophyllotoxin

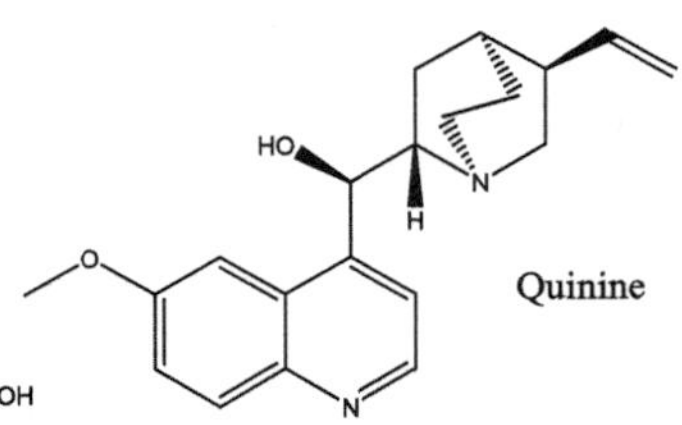

Quinine

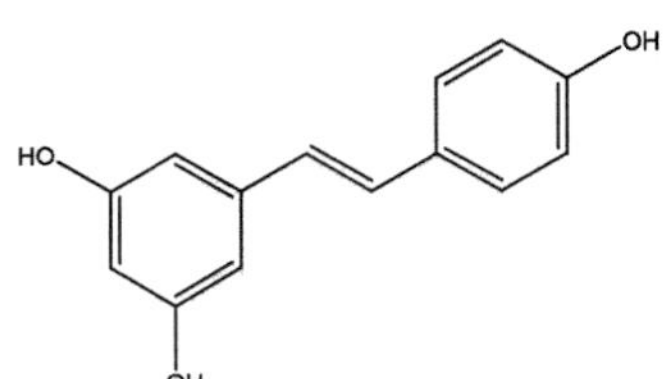

Resveratrol

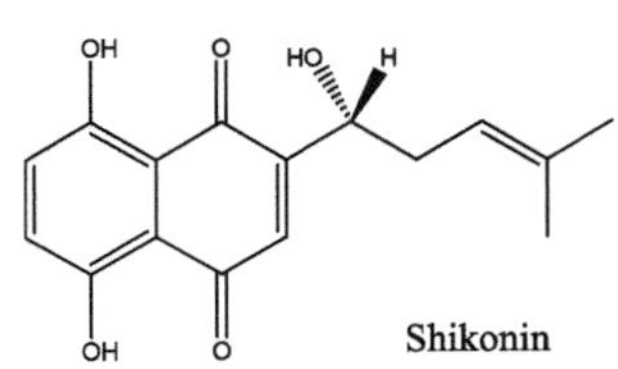

Shikonin

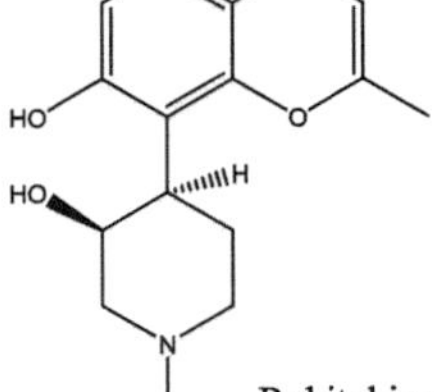

Rohitukine

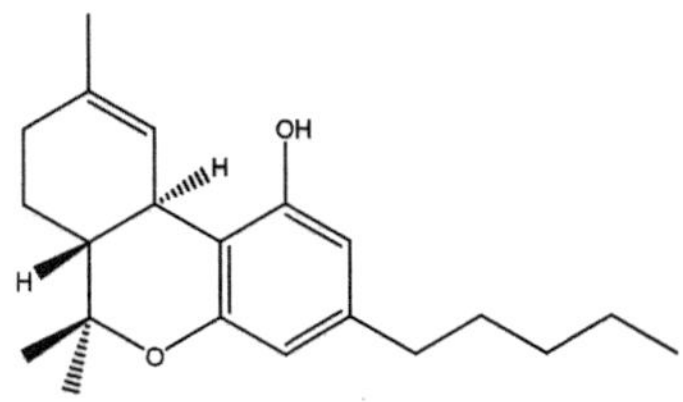

Tetrahydrocannabinol

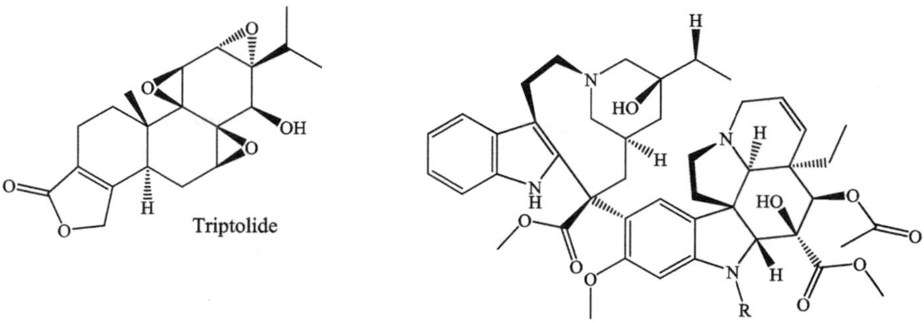

R = CH_3, Vinblastine
R = CHO, Vincristine

Fig. 8.1 Structures of PNPs discussed in this article (in alphabetical order). Abbreviations: *Glc* glucose

various forms of cancer (more details in). Following the acquisition of a manufacturing plant near Hamburg (Germany), the capacity increased to 75,000 L fermenters, which are still the largest vessels for plant tissue culture today. The plant cell fermentation technology was licensed to Bristol-Myers Squibb in 1995 to jointly commercialize paclitaxel production. In 2004, Bristol-Myers Squibb received the prestigious Presidential Green Chemistry Challenge Award of the American Chemical Society, following a transition, in 2002, to plant tissue culture as the sole source for its blockbuster anti-cancer treatment [10]. Patent protection for taxol® ended in 2000 and generic competition emerged, including a new tissue culture and formulation process for paclitaxel (and related taxanes) by Samyang Biopharm, which operates 35,000 L fermenters on its premises in Daejeon, South Korea [11].

Triterpene Saponin-Containing Cells In the mid-1980s, Nitto Denko Corp. (Ibaraki, Japan) began developing an industrial-scale process for the cultivation of ginseng cells (*Panax ginseng*). The product was approved for commercialization in Japan in 1988. The ginseng cells were shown to have contents of triterpene saponins (more specifically, ginsenosides (Fig. 8.1), which are considered to be the active principles of ginseng extracts [12]), that were very similar to those of field-cultivated ginseng [13]. The powder and extracts from the ginseng cell cultures are used to produce additives for foods, drinks and cosmetics [14]. Nitto Denko is employing bioreactors of up to 25,000 L in a two-stage production process, achieving a biomass productivity of 20 g dry weight per liter in 4 weeks of culture [15]. More recently, a collaborative team of scientists at Unhwa Corp. (Jeonju, South Korea) and the University of Edinburgh (United Kingdom) introduced cambial meristematic cells as a potentially cost-effective and reliable source of undifferentiated cells [16]. Based on information on the company website (www.unhwa.com), this technology has been adapted to generate wild ginseng cambial meristematic cells and a skincare product containing these cells (Ddobyul®) is now available commercially. More detailed information about the process (conditions and scale) is not available in the public domain at this time.

Polyphenol-Containing Cells DianaPlantSciences (Portland, OR; acquired by Symrise AG in 2014) has been developing processes for employing cell suspension cultures rich in polyphenols, which are formulated into cosmetics and dietary supplements. Cocovanol™ (launched in 2013) is a freeze-dried powder of cocoa suspension cells that, according to information on the company website (http://www.diana-group.com/), delivers high polyphenolic content without solvent extraction and contains only trace amounts of caffeine and theobromine (the bitter alkaloids of the cacao plant) (**see also Chap. 6 of this book**). Another polyphenol-containing product, ActivBerry™, which is based on bilberry suspension cells, will purportedly be launched in the near future (based on information on company website). Further information can be extracted from the patent literature but very little technical details are given in these documents. Unfortunately, no peer-reviewed publications are available on the topic.

8.3 Considerations for Commercial Targets

The commercial advances with plant tissue culture thus far have been confined to the production of (i) structurally complex pharmaceuticals with low natural abundance and (ii) cell biomass (without prior extraction) containing PNPs with presumed health-promoting properties. A combination of factors such as production cost and reliability, market scale, regulatory burden, and consumer perception determine the likelihood for a successful market introduction. In the following paragraphs, I will discuss challenges and opportunities for commercial-scale plant tissue culture.

Cost of Natural Material The prices of raw materials containing desirable PNPs vary greatly from a few U.S. dollars per kg (e.g., garlic bulbs (*Allium sativum*) (contain sulfoxides such as alliin)) to several thousand U.S. dollars per kg (e.g., saffron stigmas (*Crocus sativus*) (contain crocin and related apocarotenoid glycosides)) [17]. The extraction cost is also determined by the concentration and the accessibility of the PNP(s) of interest. For example, while Bulgarian rose (*Rosa damascena*) petals are readily available, the bulk oil price is several thousand U.S. dollars per kg [17]. The concentrations of some plant PNPs are exceedingly low (e.g., the alkaloid vincristine (Fig. 8.1) (**see also Chap. 5 of this book**) occurs at only 0.001% of dry biomass in the leaves of Madagaskar periwinkle (*Catharanthus roseus*)) and the cost of formulations containing them can be extraordinarily high [18]. Most plant tissue culture efforts have focused on such high-value products and the history for one commercial example, the production of paclitaxel (taxol®) by plant tissue culture, is discussed briefly above (**see also Chap. 7 of this book**).

Structural Complexity The extraction of salicylates from various plant sources to treat symptoms such as pain, fever and inflammation dates back to antiquity [19]. However, with the advent of facile chemical routes in the late 1800s, acetylsalicylic acid (better known as Aspirin®) went on to become the first blockbuster synthetic drug [20]. A large number of commercial drugs, even those with a PNP scaffold, are obtained by chemical synthesis or semi-synthesis. In many cases, the biological material serves as source and additional chemical steps generate the desired end product. At present, a plant-based production of a particular PNP, including but not limited to the use of tissue culture, is only competitive when the target molecule is structurally complex with several chiral centers [21]. A high degree of functionalization in a PNP can also be an advantage when the extraction from a plant-based matrix competes with the production in engineered microbial hosts.

A highly publicized example is artemisinin (Fig. 8.1), a sesquiterpene lactone employed in antimalarial therapies. Amyris Inc. (Emeryville, CA, USA) and Sanofi (Paris, France) invested tens of millions of U.S. dollars in the development of an integrated synthetic biology/chemical synthesis platform for artemisinin; however, the commercial introduction in the market for malaria drugs has been very challenging [22]. One of the reasons is the cost of production, which still cannot compete with the extraction from artemisinin's natural source, sweet wormwood (*Artemisia annua*). Furthermore, while the production of terpenoid backbones (incl. that of amorphadiene, the precursor of artemisinin) has been achieved at fairly high titers in engineered microbes, the yields achieved for highly functionalized PNPs of plant origin have been substantially lower. Often, the pathways toward plant PNPs have only been partially elucidated, and not all genes required for transferring the pathway to microbes have been cloned yet (**see also Chaps. 7 and 5 of this book**). Moreover, the obtaining high activities from multiple plant enzymes produced recombinantly in microbial hosts can be quite challenging. In summary, while synthetic biology platform will continue to be improved and may eventually make a significant contribution to the production of plant PNPs, there are currently advantages for a plant-based production of high functionalized target molecules.

Market Size The substantial up-front investment in tissue culture facilities (fermenters) and maintenance (manpower, growth medium, and energy costs) has to be supported by cost savings (when compared with alternative methods for production) as well as an appropriate size of the market. While the development of high-yielding cell suspension cultures for the production of shikonin by Mitsui was a significant breakthrough at pilot scale, the product had only a fairly short period of success in the marketplace: (i) the price tag of cell culture-derived shikonin (approximately U.S. $ 4000 per kg) was only marginally below the cost for shikonin extracted from roots of *Lithospermum erythrorhizon* and (ii) the requirement of shikonin in 1988 was only about 150 kg per year, with a predicted annual market value of U.S. $ 600,000 [23]. As a comparison, the cost for the discovery and development efforts toward cell culture production of shikonin was estimated to have been in the tens of millions U.S. dollar range [23], and the product therefore did not have a longer term commercial future.

Regulatory Burden Most products from plant tissue culture require the same regulation as the corresponding products extracted from whole plants in the U.S. (assuming that these are not 'novel' foods or pharmaceutical ingredients). However, the regulatory environment for registering plant tissue culture products differs significantly in various parts of the world. For example, plant tissue culture products would be treated as 'natural' in some markets, while in others they would be labeled as 'nature-identical' [24, 25]. Different federal agencies will be involved in reviewing commercialization efforts depending on if the product requires regulation as food ingredient or additive, or as active pharmaceutical ingredient [26]. Regulatory complications arise if genetic engineering is employed in plant tissue culture. While some concerns voiced by opponents of genetically engineered crops, for example the release of genetic material with potentially undesirable environmental effects, are irrelevant in plant tissue culture (which employs a closed production environment), the public perception in many countries has been that plant biotechnology in general poses undesirable risks [27]. Interestingly, some engineered organisms developed using newer gene editing technologies (including the revolutionary CRISPR/Cas9 system (acronym for Clustered Regularly Interspaced Short Palindromic Repeats and their associated nuclease (CRISPR associated protein 9)) are currently not regulated by the U.S. Department of Agriculture [28], and it remains to be seen if federal agencies in other countries will follow suit. The landscape for the commercialization of plant tissue culture-related products continues to evolve and future regulation in the area is therefore difficult to predict.

Processing into Products As discussed above, plant tissue cultures have been commercialized for the production and isolation of high-value PNPs. A very different, but equally viable, application of tissue culture technology is the direct formulation of suspension cells into cosmetics and dietary supplements. There are several potential advantages to the approach: (i) cost savings because no further processing is required, (ii) reduction of undesirable constituents because tissue culture is optimized for the accumulation of specific products in a fairly simple matrix, and (iii) high consistency of the product due to controlled growth conditions. Commercial examples include suspension cells of ginseng (www.unhwa.com) and cocoa (http://www.diana-group.com/). Several companies have purportedly experimented with additional plant tissue cultures at pilot scale (reviewed in [5]) but, while patent applications with limited technical detail are available, very few of these efforts have been described in the peer-reviewed literature. It is therefore not possible for an industry outsider to evaluate the commercial potential of formulated plant cells.

Consumer Acceptance In many countries, the demand for food additives, nutraceuticals and consumer care products carrying a 'natural' label has been increasing consistently over the last decade [29]. Plant tissue cultures can be a source of such 'natural' extracts, for which there is a marketing advantage over synthetic products. However, plant tissue culture products have not established a significant footprint in the marketplace, where plant parts are still the primary source of 'natural' extracts. Is synthetic biology, which includes the engineering of microbes for the production of metabolites originally sourced from plants, another challenge for the commercial development of plant tissue culture-derived products? Part of the answer to this

Table 8.2 Prospects for tissue culture in the production of PNPs

Disease area	Lead metabolite(s)	Source(s)	Comments on plant tissue culture opportunities	Generic name(s)	Development status
Actinic keratosis	Ingenol 3-O-angelate (or Ingenol mebutate)	*Euphorbia* spp.	Concentration in latex up to 0.2% [49]; tissue cultures have not been optimized for production of natural product	Picato	First FDA approval 2012
Alzheimer's disease	Galantamine (or Galanthamine)	*Galanthus* and other genera	Concentration in bulbs up to 0.3% of dry weight [50]; tissue cultures have not been optimized for production of natural product	Nivalin, Razdyne, Reminyl, Lycoremine	First FDA approval in 2001
	Huperzine	*Huperzia* and *Phlegmariurus* spp.	Concentration in above-ground tissues 0.04% ((52); up to 0.06% in *in vitro* propagated plants [51]; establishment of liquid cultures has not been successful	None	First approval by China FDA in 1990s; U.S. FDA lists no ongoing trials
Cancer	Camptothecin	*Camptotheca acuminata* (and other species of the same genus)	Concentration in leaves <0.4% [52]; approved drugs produced semi-synthetically; plant tissue culture yields of <0.1% [53] are not competitive	Irinotecan, Topotecan, Karenitecin, Gimatecan, CZ48	FDA approval 1996, FDA approval 2007, Phase III, Phase II, Phase I

(continued)

Table 8.2 (continued)

Disease area	Lead metabolite(s)	Source(s)	Comments on plant tissue culture opportunities	Generic name(s)	Development status
	Combretastatins A-1 and A4	*Combretum caffrum* (and other species of the same genus)	Sourcing from stems difficult; tissue cultures have not been optimized for production of natural product	OXi4503, Fosbretabulin	Phase II/III, Phase III
	Genistein	*Glycine max* (and other plants)	Concentration up to 0.5% in seeds [54]; being evaluated to reduce side effects of chemotherapy; hairy root cultures accumulate natural product at concentrations similar to natural source [55] and are not competitive	Genistein	Phase II
	Mertansine, Maytansine	*Maytenus* spp. (natural product might be of microbial origin)	Drug is a synthetic antibody conjugate; no plant source needed	Ado-trastuzumab emtansine, Kadcyla	First FDA approval 2013
	Homoharringtonine	*Cephalotaxus* spp.	Sourcing difficult; commercial product generated by semi-synthesis; root cultures accumulate natural product at <0.001% of dry weight [56] and are not competitive	Omacetaxine mepesuccinate	First FDA approval 2012

	Paclitaxel	*Taxus* spp.	Concentration in bark <0.05% of dry weight [57]; harvest from natural source unsustainable; successful commercial production in plant suspension cultures [7]	Taxol	FDA approval 1993 (ongoing clinical trials with analogs)
	Podophyllotoxin	*Podophyllum* spp.	Concentration in leaves up to 2.5% of dry weight [58]; cross-species co-cultures accumulate natural product at <0.01% of dry weight [59] and are not competitive; commercial drugs are produced semi-synthetically	Etoposide, Teniposide	FDA approval 1983, FDA approval 1992 (ongoing clinical trials with analogs)
	Triptolide	*Tripterygium wilfordii; Tripterygium regelii*	Extremely low concentration in roots (<0.01% of dry weight) [31]; unsustainable harvest of roots; root cultures excrete natural into medium at 5 mg/L [43]; continuous harvest possible; further investigation warranted	Minnelide	Phase I

(continued)

Table 8.2 (continued)

Disease area	Lead metabolite(s)	Source(s)	Comments on plant tissue culture opportunities	Generic name(s)	Development status
	Vinblastine	*Catharanthus roseus*	Concentration in leaves <0.01% of dry weight [60]; tissue cultures only produce biosynthetic intermediates; semi-synthetic approaches involving tissue cultures were developed in 1980s [61] but were not competitive; improvements have been made over the last few years but no step change has been achieved	Velban and others	First FDA approval 1961
	Vincristine	*Catharanthus roseus*	Concentration in leaves <0.001% of dry weight [60]; tissue cultures only produce biosynthetic intermediates; semi-synthetic approaches involving tissue cultures were developed in 1980s [61] but were not competitive; improvements have been made over the last few years but no step change has been achieved	Oncovin and others	First FDA approval 1963

Diabetes (type 2)	Phlorizin (Phloridzin)	*Malus* spp.	Concentration in fruit peel <0.1% of dry weight [62]; tissue cultures have not been optimized for production of natural product; commercial drugs produced by chemical synthesis	Dapagliflozin, Canagliflozin	FDA approval 2012, FDA approval 2013 (ongoing clinical trials with analogs)
Gout	Colchicine	*Colchicum* and *Gloriosa* spp.	Concentration in leaves and corms <0.2% of dry weight [63]; concentration in root cultures <0.03% [64] and therefore not competitive	Colcrys	FDA approval 2009
Heart arrhythmia	Ajmaline	*Rauwolfia* (or *Rauvolfia*) spp.	Concentration in roots >4% of dry weight [65]; concentration in hairy root cultures >0.5% [46], which could be competitive considering the challenge with procuring source material	Many	First FDA approval 1970s

(continued)

Table 8.2 (continued)

Disease area	Lead metabolite(s)	Source(s)	Comments on plant tissue culture opportunities	Generic name(s)	Development status
Hepatitis C virus	Castanospermine	*Castanospermum australe*	Concentration in various plant parts <0.1% [66]; cell suspension culture accumulate the natural product at low levels (<0.01%) [67] and are not competitive	Celgosivir	Phase II trials completed; current status unknown
HIV/AIDS	Calanolide A	*Calophyllum lanigerum* (and other species within the same genus)	Sourcing extremely difficult due to restrictions imposed by Malaysian government; no tissue culture data published in peer-reviewed literature	None	Phase I (completed but no follow-up)
Leukemia	Rohitukine	*Dysoxylum binectariferum* (possibly produced by endophytic fungus)	Drug produced by chemical synthesis; possibly not of plant origin and plant tissue culture therefore not applicable	Flavopiridol	Phase II

Malaria	Artemisinin	*Artemisia annua*	Concentration up to 1.3% of dry weight in leaves [68]; plant grows vigorously; cost of $ 250/kg [22]; concentration in hairy root cultures (<0.1% of cell biomass or 26 mg/l) [69]) is not competitive	Many	First China FDA approval in 1985; first U.S. FDA approval in 2009
Malaria	Quinine	*Cinchona* spp.	Concentration of <1% in bark [70]; concentrations of <0.1% were reported for hairy root cultures [71]; tissue culture not competitive	Many	Used widely until 1940s; rarely recommended now
Multiple sclerosis	Andrographolide	*Andrographis paniculata*	Concentration of >2% of dry weight in leaves [72]; concentration in cell suspension cultures <0.1% [73] and therefore not competitive	None	Phase I/II

(continued)

Table 8.2 (continued)

Disease area	Lead metabolite(s)	Source(s)	Comments on plant tissue culture opportunities	Generic name(s)	Development status
Myocardial infarction	Himbacine	*Galmulimima belgraveana*	Structure of commercial product significantly different from lead; produced by chemical synthesis	Vorapaxar	First FDA approval 2014
Nicotine dependence	Cytisine	*Laburnum anagyroides* (or *Cytisus laburnum*) (and other plants)	Little quantitative information available in peer-reviewed literature; commercial drug obtained by chemical synthesis [74]	Tabex, Varenicline, Chantix	Marketed in Bulgaria since 1960s, first FDA approval 2006
Obesity	Forskolin	*Coleus forskohlii* (or *Plectranthus barbatus*)	Concentration in roots <0.1% [72]; hairy root cultures accumulate natural product in slightly higher concentrations than roots [46] but are not competitive	None	Phase III (extract)
Pain	Capsaisin	*Capsicum annuum*	Concentration varies tremendously across cultivars; up to 1.5% of fruit dry weight [75]; cell suspension cultures produce <0.1% [76] and are not competitive	Qutenza, Theragen, Rezil	First FDA approval 2009

	Codeine	*Papaver somniferum*	Concentration of up to 10% in opium (dried latex of unripe seed pods) [77]; hairy root cultures produce up to 0.3% [78] and are not competitive	Many	Approved in many countries; often regulated under narcotic control laws
Pain	Morphine	*Papaver somniferum*	Concentration of up to 9% in opium (dried latex of unripe seed pods) [77]; hairy root cultures produce up to 0.3% [78] and are not competitive	Many	Approved in many countries; often regulated under narcotic control laws
	Tetrahydrocannabinol	*Cannabis sativa*	Concentration of up to 20% of dry weight have been reported [79]; many studies on various types of tissue cultures, but quantities are not competitive [79]; only synthetic product approved by U.S. FDA	Dronabinol	First FDA approval 1992 (other products at various stages of clinical development)

(continued)

Table 8.2 (continued)

Disease area	Lead metabolite(s)	Source(s)	Comments on plant tissue culture opportunities	Generic name(s)	Development status
Diverse ailments	Atropine, Hyoscyamine, scopolamine	Many members of Solanaceae	Concentrations vary substantially across plant species and organs but can reach 0.5% of dry weight [80]; many studies with tissue cultures have been published; however, the fairly low cost and availability of natural sources are a critical deterrent for commercial development of tissue cultures [81]	Many	Several products on market and in clinical development
	Berberine	*Berberis* spp., *Coptis* spp., *Eschscholtzia* spp., and others	Concentration up to 2.4% of dry weight in roots [82]; Berberis cannot be grown commercially in U.S. (alternate host for wheat rust fungus) [47], and other species have poor agronomic characteristics; cell cultures accumulate natural product at >2.4% [48]; further evaluation warranted	None	37 clinical trials listed by clinicaltrials.gov (most advanced: Phase II)

	Curcumin	*Curcuma longa* (and other plants)	Concentration in rhizome 0.4–2.2% of dry weight [83]; FDA issued warning letters against marketers of products claiming that cucurmin provides anti-disease effects or overall health benefits; tissue culture protocols for micropropagation have been developed but no high cucurmin producers have been published	Marketed as supplement under various names	Phase I

(continued)

Table 8.2 (continued)

Disease area	Lead metabolite(s)	Source(s)	Comments on plant tissue culture opportunities	Generic name(s)	Development status
	Epigallocatechin 3-gallate (EGCG)	*Camellia sinensis*	Concentration in green tea leaves up to 4% of dry weight [84]; FDA issued warning letters against marketers of products claiming that EGCG provides anti-disease effects or overall health benefits; it should be noted that green tea extract is an FDA-approved drug for treating certain kinds of warts; EGCG contents in tissue cultures (e.g., 0.2% in adventitious roots [85]) are not competitive	None	Phase III
	Pilocarpine	*Pilocarpus* spp.	Concentration in leaves <0.6% of dry weight [86]; production levels in tissue culture are extremely low [87] and not competitive	Many	FDA-approved for ailments affecting the eye and mouth

	Resveratrol	*Vitis vinifera* (and other plants)	Concentration in fruit skin <0.001% of dry weight [88]; large body of food science and nutrition literature; clinical evidence matter of significant dispute [89]; price too low to warrant further development of plant tissue cultures	Resveratrol	Mostly pre-clinical

question likely depends on labeling requirements for ingredients derived from genetically modified organisms. The regulatory landscape is highly complex and often differs across countries [30] (**see also Chap. 11 of this book**). The current trend toward the establishment of labeling standards would appear to be favorable for plant tissue culture that does not rely on genetic engineering. However, public concern about genetically modified organisms does not appear to be directed toward pharmaceuticals in the same way as they affect food and consumer care products. In May 2012, Protalix Biotherapeutics (www.protalix.com) was approved by the U.S. Food and Drug Administration to employ plant tissue cultures for the production of recombinant taliglucerase alpha (Elelyso®), and more products using the same platform are in advanced stages of development. It is not unlikely that plant tissue culture for producing PNPs with pharmaceutical applications would be acceptable for consumers, even if genetic engineering technologies should have been brought to bear.

8.4 Future Opportunities

Based on the considerations presented above, it would seem that plant tissue cultures have commercial potential when harvested cells are directly formulated into a nutraceutical product (simplified plant matrix and uncomplicated processing) or a PNP of particularly high pharmaceutical value is accumulated in high concentrations. Generally speaking, high value PNPs will be structurally complex (with multiple chiral centers and elaborate functionalization), accumulate at low levels in the source plant, and/or occur in species where access is limited (endangered or poor agronomic characteristics). In this paragraph, I will discuss examples of individual PNPs (not extracts) that might be produced at commercial scale by plant tissue culture technology. PNPs that have fallen out of favor among clinicians (e.g., digitoxin, ouabain, sanguinarine, tubocurarine, and yohimbine) will not be covered here (**see also Chap. 5 of this book**). I will also not discuss in the narrative PNPs whose clinical efficacy has not been demonstrated conclusively (e.g., cucurmin, epigallocatechin 3-gallate, and resveratrol) (Fig. 8.1; Table 8.2) (**see also Chap. 3 of this book**).

For some PNPs that meet the high value criteria listed above, the development of optimized tissue cultures has not been attempted or was not published; among these are ingenol 3-angelate, galanthamine, huperzine, combretastatins, and calanolide A (Fig. 8.1, Table 8.2). In other cases, the pharmaceutically relevant product is structurally distinct from the PNP lead and is obtained by chemical synthesis (which means that no plant source is needed); examples include ado-trastuzumab emtansine (synthetic antibody conjugate linked to the benzoansamacrolide, maytansine), dapagliflozin (employed for treatment of type 2 diabetes; based on the dihydrochalcone glucoside, phlorizin), vorapaxar (used for treatment of patients with a history of myocardial infarction; based on the alkaloid, himbacine), and tabex (employed to aid with smoking cessation; chemical synthesis more efficient than extraction of PNP, cytisine) (Fig. 8.1, Table 8.2).

Tissue cultures have been developed for many plants that contain pharmaceutically relevant metabolites, but the concentration of the PNP of interest has mostly been equal to or below that reported for the natural source (Table 8.2). A few tissue culture resources, however, would seem to be worth further consideration.

The diterpene epoxide, **triptolide** (Fig. 8.1, Table 8.2), accumulates to only very low concentrations (<0.01% of dry weight in various organs) in members of the genus *Tripterygium* [31]. The current harvesting of roots from mature plants requires significant agronomic inputs and suffers from low efficiency [32]. Chemical derivatives of triptolide have been evaluated in phase I clinical trials [33, 34], and minnelide, a water soluble pro-drug analogue of triptolide, has shown particularly promising activity in multiple animal models of pancreatic cancer [35]. Various types of tissue cultures of *Tripterygium* producing different PNPs were developed in the 1980s and 1990s [36–39]. However, it was recognized only recently that triptolide concentrations produced by tissue cultures (up to 0.15%) [40–42] far exceed those reported for roots. In one *Tripterygium* root culture, more than 70% of the metabolites extracted from the culture medium with an organic solvent were characterized as diterpenoids (with triptolide accounting for 16% of all detected metabolites) [43]. Such an unprecedented enrichment of the target PNP, which is very difficult to obtain in sufficient quantities from natural sources, makes tissue culture an attractive alternative to the unsustainable harvest from *Tripterygium* roots.

Ajmaline (Fig. 8.1, Table 8.2) is an alkaloid that has been used since the 1970s as a treatment of heart arrhythmias (more recently, semi-synthetic derivatives have been introduced) [44]. The commercial cultivation of the medicinal plant *Rauwolfia serpentina*, which accumulates ajmaline in stem bark and roots, has met several challenges [45]. Despite decent yields from extracting the natural producer, the scarcity of the source materials has led to the high cost of treatments involving ajmaline. A tissue culture source, such as hairy roots with yields of >0.5% [46], would be a desirable alternative.

A larger number of clinical trials (>30) have been conducted with **berberine** (Fig. 8.1, Table 8.2), in particular as a treatment for type 2 diabetes (more information at www.clinicaltrials.gov) (**see also Chap. 5 of this book**). The extraction of berberine from members of the genus *Berberis* is reasonably straightforward but the plant cannot be grown commercially (by law) in several countries due to the fact that is serves as an alternate host for the wheat rust fungus [47]. Other natural producers, such as *Coptis* spp. or *Eschscholtzia* spp., have very poor agronomic performance and are not viable alternatives. Tissue cultures had been developed to produce berberine at fairly high yields (>3%) in the 1980s [48], and a reevaluation of their commercial potential would therefore be warranted.

Acknowledgements Some aspects of the author's research described in this article were funded by the National Institutes of Health (award number RC2GM092561).

References

1. Haberlandt G. Kulturversuche mit isolierten Pflanzenzellen. Sitzungsb Akad Wiss Wien. 1902;101:69–72.
2. Vasil V, Hildebrandt AC. Differentiation of tobacco plants from single, isolated cells in microcultures. Science. 1965;150:889–92.
3. Thorpe TA. History of plant tissue culture. Mol Biotechnol. 2007;37(2):169–80.
4. Barz W, Ellis BE. Plant cell cultures and their biotechnological potential. Ber Deut Bot Ges. 1981;94:1–762.
5. Wilson SA, Roberts SC. Recent advances towards development and commercialization of plant cell culture processes for the synthesis of biomolecules. Plant Biotechnol J. 2012;10(3):249–68.
6. Fujita Y. Industrial production of shikonin and berberine. Ciba Found Symp. 1988;137:228–35.
7. McElroy C, Jennewein S. Taxol® biosynthesis and production: from forests to fermenters, chap. 7. In: Schwab W, Lange BM, Wüst M, editors. Biotechnology of natural products. Cham: Springer. 2017; (in press)
8. Tschofen M, Knopp D, Hood E, Stoger E. Plant molecular farming: much more than medicines. Annu Rev Anal Chem. 2016;9:271–94.
9. Malik S, Bhushan S, Sharma M, Ahuja PS. Biotechnological approaches to the production of shikonins: a critical review with recent updates. Crit Rev Biotechnol. 2016;36(2):327–40.
10. Ritter SK. Green innovations. Chem Eng News. 2004;82(28):25–30.
11. Choi HK, Son JS, Na GH, Hong SS, Park YS, Song JY. Mass production of paclitaxel in plant cell culture. J Plant Biotechnol. 2002;29:59–62.
12. Wong AST, Che CM, Leung KW. Recent advances in ginseng as cancer therapeutics: a functional and mechanistic overview. Nat Prod Rep. 2015;32:256–72.
13. Furuya T, Yoshikawa T, Orihara Y, Oda H. Saponin production in cell suspension cultures of Panax ginseng. Planta Med. 1983;48:83–7.
14. JY W, Zhong JJ. Production of ginseng and its bioactive components in plant cell culture: current technological and applied aspects. J Biotechnol. 1999;68(2–3):89–99.
15. Hibino K, Ushiyama K. Commercial production of ginseng by plant tissue culture technology. In: Fu TJ, Singh G, Curtis WR, editors. Plant cell and tissue culture for the production of food ingredients. New York: Plenum Publishers; 1999. p. 215–24.
16. Lee EK, Jin YW, Park JH, Yoo YM, Hong SM, Amir R, et al. Cultured cambial meristematic cells as a source of plant natural products. Nat Biotechnol. 2010;28(11):1213–7.
17. Lubbe A, Verpoorte R. Cultivation of medicinal and aromatic plants for specialty industrial materials. Ind Crop Prod. 2011;34(1):785–801.
18. Hornberger J, Reyes C, Lubeck D, Valente N. Economic evaluation of rituximab plus cyclophosphamide, vincristine and prednisolone for advanced follicular lymphoma. Leuk Lymphoma. 2008;49(2):227–36.
19. Sumner J. The natural history of medicinal plants. Portland, Timber Press.
20. Jack DB. One hundred years of aspirin. Lancet. 1997;350:437–9.
21. Newman DJ, Cragg GM. Natural products as sources of new drugs from 1981 to 2014. J Nat Prod. 2016;79(3):629–61.
22. Peplow M. Synthetic malaria drug meets market resistance. Nature. 2016;530(7591):389–90.
23. Neumann KH, Kumar A, Imani J. Secondary metabolism. In: Plant cell and tissue culture – a tool in biotechnology: basics and application. Berlin/Heidelberg: Springer; 2009. p. 181–226.
24. Sigurdson GT, Tang P, Giusti MM. Natural colorants food colorants from natural sources. Annu Rev Food Sci Technol. 2017;8:261–80.
25. Davies KM, Deroles SC. Prospects for the use of plant cell cultures in food biotechnology. Curr Op Biotechnol. 2014;26:133–40.

26. Sparrow PAC, Irwin JA, Dale PJ, Twyman RM, Ma JKC. Pharma-planta: road testing the developing regulatory guidelines for plant-made pharmaceuticals. Transgenic Res. 2007;16(2):147–61.
27. Frewer LJ, van der Lans IA, Fischer ARH, Reinders MJ, Menozzi D, Zhang XY, et al. Public perceptions of agri-food applications of genetic modification – a systematic review and meta-analysis. Trends Food Sci Technol. 2013;30(2):142–52.
28. Waltz E. CRISPR-edited crops free to enter market, skip regulation. Nat Biotechnol. 2016;34(6):582.
29. Carocho M, Barreiro MF, Morales P, Ferreira ICFR. Adding molecules to food, pros and cons: a review on synthetic and natural food additives. Compr Rev Food Sci. 2014;13(4):377–99.
30. Vigani M, Olper A. GMO standards, endogenous policy and the market for information. Food Policy. 2013;43:32–43.
31. Lange BM, Fischedick JT, Lange MF, Srividya N, Šamec D, Poirier BC. Integrative approaches for the identification and localization of specialized metabolites in Tripterygium roots. Plant Physiol. 2017;173:456–69.
32. Helmstädter A. Tripterygium wilfordii Hook. f. – How a traditional Taiwanese medicinal plant found its way to the west. Pharmazie. 2013;68(7):643–6.
33. Meng CC, Zhu HC, Song HM, Wang ZM, Huang GH, Li DF, et al. Targets and molecular mechanisms of triptolide in cancer therapy. Chin J Cancer Res. 2014;26(5):622–6.
34. Zhou ZL, Yang YX, Ding J, Li YC, Miao ZH. Triptolide: structural modifications, structure-activity relationships, bioactivities, clinical development and mechanisms. Nat Prod Rep. 2012;29(4):457–75.
35. Chugh R, Sangwan V, Patil SP, Dudeja V, Dawra RK, Banerjee S, et al. A preclinical evaluation of minnelide as a therapeutic agent against pancreatic cancer. Sci Transl Med. 2012;4(156):156ra13.
36. Kutney JP, Beale MH, Salisbury PJ, Sindelar RD, Stuart KL, Worth BR, et al. Tripdiolide from tissue culture of Tripterygium wilfordii. Heterocycles. 1980;14(10):1465–7.
37. Kutney JP, Han K, KuriBrena F, Milanova RK, Roberts M. Studies with plant cell cultures of the Chinese herbal plant, Tripterygium wilfordii. Synthesis and biotransformation of diterpene analogues. Heterocycles. 1997;44:95–104.
38. Nakano K, Oose Y, Masuda Y, Kamada H, Takaishi Y. A diterpenoid and triterpenes from tissue cultures of Tripterygium wilfordii. Phytochemistry. 1997;45(2):293–6.
39. Nakano K, Yoshida C, Furukawa W, Takaishi Y, Shishido K. Terpenoids in transformed root culture of Tripterygium wilfordii. Phytochemistry. 1998;49(6):1821–4.
40. Miao GP, Zhu CS, Feng JT, Han J, Song XW, Zhang X. Aggregate cell suspension cultures of Tripterygium wilfordii Hook. f. For triptolide, wilforgine, and wilforine production. Plant Cell Tissue Organ Cult. 2013;112(1):109–16.
41. Miao GP, Zhu CS, Yang YQ, Feng MX, Ma ZQ, Feng JT, et al. Elicitation and in situ adsorption enhanced secondary metabolites production of Tripterygium wilfordii Hook. f. adventitious root fragment liquid cultures in shake flask and a modified bubble column bioreactor. Bioprocess Biosyst Eng. 2014;37(4):641–50.
42. Zhu CS, Miao GP, Guo J, Huo YB, Zhang X, Xie JH, et al. Establishment of Tripterygium wilfordii Hook. f. hairy root culture and optimization of its culture conditions for the production of triptolide and wilforine. J Microbiol Biotechnol. 2014;24(6):823–34.
43. Inabuy FS, Fischedick JT, Lange I, Hartmann M, Srividya N, Parrish AN, Xu M, Peters RJ, Lange BM. Biosynthesis of diterpenoids in Tripterygium adventitious root cultures. Plant Physiol. 2017;175(1):92–103.
44. Postema PG, Wolpert C, Amin AS, Probst V, Borggrefe M, Roden DM, et al. Drugs and Brugada syndrome patients: review of the literature, recommendations, and an up-to-date website. Heart Rhythm. 2009;6(9):1335–41.
45. Goel MK, Mehrotra S, Kukreja AK, Shanker K, Khanuja SPS. In vitro propagation of Rauwolfia serpentina using liquid medium, assessment of genetic fidelity of micropropagated plants, and simultaneous quantitation of reserpine, ajmaline, and ajmalicine. Methods Mol Biol. 2009;547:17–33.

46. Pandey P, Kaur R, Singh S, Chattopadhyay SK, Srivastava SK, Banerjee S. Long-term stability in biomass and production of terpene indole alkaloids by hairy root culture of Rauvolfia serpentina and cost approximation to endorse commercial realism. Biotechnol Lett. 2014;36(7):1523–8.
47. Jin Y, Szabo LJ, Carson M. Century-old mystery of Puccinia striiformis life history solved with the identification of Berberis as an alternate host. Phytopathology. 2010;100(5):432–5.
48. Sato F, Yamada Y. High berberine-producing cultures of Coptis japonica cells. Phytochemistry. 1984;23(2):281–5.
49. Vasas A, Hohmann J. Euphorbia diterpenes: isolation, structure, biological activity, and synthesis (2008–2012). Chem Rev. 2014;114(17):8579–612.
50. Lubbe A, Pomahacova B, Choi YH, Verpoorte R. Analysis of metabolic variation and galanthamine content in Narcissus bulbs by ^{1}H-NMR. Phytochem Anal. 2010;21(1):66–72.
51. Ma XQ, Gang DR. In vitro production of huperzine a, a promising drug candidate for Alzheimer's disease. Phytochemistry. 2008;69(10):2022–8.
52. Montoro P, Maldini M, Piacente S, Macchia M, Pizza C. Metabolite fingerprinting of Camptotheca acuminata and the HPLC-ESI-MS/MS analysis of camptothecin and related alkaloids. J Pharm Biomed. 2010;51(2):405–15.
53. Kai GY, Wu C, Gen LY, Zhang LQ, Cui LJ, Ni XL. Biosynthesis and biotechnological production of anti-cancer drug camptothecin. Phytochem Rev. 2015;14(3):525–39.
54. Laurenz RT, Pavani NS, Thelen KD. Determination of isoflavone (genistein and daidzein) concentration of soybean seed as affected by environment and management inputs. J Sci Food Agric. 2017;97(10):3342–47.
55. Chandra S, Chandra R. Engineering secondary metabolite production in hairy roots. Phytochem Rev. 2011;10(3):371–95.
56. Wickremesinhe ERM, Arteca RN. HPLC separation of cephalotaxine, harringtonine and homoharringtonine from callus and root cultures of Cephalotaxus harringtonia. J Liq Chromatogr. 1996;19(6):889–97.
57. vanRozendaal ELM, Lelyveld GP, vanBeek TAA. Simplified method for the determination of taxanes in yew needles by reversed-phase (C-18) high pressure liquid chromatography. Phytochem Anal. 1997;8(6):286–93.
58. Zheljazkov VD, Cantrell CL, Astatkie T. Variation in podophyllotoxin concentration in leaves and rhizomes of American mayapple (Podophyllum peltatum L.). Ind Crop Prod. 2011;33(3):633–7.
59. Lin HW, Kwok KH, Doran PM. Production of podophyllotoxin using cross-species coculture of Linum flavum hairy roots and Podophyllum hexandrum cell suspensions. Biotechnol Prog. 2003;19(5):1417–26.
60. Pan QF, Mustafa NR, Tang KX, Choi YH, Verpoorte R. Monoterpenoid indole alkaloids biosynthesis and its regulation in Catharanthus roseus: a literature review from genes to metabolites. Phytochem Rev. 2016;15(2):221–50.
61. Fujita Y, Hara Y, Morimoto T, Misawa M. Semisynthetic production of vinblastine involving cell cultures of Catharanthus roseus and chemical reaction. Curr Plant Sci Biot. 1990;9:738–43.
62. Ehrenkranz JRL, Lewis NG, Kahn CR, Roth J. Phlorizin: a review. Diabet Metab Res. 2005;21(1):31–8.
63. Alali F, Tawaha K, Qasaymeh RM. Determination of colchicine in Colchicum steveni and Chierosolymitanum (Colchicaceae): comparison between two analytical methods. Phytochem Anal. 2004;15(1):27–9.
64. Ghosh B, Mukherjee S, Jha TB, Jha S. Enhanced colchicine production in root cultures of Gloriosa superba by direct and indirect precursors of the biosynthetic pathway. Biotechnol Lett. 2002;24(3):231–4.
65. Kumar S, Singh A, Bajpai V, Srivastava M, Singh BP, Ojha S, et al. Simultaneous determination of bioactive monoterpene indole alkaloids in ethanolic extract of seven Rauvolfia species using UHPLC with hybrid triple quadrupole linear ion trap mass spectrometry. Phytochem Anal. 2016;27(5):296–303.

66. Abu-Zeyad R, Khan AG, Khoo C. Occurrence of arbuscular mycorrhiza in Castanospermum australe A. Cunn. & C. Fraser and effects on growth and production of castanospermine. Mycorrhiza. 1999;9(2):111–7.
67. Roja G, Heble MR. Castanospermine, an HIV inhibitor from tissue cultures of Castanospermum australe. Phytother Res. 1995;9(7):540–2.
68. Suberu J, Song LJ, Slade S, Sullivan N, Barker G, Lapkin AA. A rapid method for the determination of artemisinin and its biosynthetic precursors in Artemisia annua L. crude extracts. J Pharmaceut Biomed. 2013;84:269–77.
69. Patra N, Srivastava AK. Artemisinin production by plant hairy root cultures in gas- and liquid-phase bioreactors. Plant Cell Rep. 2016;35(1):143–53.
70. Fabiano-Tixier AS, Elomri A, Blanckaert A, Seguin E, Petitcolas E, Chemat F. Rapid and green analytical method for the determination of quinoline alkaloids from Cinchona succirubra based on microwave-integrated extraction and leaching (MIEL) prior to high performance liquid chromatography. Int J Mol Sci. 2011;12(11):7846–60.
71. Geerlings A, Hallard D, Caballero AM, Cardoso IL, van der Heijden R, Verpoorte R. Alkaloid production by a Cinchona officinalis 'Ledgeriana' hairy root culture containing constitutive expression constructs of tryptophan decarboxylase and strictosidine synthase cDNAs from Catharanthus roseus. Plant Cell Rep. 1999;19(2):191–6.
72. Cuthbertson DJ, Johnson SR, Piljac-Zegarac J, Kappel J, Schäfer S, Wüst M, et al. Accurate mass-time tag library for LC/MS-based metabolite profiling of medicinal plants. Phytochemistry. 2013;91:187–97.
73. Vakil MMA, Mendhulkar VD. Enhanced synthesis of andrographolide by Aspergillus niger and Penicillium expansum elicitors in cell suspension culture of Andrographis paniculata (Burm. f.) Nees. Bot Stud. 2013;54:49.
74. Rouden J, Lasne MC, Blanchet J, Baudoux J. (−)-Cytisine and derivatives: synthesis, reactivity, and applications. Chem Rev. 2014;114(1):712–78.
75. Welch CJ, Regalado EL, Welch EC, Eckert IMK, Kraml C. Evaluation of capsaicin in chili peppers and hot sauces by MISER HPLC-ESIMS. Anal Methods. 2014;6(3):857–62.
76. Kehie M, Kumaria S, Tandon P. Biotechnological enhancement of capsaicin biosynthesis in cell suspension cultures of Naga King Chili (Capsicum chinense Jacq.). Bioprocess Biosyst Eng. 2016;39(1):205–10.
77. Fakhari AR, Nojavan S, Ebrahimi SN, Evenhuis CJ. Optimized ultrasound-assisted extraction procedure for the analysis of opium alkaloids in papaver plants by cyclodextrin-modified capillary electrophoresis. J Sep Sci. 2010;33(14):2153–9.
78. Hashemi SM, Naghavi MR. Production and gene expression of morphinan alkaloids in hairy root culture of Papaver orientale L. using abiotic elicitors. Plant Cell Tissue Org Cult. 2016;125(1):31–41.
79. Andre CM, Hausman JF, Guerriero G. Cannabis sativa: the plant of the thousand and one molecules. Front Plant Sci. 2016;7:19.
80. Jakabova S, Vincze L, Farkas A, Kilar F, Boros B, Felinger A. Determination of tropane alkaloids atropine and scopolamine by liquid chromatography-mass spectrometry in plant organs of Datura species. J Chromatogr A. 2012;1232:295–301.
81. Ahmad S, Garg M, Tamboli E, Abdin M, Ansari S. In vitro production of alkaloids: factors, approaches, challenges and prospects. Pharmacogn Rev. 2013;7(13):27–33.
82. Wu JK, Yu D, Sun HF, Zhang Y, Zhang WW, Meng FJ, et al. Optimizing the extraction of anti-tumor alkaloids from the stem of Berberis amurensis by response surface methodology. Ind Crop Prod. 2015;69:68–75.
83. Gokhul V, Yuvapriya S, Chandramohan M, Muthukumaran P. Isolation and extraction of curcumin from three different varieties of Curcuma Longa L – a comparative study. Int J Pharm Res Alli. 2015;4(2):79–84.
84. Rahim AA, Nofrizal S, Saad B. Rapid tea catechins and caffeine determination by HPLC using microwave-assisted extraction and silica monolithic column. Food Chem. 2014;147:262–8.
85. Kim Y, Kim HG, Sim SJ, Kim JC, Min JY, Hwang JG, Kang SM, Moon HS, Kim JK, Choi MS. Effects of culture media on catechins and caffeine production in adventitious roots of tea tree (Camellia sinensis L.). J Agric Life Sci. 2013;47:11–20.

86. Sawaya ACHF, Vaz BG, Eberlin MN, Mazzafera P. Screening species of Pilocarpus (Rutaceae) as sources of pilocarpine and other imidazole alkaloids. Genet Res Crop Ev. 2011;58(3):471–80.
87. Abreu IN, Mazzafera P, Eberlin MN, Zullo MAT, Sawaya ACHF. Characterization of the variation in the imidazole alkaloid profile of Pilocarpus microphyllus in different seasons and parts of the plant by electrospray ionization mass spectrometry fingerprinting and identification of novel alkaloids by tandem mass spectrometry. Rapid Commun Mass Spectrom. 2007;21(7):1205–13.
88. Iacopini P, Baldi M, Storchi P, Sebastiani L. Catechin, epicatechin, quercetin, rutin and resveratrol in red grape: content, in vitro antioxidant activity and interactions. J Food Compos Anal. 2008;21(8):589–98.
89. Tome-Carneiro J, Larrosa M, Gonzalez-Sarrias A, Tomas-Barberan FA, Garcia-Conesa MT, Espin JC. Resveratrol and clinical trials: the crossroad from in vitro studies to human evidence. Curr Pharm Design. 2013;19(34):6064–93.

Tailoring Natural Products with Glycosyltransferases

9

Katja Härtl, Kate McGraphery, Julian Rüdiger, and Wilfried Schwab

Abbreviations

ABA	Abscisic acid
ABA-GE	ABA-glucose ester
ABC	ATP-binding cassette
AG	Apocarotenoid glycoside
AVI	Anthocyanin vacuolar inclusion
CAZymes	Carbohydrate-Active enZymes
CRC	Colorectal cancer
E. coli	*Escherichia coli*
ER	Endoplasmatic reticulum
GALNT12	Polypeptide *N*-acetylgalactosaminyltransferase 12
galU	Glucose-1-phosphate uridylyltransferase
GST	Glutathione S-transferase
GT	Glycosyltransferase
KAH	Kaurenoic acid 13-hydroxylase
KO	Kaurine oxidase
LC-MS	Liquid chromatography mass-spectrometry
MATE	Multidrug and toxic extrusion
MEP/DOXP	2-*C*-methyl-D-erythritol 4-phosphate/1-deoxy-D-xylulose 5-phosphate
MXT	Mitoxantrole
pgi	Phosphoglucose isomerase
PSPG box	Plant secondary product glycosyltransferase box

K. Härtl • K. McGraphery • J. Rüdiger • W. Schwab (✉)
Biotechnology of Natural Products, Technische Universität München,
Liesel-Beckmann-Str. 1, 85354 Freising, Germany
e-mail: katja.haertl@tum.de; kate.mcgraphery@tum.de; julian.ruediger@tum.de;
wilfried.schwab@tum.de

W. Schwab et al. (eds.), *Biotechnology of Natural Products*,
https://doi.org/10.1007/978-3-319-67903-7_9

SG	Steviol glycoside
ushA	UDP-glucose hydrolase
zwf	D-glucose-6-phosphate dehydrogenase

9.1 Introduction

In the biosynthesis of plant natural products, modification of their chemical structure by tailoring enzymes is pivotal. These reactions include transfer of sugar units catalyzed by carbohydrate-active enzymes (CAZymes; http://www.cazy.org/) which are proteins whose functional domains are able to form, degrade or modify glycosidic bonds [1]. Their general abundance is in accordance with their high biological significance in Nature. In fact, most of the dry weight of plant biomass on earth exists as carbohydrate polymers, which emerged from glycosyl transfer reactions [2]. Some end products of primary plant metabolism are the major base of the food and the feed industry (for example, starch in grain and tuber crops). Glycosides and glucose esters of secondary metabolites are in general less abundant, but nevertheless of equal importance. On a cellular level, glycosylation functions, among other roles, in energy metabolism, pathogen virulence, molecular defense, and storage information signaling [3]. Addition of a sugar residue also increases water solubility and stability, and thus affects the bioactivity of acceptor molecules [4]. Therefore, the interest of the pharmaceutical, cosmetics and food industry in these compounds has increased continuously. At present, glycosides are used as therapeutic drugs, functional ingredients, and dietary supplements. They can be produced either chemically, for example by the Koenigs-Knorr reaction, or biochemically by the action of glycosidases, transglycosidases, glycoside phosphorylases and glycosyltransferases [5].

Glycosyltransferases (GTs) form a vast class of CAZymes that transfer sugar moieties from activated donors (usually nucleoside diphosphate (NDP) activated monosaccharides) to specific acceptor molecules (generally alcohols and carboxylic acids) (Fig. 9.1). However, transfer to N-, S-, and C-atoms, resulting in the formation of glycosylamines, thioglycosides, and *C*-glycosides, respectively, have also frequently been observed [6–8]. Recently, family 1 GT enzymes have drawn special interest from industry, because they change the physicochemical properties and consequently the bioactivity of lipophilic small molecules, thus offering new and exciting possibilities for the biotechnological production of bioactive glycosides [4, 9]. An advantage of family 1 GTs in this context is due to their ability to catalyze glycosidic bond formation regio- and stereoselectively with high yield [10, 11]. Hence, the property of the molecule of interest can be adapted on demand. This is especially important for the application of carbohydrate-tailored drugs, where the precise identity and position of the glycosyl residue decides about the mechanism of action and the bioavailability [12, 13]. The rapidly growing number of publicly available genomes has enabled the identification of new GT sequences that encode proteins with novel activities, and the first GT enzymes have now been successfully applied in bio-catalytic processes [14]. The following chapter will cover the last

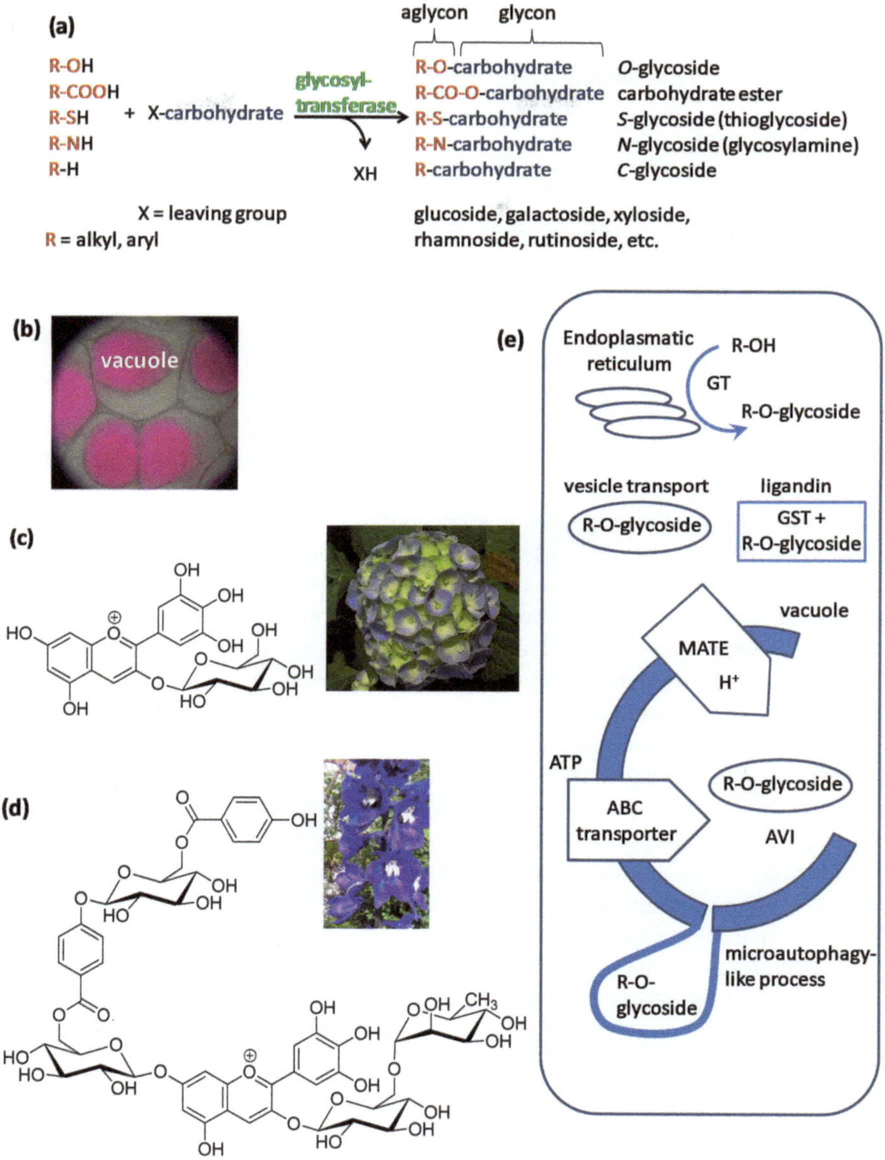

Fig. 9.1 Reactions catalyzed by glycosyltransferase (**a**). Anthocyanin-storing vacuoles of *Rhoeo spathacea* (**b**). Cells have been plasmolyzed. Picture was adapted from Wikipedia (https://en.wikipedia.org/wiki/Vacuole; last accessed April, 2017). *Hydrangea* variety producing the anthocyanin myrtillin (delphinidin-3-*O*-glucoside) (**c**), and *Delphinium* variety accumulating the anthocyanin violdelphin (delphinidin 3-*O*-rutinoside-7-*O*-(6-*O*-(4-*O*-(6-*O*-(*p*-hydroxybenzoyl)-glucosyl)-oxybenzoyl)-glucoside) (**d**). Transport of glycosides into the vacuole (**e**). ABC ATP-binding cassette transporter, AVI anthocyanin vacuolar inclusions, GST glutathion S-transferase, GT glycosyltransferase, MATE multidrug and toxic extrusion protein, R alkyl or aryl

10 years of research on family 1 GT enzymes with focus on the biotechnological production of glycosides/glucose esters and special emphasis upon production in whole cell biocatalysts and applications in industry.

9.2 The Significance of Glycosylation in Plants

Glycosylation is one of the most important tailoring mechanisms of bioactive compounds in plants [15]. The addition of a glycon (sugar unit) fundamentally changes the physicochemical properties of the acceptor aglycon (non-sugar component). Through glycosylation by GTs, plants can modulate the structure and function of secondary metabolites.

9.2.1 Glycosylation Increases Solubility

Glycosides and glucose esters of natural products are less hydrophobic than the aglycon on its own [16]. Thus, upon glycosylation of the aglycon the water solubility increases. In particular, this is evident for hydrophobic metabolites like flavonoids that feature a complex phenolic ring structure [17]. Flavonoids are major secondary metabolites in many fruits and vegetables, where they act among others inter alia as colored pigments. Due to enhanced solubility by glycosylation, plants can accumulate glycosylated flavonoids and related anthocyanins in the lumen of the vacuole in high concentrations [18] (Fig. 9.1) (**see also Chap. 3 of this book**). Glycosylation of flavonoids takes place in the cytosol right after the formation of the aglycons, since some of the non-glycosylated precursor molecules are unstable under physiological conditions. Utilizing a GT enzyme from apple, the 3,5-β-D-glucoside of resveratrol was produced, thereby increasing the water-solubility of an otherwise hydrophobic compound by 1700-fold [19] (**see also Chap. 3 of this book**). As a currently applied dietary supplement resveratrol is administered orally to patients. However, its medical exploitation is strongly limited by the low water-solubility. Thus, the future usage of resveratrol glucosides as therapeutic agent is envisaged [20, 21]. Similarly, the biological availability of the isoflavone gentistin is limited by its insolubility in water. Consequently, water-soluble gentisin glycosides were produced whereby solubility was increased 1000–10,000-fold and antioxidant activity was maintained [22]. Likewise, quercetin glycosides were developed to improve water solubility of the flavonoid for food and other applications [23].

9.2.2 Glycosylation Increases Stability

Glycosides and glucose esters exhibit not only a higher hydrophilicity but also an enhanced stability. Glycosylation stabilizes and intensifies the color of plant anthocyanidins – secondary metabolites that are responsible for the red, blue, and purple pigmentation of flowers and fruits in various plant species (**see also Chap. 4 of this**

book). This stabilization is probably further enhanced by acylation of the anthocyanin core with phenolic acids, thereby promoting intermolecular sandwich type stacking of the aromatic nuclei of the molecules [24]. However, the substitutions of the aromatic acyl groups occur at glycosyl residues. Thus, without glycosylation, a subsequent acylation with phenolic acids would be limited to the available hydroxyl groups of the aglycon. The poly-acylated anthocyanin glycoside violdelphin, one of the main color pigments of *Delphinium* flowers [25], is more stable and shows a stronger blue color than its less modified form myrtillin (Fig. 9.1), which is found in varieties of the genus *Hydrangea* [26]. Similarly, the *O*-glucosides of hydroxycinnamic acids are more stable than the phenolic acids, of which 20–40% degrade at room temperature during one year of storage [27]. Furthermore, the half-life of furaneol-glucoside exceeds that of the free flavor molecule by 35-fold [28].

9.2.3 Glycosylation Controls Sequestration/ Compartmentalization

Anthocyanins are believed to be synthesized by a multi-enzyme complex located on the cytosolic side of the endoplasmic reticulum [29]. The aglycons are promptly glycosylated, and the glycosides are administered from the cytosol to the vacuole, where they accumulate to high levels [30] (**see also Chap. 4 of this book**). The low pH value of the vacuole suppresses oxidation reactions, and ensures anthocyanin stability. Consequently, a specific transport mechanism is required that permits the translocation of sugar-bound metabolites across the cytoplasm and through the tonoplast [31].

For transport across the cytoplasm, two main models have been proposed: the ligandin model and the vesicle-mediated transport model, both being not mutually exclusive (Fig. 9.1). The ligandin model suggests binding of cytoplasmic anthocyanins to glutathione S-transferase (GST) proteins, which function as carriers that escort/stabilize anthocyanins until they are taken up into the vacuole [32–34]. As an example, two GST proteins from *Vitis vinifera* were associated to this mode of transport as their overexpression correlated with increased anthocyanin content in grape suspension cells [33]. Furthermore, it was shown that *Arabidopsis* mutant seedlings deficient in expression of the GST Transparent Testa 19 (TT19) barely accumulated anthocyanins [32, 34], whereas by re-introduction of a functional TT19 the mutant phenotype could be rescued [34]. Similarly, a GST seems to be critical for anthocyanin formation in strawberry as comparative transcriptome analyses of red- and natural white-fruited strawberry genotypes uncovered *GST* as highly differentially expressed gene [35]. According to the vesicular transport model, anthocyanins enter the ER lumen and are subsequently transported to the vacuole in vesicles, which could be detected microscopically due to the fact that the contained anthocyanins possess autofluorescence [36]. This process leads to the vacuolar accumulation of the glycosides in anthocyanic vascular inclusions (AVIs). AVIs have been described in more than 70 anthocyanin-producing species [31]. Since vanadate, a general inhibitor of ABC transporters, induced a dramatic increase of anthocyanin-containing sub-vacuolar structures it was suggested that cells utilize components of the protein secretory trafficking pathway for the direct transport of anthocyanins from the endoplasmic reticulum to the vacuole [37].

For transport into the vacuole two major mechanisms were suggested: On one hand, active transport was proposed that is mediated by directly energized ATP-binding cassette (ABC) transporters, such as the multidrug resistance-associated proteins (MRPs) ZmMRP3 and 4 in maize (*Zea mays*) [38]. Genetic loss of ZmMrp3 function in mutant plants led to a distinct pigmentation pattern, resulting from mislocalized and significantly reduced anthocyanin levels. Furthermore, yeast microsomes expressing the ZmMrp3 homologue ABCC1 from *V. vinifera* demonstrated the transport of anthocyanins in the presence of GSH [39]. On the other hand, a second transport involving Multidrug And Toxic Extrusion (MATE) family proteins was proposed that depends on a pre-existing, vacuolar membrane spanning H^+ gradient [40–42]. Characterization of MATE2 from *Medicago truncatula* [41] and anthoMATE1 and 3 from *V. vinifera* [42] revealed the translocation of glycosylated pigments that have been further decorated by acylation or malonylation. However, both transport mechanisms are presumably not mutually exclusive and seem to strongly depend on the type and structure of the transported secondary metabolite and may require other, not yet known factors. As an example, abscisic acid glucosyl ester is transported by both, proton-antiport and ABC-binding cassette mechanisms [43]. Only recently, a novel way of vacuolar transport has been put forward (Fig. 9.1). The authors observed by confocal microscopy that cytoplasmic anthocyanins aggregate in AVIs in close proximity to the vacuolar surface, and are directly enclosed by the tonoplast in a microautophagy-like process [44]. The endoplasmic reticulum-to-vacuole vesicular transport of anthocyanins mediated by a *trans* Golgi network-independent mechanism presumably contributes to the formation of AVIs [37].

A new, putative flavonoid carrier has been found in epidermal tissues of carnation petals [45]. The amino acid sequence is similar to mammalian bilitranslocase, a plasma membrane transporter found in liver and gastric mucosa. There, the protein mediates the uptake of the pigment bilirubin, dietary anthocyanins and nicotinic acid.

9.2.4 Glycosylation Affects Bioactivity and -Availability

Plants can regulate the level of bioactive secondary metabolites by linking them to sugar units. The plant hormone abscisic acid (ABA), which has important functions in plant responses to abiotic stresses, seed development, and germination [46], is inactivated by esterification with glucose [47]. While "free", unbound ABA can trigger stomatal closure and is mainly found in extra-vacuolar compartments, the ABA-glucose-ester (ABA-GE) is non-reactive and stored in the vacuole [48]. A second, oxidative mechanism of ABA inactivation has been proposed [49]. However, glucose conjugation has one decisive advantage to other modifications – it is reversible. ABA-GE is hydrolyzed by vacuolar β-glucosidases in a fast, one-step reaction to generate free cytosolic ABA [50, 51]. Therefore, ABA-GE is thought to act as an easily accessible ABA reservoir that can be tapped by plants to regulate ABA homeostasis [43]. Another example is the recent identification of xanthophyll-derived apocarotenoid glycosides (AGs) in leaves of *Arabidopsis thaliana* [52]. Increased carotenoid pathway flux in leaves of transgenic plants resulted in higher levels of AGs. Accordingly, the authors hypothesized that formation of AGs regulates the cellular level of carotenoids. Similarly, it has been shown that glycosylation of monoterpenes reduces the concentrations of free, aroma-active terpenols in grapes, which however, can be released again during vinification [53].

Furthermore, glycosides and glucose esters act as precursors for various biosynthetic pathways, because the glycon features several functional hydroxyl groups that enable additional metabolic reactions [25]. For example, galloylglucose ester is generally considered as precursor of ellagitannins and ellagic acid, polyphenolic antioxidants in strawberry [54], raspberry [54], grapevine [55], pedunculate oak [56], and pomegranate (*Punica granatum*) [57]. It is hypothesized that the galloylglucose ester functions as gallic acid donor as well as acceptor and thus, enables dimerization of the phenolic acid in the ellagitannin pathway [54].

9.2.5 Glycosylation Reduces Toxicity

As sedentary organisms, plants have developed strategies to mitigate the poisonous effects of toxic chemicals they are confronted with. One possible detoxification process is the conjugation of toxins to glucose molecules, making them more water soluble and enabling transport into the vacuole. Spotted knapweed (*Centaurea maculosa*) is able to metabolize maculosin – a host-specific toxin produced by the fungus *Alternaria alternata* – to the corresponding *β-O*-glucoside, which lacks the phytotoxicity of the aglycon [58]. Moreover, it was shown that GT enzymes are able to diminish the harmful effects of the mycotoxins zearalenone [59] and deoxynivalenol [60]. Thus, the toxins become masked but remain present in the plant tissue. Toxicological data are scarce, but several studies revealed the potential threat to consumer safety from these substances due to possible hydrolysis during mammalian digestion [61].

An interesting observation has been made by Australian wine producers. After a number of vineyards were exposed to smoke from bushfires and the berries were processed, the resulting wines exhibited a strong off-flavor [62]. Strong "smoky", "burnt" and "ashtray" characters were reported [62, 63]. Subsequently, the presence of the *β*-D-glucosides of guaiacol, its conjugates, and other related phenolic smoky compounds was confirmed [64, 65]. Apparently, the plants tailor the phenols by glycosylation to cope with the toxic effect. However, during the winemaking process (e.g. fermentation) and consumption of the wine, the glycosidic bonds are hydrolyzed and the smoky compounds are released, thus causing an unpleasant taste [66, 67]. The wines are disliked due to the off-flavor, which eventually leads to a loss of income for the wine producers.

9.2.6 Glycosylation Affects Perception

The elucidation of the cause of the smoke-tainted wines prompted further analyses of the in-mouth hydrolysis of glycosidically bound flavor compounds [67]. It was shown that enzymes of the human saliva are able to release the volatile aglycones from their natural glycoconjugates even under low pH and elevated ethanol conditions, confirming the in-mouth breakdown of monosaccharide and disaccharide glycosides. Thus, the long, lingering aftertaste of wines, desirable or undesirable, may be due to retro-nasal perception of aromas released from glycosides, which occur naturally in grapes [67, 68].

Steviol glycosides, a mixture of glycosides of the diterpene steviol, have recently been approved as sweetener in the EU [69]. Stevioside and rebaudioside A are the main diterpene glycosides present in leaf tissues, but only rebaudioside A imparts a desirable sweet taste, while stevioside produces a residual bitter aftertaste (Fig. 9.2). Glycosylation of stevioside yields rebaudioside A and can increase the ratio of rebaudioside A to stevioside in steviol glycoside products, providing a conceivable strategy to improve the organoleptic properties of steviol glycoside products. Hence, several enzymatic processes have been recently suggested to produce rebaudioside A by glucosylation of stevioside [70–74].

9.3 Glucoside/Glucose Ester Synthesis

Glycosidic and glucose ester bond formation is mostly achieved chemically by a series of steps involving the protection of interfering hydroxyl groups, activation of a leaving group at the anomeric carbon proceeding in an S_N1 reaction mechanism, use of a heavy metal catalyst in water free medium, exclusion of light, and eventually deprotection [75]. Although decisive progress has been made in improving the methods and techniques since the classical Koenigs-Knorr reaction was published [76], chemical glycosylation has considerable drawbacks. It still suffers from low yields, high costs, usage of toxic heavy metal catalysts, and the formation of unspecific products [9, 77].

In contrast, biocatalytic reactions are a promising alternative due to the mild reaction conditions, the regio- and stereo-selectivity of the enzymatic reactions, and the ability to accept both hydrophilic and hydrophobic substrates. Furthermore, the protection of functional groups is not required. Consequently, fewer process steps are needed, which reduces production costs and ecological damage [77, 78]. Among the glycosidic bond mediating enzymes are (i) glycoside hydrolases that also efficiently catalyze the reverse hydrolytic reaction (condensation) whereby glycosides are formed, (ii) transglycosidases that are able to catalyze the transfer of glycosidic bonds within carbohydrate molecules and between glycosides, (iii) glycoside phosphorylases that use phosphate-activated glycosyl donors, and (iv) glycosyltransferases that utilize nucleotide diphosphate activated sugar donors as co-substrates to form glycosides [5].

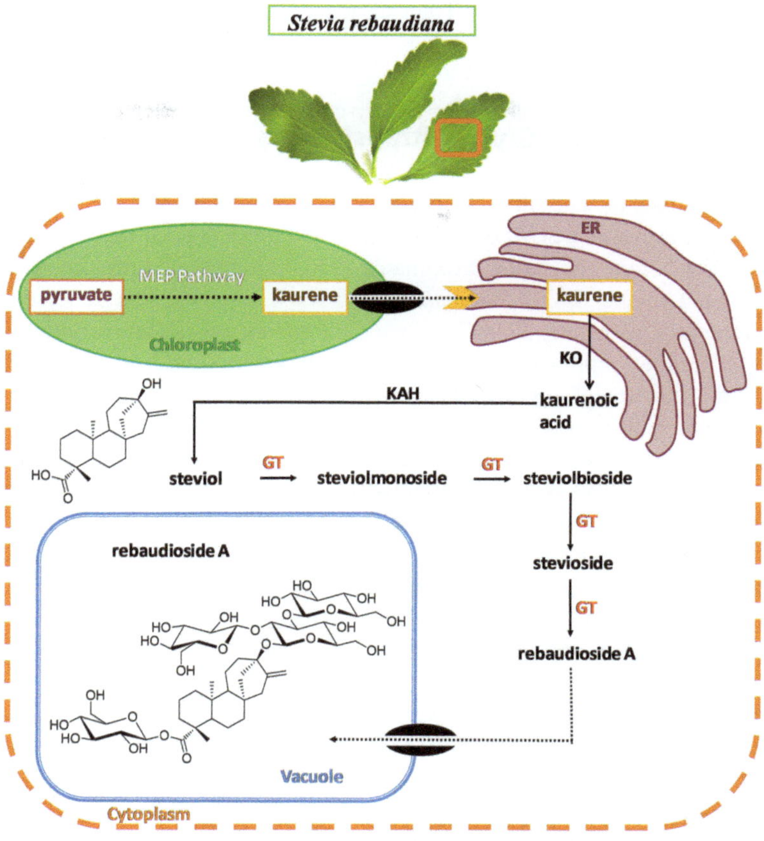

Fig. 9.2 Schematic representation of the biosynthetic pathway of steviol glycoside, rebaudioside A. This biosynthetic pathway takes place in the leaf tissue of *Stevia rebaudiana*. The multi-step methylerythritol 4-phosphate (MEP) pathway in the chloroplast stroma converts the initial precursor pyruvate to kaurene. It is then transported to the endoplasmic reticulum (ER) where it is oxidized by kaurene oxidase (KO) and kaurenoic acid hydroxylase (KAH) to form steviol. Subsequently, steviol is glycosylated by multiple glycosyltransferases (GT) to form rebaudioside A – the sweetest and least bitter tasting glycoside – which is transported into the vacuole for storage

However, successful application of these CAZymes in industrial processes is limited for various reasons, which also depend on the enzyme class [9]. Glycosidases show high promiscuity regarding their acceptor substrate specificity, yet production outcome is poor, and conversion of acceptors with multiple hydroxyl groups often results in an isomeric product mixture. Thus, expenses for the purification of the product make a considerable contribution to the overall costs. Transglycosidases and glycoside phosphorylases accept only a limited number of acceptors and exhibit poor regioselectivity, similar to glycosidases. GT enzymes on the other hand accept a wide range of hydrophobic and hydrophilic acceptors, while showing regio- and stereoselective product formation. Their disadvantage is the requirement of expensive co-substrates.

9.4 Family 1 Plant Glycosyltransferases

GT enzymes can be classified in families based on their reaction mechanism, sequence similarity, as well as donor, acceptor and product specificity [79–82]. Up to date, 103 GT families and approximately 316,000 annotated protein sequences are contained in the CAZY database (http://www.cazy.org, last accessed April 2017). Recently, a specific PlantCAZyme database was established, giving credit to the vast number of plant GTs [83]. Plant GT enzymes, similar to all GT proteins, catalyze the transfer of sugar units from activated nucleotide diphosphate sugar donors. Plant GTs mostly use UDP-glucose, although UDP-rhamnose, UDP-galactose, UDP-xylose, UDP-arabinose, and UDP-glucuronic acid have also been reported as donors [84–87]. Additionally, glycosides and glucose esters can be distinguished by the type of bond that is formed. Although *O*-glycosylation is the most common modification, *N*-, *S*-, and *C*-glycosides have also been described [3]. The sequence identity of plant GTs may vary, but they share distinctive characteristics, such as structural folds, stereo-chemical mechanism of glycosidic bond formation, and a conserved motif called plant secondary product glycosyltransferase (PSPG) box.

Two structural folds (Fig. 9.3) have been characterized intensively, the GT-A fold and the GT-B fold [92]. GT-A folded GTs possess a single Rossmann fold and a conserved metal-binding motif [93, 94]. In contrast, GT-B enzymes do not require metal ions and contain two Rossmann folds. These Rossmann folds are linked, facing one another, and form an active cleft between them [95, 96]. A third fold, named GT-C fold, has been proposed along with the other two folds but the distinctiveness of the GT-C fold remains controversial [97–99]. Recently, a new fold was reported for a bacterial GT involved in the glycosylation of serine-rich repeat streptococcal adhesins [91]. After X-ray crystallography the authors were able to identify a distinct structure, different from all known GT folds, and a new metal-binding site. Consequently, this new structure was called GT-D fold.

GT enzymes can also be classified according to the anomeric configuration of the product. Enzymes that retain the stereo-chemistry at the anomeric center of the donor substrates are called "retaining" GTs. In contrast, enzymes that invert the stereochemistry are called "inverting" GTs [82]. Family 1 GT proteins are inverting enzymes that adopt the GT-B fold, and tailor lipophilic small-molecule acceptors [100]. Additionally, most of the family 1 GT enzymes feature the so-called PSPG box, a conserved C-terminal motif responsible for the interaction with the sugar donor [101]. An interaction between the highly conserved HCGWNS motif and UDP-glucose has been revealed [102], and the last amino acid of the PSPG box probably controls the selection of the sugar donor [103].

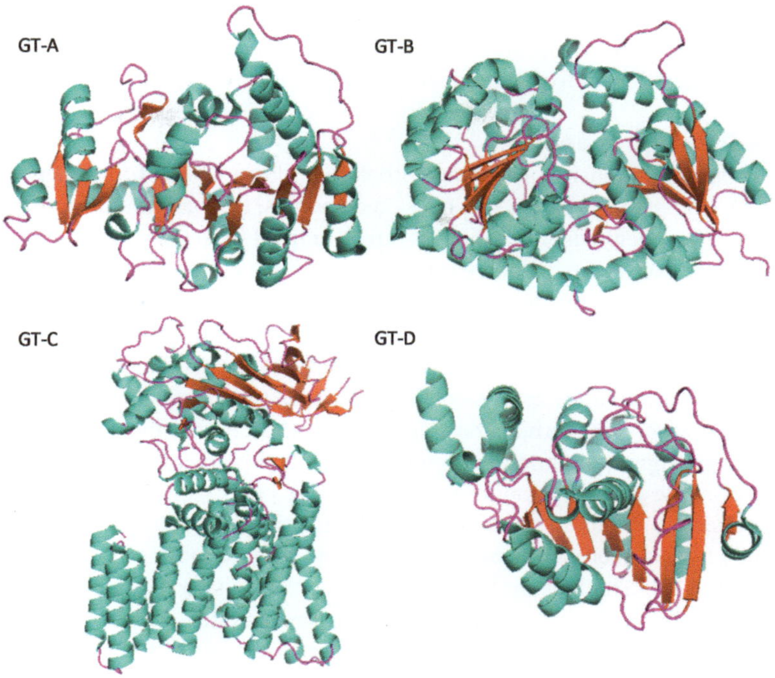

Fig. 9.3 Representative examples of the four possible structural GT folds. GT-A: PDB 1FOA [88]; GT-B: PDB 4REL [89]; GT-C: PDB 3RCE [90], and GT-D: 4PHR [91]. The pdb-files were downloaded from RCSB Protein Data Base (http://www.rcsb.org/pdb/home/home.do; last accessed May, 2016), and visualized by the PyMOL v.1.7.4 software

9.5 Substrates of Family 1 Plant Glycosyltransferases

Plants produce a tremendous diversity of low molecular weight metabolites, many of them being of commercial importance. The variety is mostly ensured by decoration of a limited number of common skeletons by hydroxylation, methylation, acylation and glycosylation. Glycosylation is one of the most widespread modifications.

9.5.1 Secondary Metabolites

Although secondary metabolites may not be essential for plant growth and development under artificial growth conditions, they are of high relevance for survival in natural environments. During evolution, when plants became sedentary, their inability to move forced them to be able to adapt to constantly changing environmental conditions. At times they are exposed to changes in temperature, UV radiation, salinity, water status, and pathogen pressure. This has led to the development of a wide variety of secondary metabolites that are constantly modified to meet the plant's requirements. Many metabolites acting in defense responses are glycosylated for stabilization, detoxification, and sequestration.

9.5.1.1 Phenylpropanoids

Phenylpropanoids consist of a three carbon side chain linked to a phenyl group (Fig. 9.4). Like most phenolic secondary metabolites, their aromatic core structure is derived from phenylalanine, which in turn is derived from the shikimate pathway. Phenylalanine is converted in three reaction steps to the activated 4-coumaroyl-CoA thioester, which gives rise to simple (e.g. hydroxycinnamic acids) and more complex (e.g. flavonoids, stilbenes, lignins) phenolics; this biosynthetic pathway has been well studied and reviewed [104, 105]. The phenylpropanoid biosynthetic pathway provides metabolites acting as UV protectants (flavones) or as chemo-attractants for *Rhizobia* – mycorrhiza forming bacteria [106]. Furthermore, the precursors for polymeric lignin, antimicrobial phytoalexins, and pigments (anthocyanins) are derived from the general phenylpropanoid pathway [107] (**see also Chap. 4 of this book**). Phenylpropanoids have also drawn attention for their health-promoting properties. Anticancerogenic, antioxidative, and antimicrobial effects have been noted [108–110]. The already high structural diversity is further increased by the action of family 1 GTs through addition of sugar residues.

Caffeic acid (3,4-dihydroxycinnamic acid) is an example of a simple phenylpropanoid. Its esterification with glucose by GTs has been described frequently [10, 111], most recently by enzymes from the tea plant (*Camellia sinensis*) [112]. Strawberry, raspberry and grape GTs that produce phenylpropanoyl glucose esters show substrate promiscuity as they form glucose esters of a variety of (hydroxyl) cinnamic acids and (hydroxyl)benzoic acids, including gallic acid, and might also to be involved in the biosynthesis of ellagitannins [54].

Monolignols are the precursors of lignins, which are integrated into the cell walls of higher plants to enhance shoot stabilization. They are converted to glycosides by GTs, as has been shown for UGT72B1 from *Arabidopsis*, which catalyzes the glucosylation of coniferyl alcohol and coniferyl aldehyde [113]. Over-expression of PtGT1 from poplar (*Populus tomentosa*) in tobacco plants led to increased lignin content and an early flowering phenotype [114], indicating that monolignol glucosides might play a role in the formation of lignin in plant cell walls (although their specific functions remain elusive) [115].

Coumarins are yet another group of phenylpropanoids relevant for industry. Coumarin (1-benzopyran-2-one), the eponym of this class of metabolites, is commonly used in perfumes and fragrances because of its vanilla-like odor. *In planta*, coumarins are hydroxylated, which enables glycosylation of these functional groups, and these metabolites rarely occur in aglycon form. Two GTs from tobacco were found to glucosylate scopoletin (6-methoxy-7-hydroxycoumarin), which is considered a phytoalexin because it possesses antimicrobial properties and is accumulated by plant tissues upon pathogen infection [116]. Glycosylation of umbelliferone (7-hydroxycoumarin), another hydroxycoumarin with phytoalexin activity, was recently reported to be mediated by PNgt1 and 2, two GTs from Ipomoea morning glory (*Pharbitis nil*) [117].

9.5.1.2 Flavonoids, Anthocyanins

Flavonoids, which arise from phenylpropanoids by condensation with three molecules of malonyl CoA, are characterized by a 15-carbon flavan skeleton (Fig. 9.4) (**see also Chap. 4 of this book**). Among them are anthocyanins – glycosylated plant pigments responsible for fruit color [118]. The first plant GT gene was identified in a study on the genetic instability of transponsons in maize. *Bronze 1*, the gene responsible for the dark pigmentation of maize grains later turned out to encode an anthocyanidin GT [101]. In most cases, UDP-D-glucose serves as the preferred sugar donor for plant GTs, which is the reason why glucosides are so abundant in the plant kingdom. However, in red-fleshed kiwifruit (*Actinidia chinensis*) cyanidin 3′-*O*-xylo-3-*O*-galactoside is the main anthocyanin. It is formed by two sequential glycosylation steps. The first sugar (galactose) is transferred by F3GT1, whereas xylose is transferred by F3GGT1 [87]. Similarly, malvidin-3,5-*O*-bis-glucoside is synthesized by two sequential glycosylation reactions, but the monosaccharides are transferred to different positions of the anthocyanidin skeleton. The second glycosylation step is catalyzed by 5-*O*-glucosyltransferase Va5GT from grape (*Vitis amurensis*), which converts malvidin-3-*O*-glucoside to the corresponding 3,5-*O*-bis-glucoside [119, 120]. RhGT1, a dual function anthocyanidin 3,5-*O*-glucosyltransferase from *Rosa hybrida*, produces the 3,5-*O*-bis-glucoside directly from anthocyanidins and is able to glycosylate a wider spectrum of flavonoid metabolites including apigenin (flavone) and galangin (flavonol) [121]. The diversity of plant natural products originates from such combinatorial modifications.

9.5.1.3 Dihydrochalcones, Acylphloroglucinol, Stilbenes, Curcumin

Phloretin derivatives are classified as dihydrochalcones and belong to the group of acylphloroglucinols (Fig. 9.4). They are the main phenolic metabolites in apple (*Malus* × *domestica*) and pear (*Pyrus communis*) leaves, where they are thought to act in pathogen resistance [122]. They mainly occur in glycosylated form, with phloridzin (phloretin 2′-*O*-glucoside) being the most abundant product [123]. Phloridzin is produced by a GT in apple and pear that is specific for the acceptor substrate phloretin but shows relaxed activity for the donor substrate. MdPGT1 can glycosylate phloretin in the presence of three sugar donors: UDP-glucose, UDP-xylose and UDP-galactose [124–126]. Trilobatin (phloretin-4′-*O*-glucoside) is formed by MdPh-4′-OGT in apple [127].

Only recently, acylphloroglucinols have also been discovered in strawberry fruit [128]. They are synthesized by a dual functional chalcone/valerophenone synthase, which readily catalyzes the condensation of two intermediates in the branched-chain amino acid metabolism, isovaleryl-Coenzyme A (CoA) and isobutyryl-CoA, with three molecules of malonyl-CoA to form phlorisovalerophenone and phlorisobutyrophenone, respectively. Glucosylation is finally catalyzed by the promiscuous

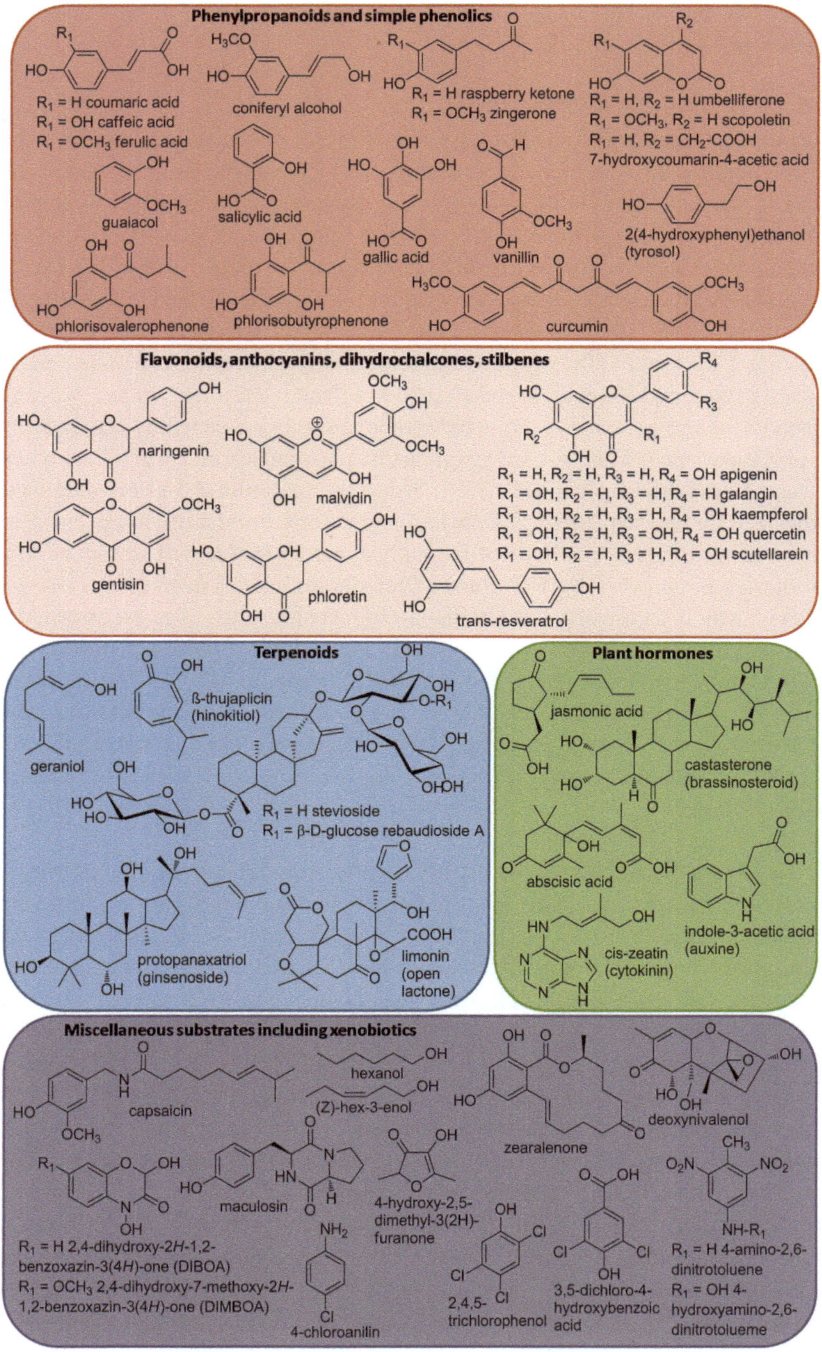

Fig. 9.4 Chemical structures of GT acceptor molecules

UGT71K3 enzyme [129], which also participates in the glucosylation of the key strawberry flavor compound 4-hydroxy-2,5-dimethyl-3(2H)-furanone [130].

Similar to coumarins, stilbenes are considered as phytoalexins but are also thought to act in defense response signaling and protection against UV-radiation [131]. *Trans*-resveratrol (*trans*-3,5,4′-trihydroxystilbene) has received attention lately, as it lowered blood pressure and insulin resistance when administered to rodents [132]. Glucosylation of resveratrol significantly increases its solubility and is catalyzed by PaGT3 (in *Phytolacca americana)* [133] or a bi-functional resveratrol/hydroxycinnamic acid glucosyltransferase (in *Vitis labrusca*) [134]. Four different glucosides of resveratrol, namely resveratrol 3-*O*-β-D-glucoside, resveratrol 4′-*O*-β-D-glucoside, resveratrol 3,5-*O*-β-D-diglucoside, and resveratrol 3,5,4′-*O*-β-D-triglucoside are produced by an UDP glucosyltransferase from *Bacillus licheniformis* [21] (**see also Chap. 3 of this book**).

Curcumin is the yellow pigment of turmeric, the dried rhizome of *Curcuma longa.* It is used primarily as food colorant, but also supposedly possesses pharmacological activity. However, its low water solubility limits further pharmacological exploration and practical application. Thus, glucosylation of curcumin was analyzed in cultured *Catharanthus roseus* cells and two GTs were characterized. CaUGT2 catalyzed the formation of curcumin monoglucoside from curcumin and also the conversion of curcumin monoglucoside to curcumin diglucoside [135], whereas UCGGT catalyzed the 1,6-glucosylation of curcumin 4′-*O*-glucoside to yield curcumin 4′-*O*-gentiobioside [136].

9.5.1.4 Terpenoids

Terpenoids are synthesized *in-vivo* via two biosynthetic pathways, either by the mevalonate or the 2-*C*-methyl-D-erythritol 4-phosphate/1-deoxy-D-xylulose 5-phosphate (MEP/DOXP) pathway [137]. In plants, terpenoids are involved in defense and stress response, and plant insect interaction [138]. Many terpenoids are fragrant volatile metabolites used in many different industrial applications, primarily in the fragrance and food sectors. The monoterpene alcohol geraniol is a commercially highly relevant terpenoid, as it possesses an odor commonly associated to rose flowers [139]. In addition, it is also a crucial flavor contributor in various grapevine varieties, and several monoterpenol GTs mediating gluco-conjugation of geraniol among other terpenoid alcohols have been reported recently [53, 140]. Similarly, AdGT4 from the kiwi plant (*Actinidia deliciosa*) glycosylates a range of terpenes and primary alcohols which are found as glycosides in ripe kiwifruit. Two of the enzyme's preferred primary alcohol aglycones, hexanol and (Z)-hex-3-enol, contribute strongly to the 'grassy-green' aroma notes of ripe kiwifruit [141]. In tea plants (*C. sinensis*) terpenoid alcohols are stored as β-primeverosides, diglycosides formed from glucose and xylose, by the sequential action of CsGT1 and CsGT2 [142].

9.5.2 Plant Hormones

Phytohormones are bioactive compounds of plant origin that function as messenger molecules and play essential roles in germination, growth, development, and defense reactions. They are thought to be glycosylated by GTs to sustain cellular homeostasis, in other words to maintain an equilibrium of active and inactive (non-glycosylated and glycosylated) products [143]. Brassinosteroids, a family of steroid hormones, regulate cell division and elongation. In *A. thaliana* it was shown that homologous UGT73C5 and UGT73C6 catalyze the glucosylation of the brassinosteroid castasterone, thereby causing its inactivation [144, 145]. Likewise, cytokinins are involved in plant growth but also environmental responses, and GT activity on cytokinins such as *cis*-zeatin has been verified [146]. Similarly, GT mediated glucose esterification of auxins [147] and abscisic acid [148] has been shown, and a promiscuous GT enzyme from the immature seeds of morning glory (*Ipomoea nil*) glucosylates 2-*trans*-abscisic acid, indole-3-acetic acid, salicylic acid (SA) and (+/−)-jasmonic acid [149].

9.5.3 Miscellaneous Substrates (Alkaloids, Benzoxazinoids, Furanones, and Xenobiotics)

Alkaloids are bioactive nitrogen-containing plant metabolites derived from amino acids (**see also Chap. 5 of this book**). Capsaicinoids are branched or straight-chain alkylvanillylamides produced by species of the genus *Capsicum*, which are important sources of foods, spices, and medicines. Capsaicin (8-methyl-*N*-vanillyl-6-nonenamide) is the most poignant compound among naturally-occurring capsaicinoids and shows interesting bioactivity. However, capsaicinoids exhibit low water-solubility and are consequently only poorly absorbed after oral administration [150]. The search for capsaicinoid GTs yielded PaGT3, which was isolated from pokeweed (*P. americana*) and converted capsaicin and 8-nordihydrocapsaicin to their corresponding glucosides [150].

Benzoxazinoids (Bx) are defensive metabolites in various species of the *Poaceae* and are derived from tryptophan. DIBOA (2,4-dihydroxy-2H-1,4-benzoxazin-3(4H)-one) and its *C*-7-methoxy derivative DIMBOA are the predominant benzoxazinoids in maize, and were reported to be glucosylated by the GT Co-BX8 from larkspur (*Consolida orientalis*) [151]. The Bx-glucosides possess reduced toxicity compared to the aglycons and are stored in the vacuole. When the cells are disrupted upon wounding and/or infection, the toxic aglycons are released by a pre-existing β-glucosidase (Glu) that accumulates in plastids, defining them as phytoanticipins [152].

Furanones such as the key strawberry flavor compound 4-hydroxy-2,5-dimethyl-3(2H)-furanone (HDMF, furaneol®) are downstream metabolites of fructose-1,6-diphosphate [153]. HDMF is synthesized during strawberry development but becomes glucosylated by UGT71K3a and UGT71K3b, rendering the volatile odorless [130]. UGT71K3 isozymes are promiscuous and also accept acylphloroglucinols and vanillin as acceptor molecules, probably due to structural similarities with HDMF [129].

Xenobiotics are man-made chemical substances, such as pesticides and industrial chemicals that form residues in plants. Some of these have cytotoxic effects, but plants can prevent these by conjugating xenobiotic breakdown products to sugar residues. The explosive 2,4,6-trinitrotoluene (TNT) is one such chemical contaminating soil and groundwater today. Six GTs from *A. thaliana* were found to glucosylate hydroxylaminodinitrotoluenes and aminodinitrotoluenes, degradation products of the TNT-metabolism [154], while 44 GTs catalyzed the *O*-glucosylation of chlorinated phenols, but only one, UGT72B1, showed appreciable *N*-glucosylating activity toward chloroanilines [155]. Similarly, FaGT2, a multifunctional GT enzyme from strawberry fruits, is involved in the metabolism of natural and xenobiotic compounds, such as the herbicide 2,4,5-trichlorophenol and the herbicide analogue 3,5-dichloro-4-hydroxybenzoic acid [156].

9.6 Glucoside Production by Whole Cell Biocatalysts

Natural product glycosides/carbohydrate esters are new and promising bioactive substances with prospective in functional foods, drug development, cosmetics, and many other applications [157, 158]. The bio-catalytic synthesis of glucosides using GT enzymes has many advantages over classic chemical approaches [159]. The production can be carried out either *in-vitro,* employing purified, heterologously produced enzyme [160–162], or *in-vivo,* utilizing a whole-cell biotransformation system [74, 133, 163, 164]. Production of glycosides using living cells provides major advantages (Fig. 9.5) [77, 165, 166], as the cells can be grown to high density and cultivation is cost-efficient compared to chemical approaches or *in-vitro* systems. Furthermore, if supplied with appropriate nutrients, the cells will be able to take up the substrates by diffusion or active transport. Consequently, endogenous or exogenous GTs glycosylate the substrates and the products are excreted into the medium, where they can be easily collected as the biocatalyst can readily be removed. Moreover, addition of co-factors is not required, since the cells provide their own recycling machinery [162]. Since many plant genomes and consequently GT sequences are publicly available, the next step in the development of whole-cell biocatalysts, currently applied in newest research, is the creation of transgenic cells overexpressing the genes of interest (Table 9.1). However, the whole-cell approach has also some limitations [189]. Often the availability of the substrate is restricted by its solubility, and too high concentrations can be toxic for the cells. Furthermore, some substrates are not able to enter the cells [159].

9.6.1 Production System

Glycoside production using whole-cell biocatalysts can be performed in two technical production systems; shaking flasks and stirred-tank reactors.

Production in partitioned shaking flasks, as described for the production of β-glucosides employing plant GTs, consists of three sequential steps [162, 166]. (I)

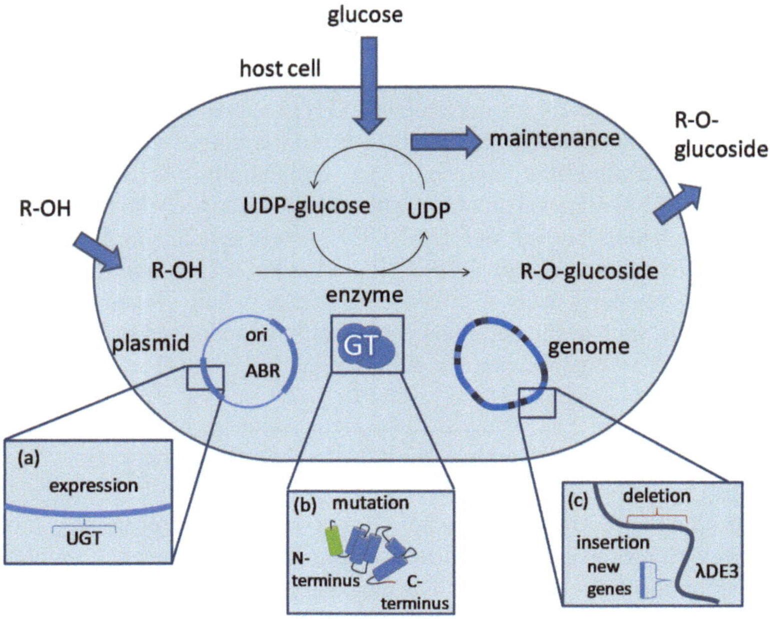

Fig. 9.5 Possible starting points for optimization of the whole cell biocatalyst (host cell). Utilization of improved plasmids to regulate heterologous GT expression (**a**) (ori: origin of replication, ABR: antibiotic resistance). Amino acid sequence mutation for enhanced protein expression, increased enzyme stability, modified substrate specificity and enzymatic activity (**b**). Host improvement through deletions and insertions/additions to the host genome, including metabolic engineering to improve the metabolic flux towards glycoside formation (**c**). R alkyl or aryl

Biomass production: Recombinant cells are grown to a high cell density using a complex medium. (II) GT production: The culture medium is changed to a defined minimal medium. This is done to reduce unwanted byproducts and to simplify the downstream applications. Subsequently, the expression of the recombinant GT is actively induced. (III) Glycosylation: The aglycon is added to the medium and the conversion takes place. Although the use of shaking flasks is simple, it offers very little process control and is not suitable for up-scaling.

The stirred-tank reactor on the other hand offers superior aeration and can be scaled up [166]. Furthermore, it provides online data like, among others, pH value, precise temperature, and dissolved O_2 concentration. This enables a more precise control of the process by, for example, using intelligent software and protocols [190]. The production of 3-*O*-xylosyl quercetin using an engineered *E. coli* strain was improved through up-scaling. By changing the cultivation vessel from 5-mL culture tubes to a 3-L bioreactor, the production of quercetin 3-*O*-xylose was boosted from 55–98% [170].

Table 9.1 Functions of GTs that have been explored in biotechnological efforts *in vivo*

Enzyme	Genetic source	Host	Sugar donor	Acceptor	Product	References
3GT	*Arabidopsis thaliana*	*E. coli*	UDP-glucose	Cyanidin	Cyanidin 3-*O*-glucoside	[167]
AGT (At4g09500)	*A. thaliana*	*E. coli*	UDP-glucose	Apigenin, baicalein	Apigenin 7-*O*-β-D-glucoside, baicalein 7-*O*-β-D-glucoside	[168]
AmUGT10	*Antirrhinum majus*	*E. coli*	UDP-glucuronic acid	Luteolin	Luteolin 7-*O*-glucuronide, luteolin 7-*O*-glucoside	[169]
arGt-3	*A. thaliana*	*E. coli*	UDP-xylose	Quercetin	Quercetin 3-*O*-xylose	[170]
AtUGT73B3	*A. thaliana*	*E. coli*	UDP-glucose	Kaempferol	Kaempferol 3-*O*-glucoside	[171]
AtUGT78D1	*A. thaliana*	*E. coli*	UDP-rhamnose, dTDP-6-deoxytalose	Quercetin, kaempferol	Quercetin 3-*O*-rhamnoside, kaempferol 3-*O*-rhamnoside, kaempferol 3-*O*-glucoside, quercetin 3-*O*-(6-deoxytalose)	[171–173]
AtUGT78D2 (At5g17050)	*A. thaliana*	*E. coli*	UDP-*N*-acetylglucosamine, UDP-glucose, UDP-xylose	Quercetin, kaempferol, pelargonidin	Quercetin 3-*O*-*N*-acetylglucosamine, quercetin 3-*O*-glucoside, kaempferol 3-*O*-glucoside, quercetin 3-*O*-xyloside, pelargonidin 3-*O*-glucoside	[86, 171, 174–176]
AtUGT78D3	*A. thaliana*	*E. coli*	UDP-glucose, UDP-xylose, UDP-arabinose	Quercetin	Quercetin 3-*O*-xyloside, quercetin 3-*O*-glucoside, quercetin 3-*O*-arabinoside	[86]
DicGT4	*Dianthus caryophyllus*	*S. cerevisiae*	UDP-glucose	Naringenin, phloretin	Naringenin 7-*O*-glucoside, naringenin 4′-*O*-glucoside, phloretin 2′-*O*-glucoside	[177]

(continued)

Table 9.1 (continued)

Enzyme	Genetic source	Host	Sugar donor	Acceptor	Product	References
F3GT	*Petunia hybrida*	*E. coli*	UDP-galactose	Quercetin, kaempferol, myricetin, morin, fisetin	3-*O*-galactosides of quercetin, kaempferol, myricetin, morin, and fisetin	[163]
F7GT	*Scutellaria biacalensis*	*E. coli*	UDP-glucose	Apigenin	Apigenin 7-*O*-glucoside, apigenin 7,4′-*O*-diglucoside, apigenin 5,7-*O*-diglucoside, apigenin 5,7,4′-*O*-triglucoside	[160, 161]
FeCGTa (UGT708C1)	*Fagopyrum esculentum*	*E. coli*	UDP-glucose	Phloretin, 2-hydroxynaringenin, 2-hydroxypinocembrin, 2-phenyl-2′,4′,6′-trihydroxyacetophenone, 2′,4′,6′-trihydroxyaceto phenone	Ploretin 3-*C*-glucoside, *C*-glucosides of 2-hydroxynaringenin; 2-hydroxypinocembrin, 2-phenyl-2′,4′,6′-trihydroxyacetophenone, and 2′,4′,6′-trihydroxyacetophenone	[178]
GT73C5	*A. thaliana*	*E. coli*	UDP-glucose	Geraniol, citronellol, farnesol, terpineol, perillyl alcohol, linalool, menthol	Geranyl glucoside, citronellyl glucoside, farnesyl glucoside, terpineoyl glucoside, perillyl glucoside, linalyl glucoside, menthyl glucoside	[166]
GT75D1	*A. thaliana*	*E. coli*	UDP-glucose	Artemisinic acid	Artemisinic acid glucose ester	[166]
PaGT3	*Phytolacca americana*	*E. coli / P. americana cell suspension culture*	UDP-glucose	*Trans*-resveratrol	Trans-resveratrol 3-*O*-β-glucoside, trans-resveratrol 4′-*O*-β-glucoside	[133]
PGT8	*P. hybrida*	*E. coli*	UDP-glucose	Pelargonidin, cyanidin	Pelargonidin 3-*O*-glucoside, cyanidin 3-*O*-glucoside	[176, 179]
PhUGT	*P. hybrida*	*E. coli*	UDP-galactose	Quercetin	Quercetin 3-*O*-galactoside	[169]

RhaGT	*A. thaliana*	*E. coli*	UDP-rhamnose, UDP-glucose	Quercetin, kaempferol, myricetin, morin, fisetin	3-*O*-rhamnosides and 3-*O*-glucosides of quercetin, kaempferol, myricetin, morin, and fisetin	[163]
rUGT1A7	*Rattus norvegicus*	*S. pombe*	UDP-glucuronic acid	4-methylumbelliferone	4-Methylumbelliferone-β-D-glucuronide	[180]
SbGT30, SbGT34, SbGT56	*S. biacalensis*	*E. coli / S. cerevisiae*	UDP-glucose	Scutellarein	Scutellarein 7-*O*-glucoside	[164]
SbUGTA1	*Starmerella bombicola*	*E. coli*	UDP-glucose	16-hydroxy-palmitic acid	16-*O*-(β-D-glucopyranosyl)-octadecenoic acid	[162]
UGT1A1, UGT1A6–10, UGT2A1, UGT2B15	*Homo sapiens*	*Schizosaccharomyces pombe*	UDP-glucuronic acid	4-methylumbelliferone	4-Methylumbelliferone-β-D-glucuronide	[180]
UGT2B17	*Homo sapiens*	*S. pombe*	UDP-glucuronic acid	4-methylumbelliferone, testosterone	4-Methylumbelliferone-β-D-glucuronide, testosterone glucuronide	[180]
UGT71C1	*A. thaliana*	*E. coli*	UDP-glucose	Quercetin	Quercetin 3′-*O*-glucoside, quercetin 7-*O*-glucoside, quercetin 7,3′-*O*-diglucoside	[161]
UGT71C2	*A. thaliana*	*S. pombe*	UDP-glucose	Vanillin	Vanillin β-D-glucoside	[181]
UGT71C3	*A. thaliana*	*E. coli*	UDP-glucose	Quercetin	Quercetin 3-O-glucoside, quercetin 3,3′-*O*-diglucoside, quercetin 3,4′-*O*-diglucoside, quercetin 3,7-*O*-diglucoside	[161]
UGT72B1	*A. thaliana*	*S. pombe*	UDP-glucose	Vanillin	Vanillin β-D-glucoside	[181]

(continued)

Table 9.1 (continued)

Enzyme	Genetic source	Host	Sugar donor	Acceptor	Product	References
UGT72B14	*Rhodiola sachalinensis*	*E. coli*	UDP-glucose	Tyrosol	Tyrosol glucoside	[182]
UGT72E2	*A. thaliana*	*S. pombe*	UDP-glucose	Vanillin	Vanillin β-D-glucoside	[14, 181]
UGT73A15 (CaUGT2)	*Catharanthus roseus*	*E. coli*	UDP-glucose	Curcumin, quercetin, trans-resveratrol, carvacrol	Curcumin monoglucoside, curcumin diglucoside, quercetin 4′-*O*-glucoside, quercetin 3-*O*-glucoside, quercetin 3,4′-O-diglucoside, quercetin 3,3′-*O*-diglucoside, quercetin 7,4′-*O*-diglucoside, quercetin 3,7-*O*-diglucoside, quercetin 3,7,3′-*O*-triglucoside, quercetin 3,7,4′-*O*-triglucoside, resveratrol 3-*O*-glucoside and resveratrol 4′-*O*-glucoside, carvacrol glucoside	[160–162]
UGT73C6	*A. thaliana*	*S. cerevisiae*	UDP-glucose	Zearalenone	α- and β-zearalenone 4-*O*-glucoside	[60]
UGT76G1	*Stevia rebaudiana*	*S. cerevisiae*	UDP-glucose	Stevioside	Rebaudioside A	[74]
UGT78K1	*Glycine max*	*E. coli*	UDP-glucose, UDP-rhamnose	Fisetin, kaempferol, myricetin, morin, quercetin	Glucosides and rhamnosides of fisetin, kaempferol, myricetin, morin, quercetin	[175, 183]
UGTPg1	*Panax ginseng*	*S. cerevisiae*	UDP-glucose	Protopanaxatriol	Ginsenosides F1 and Rh1	[184]
UGTPG100	*P. ginseng*	*S. cerevisiae*	UDP-glucose	Protopanaxatriol	Ginsenosides F1 and Rh1	[184]
UGTPg29	*P. ginseng*	*S. cerevisiae*	UDP-glucose	Ginsenoside Rh2	Ginsenoside Rg3	[185]

UGTPg45	*P. ginseng*	*S. cerevisiae*	UDP-glucose	Protopanaxadiol, ginsenoside Rh2	Ginsenoside Rh2, ginsenoside Rg3	[185]
VvGT14a	*Vitis vinifera*	*E. coli*	UDP-glucose	Eugenol	Eugenol glucoside	[162]
VvGT15c	*V. vinifera*	*E. coli*	UDP-glucose	Geraniol	Geranyl glucoside	[162]
VvGT2	*V. vinifera*	*E. coli*	UDP-glucose	Gallic acid, sinapic acid, protocatechuic acid, *p*-coumaric acid, vanillic acid, nicotinic acid, *p*-aminobenzoic acid, ferulic acid, cinnamic acid, isophthalic acid, terephthalic acid, cyclohexanecarboxylic acid, syringic acid, salicylic acid, *p*-hydroxybenzoic acid	Glucose esters of all acceptors	[186]
VvUGT	*V. vinifera*	*E. coli*	UDP-glucuronic acid	Quercetin	Quercetin 3-*O*-glucuronide	[169]
YjiC (AAU40842)	*Bacillus licheniformis*	*E. coli*	UDP-glucose	Phloretin, genistein, biochanin A, daidzein, formononetin	Phloretin 4′-*O*-glucoside, phloretin 2′-*O*-glucoside, phloretin 4′,4-*O*-diglucoside, phloretin 2′,4′,4-*O*-triglucoside, genistein 4′-*O*-β-D-glucoside, genistein 7-*O*-β-D-glucoside, genistein 4′,7-*O*-β-D-diglucoside, biochanin A 7-*O*-β-D-glucoside, daidzein 4′-*O*-β-D-glucoside, daidzein 7-*O*-β-D-glucoside, daidzein 4′, 7-*O*-β-D-diglucoside, formononetin 7-*O*-β-D-glucoside	[187, 188]

9.6.2 Types of Whole-Cell Biocatalysts

Presently, *E. coli* is the biocatalyst of choice in most natural product glycol-diversification approaches, as it is a well-studied bacterium with many genetic tools already available [191] (Table 9.1). A broad range of natural plant products has already been successfully glycosylated in transgenic *E. coli* cells expressing GTs [192]. The flavonol quercetin was transformed to the 3-*O*-rhamoside and 3-*O*-*N*-acetylglucosamine by *E. coli* cells expressing GTs from *A. thaliana* [172, 174], and anthocyanidins [167], kaempferol [171] and 2(4-hydroxyphenyl)ethanol [182] were glycosylated employing various GT enzymes in transgenic *E. coli* cells (**see also Chap. 4 of this book**). *C*-glucosides of flavonoids and related compounds (2-hydroxyflavanone, dihydrochalcone, and trihydroxyacetophenone) were produced by *E. coli* expressing a buckwheat *C*-glucosyltransferase [178].

However, depending on the substrate compatibility or the final application, a different host organism might be of advantage. For example, *Saccharomyces cerevisiae* cells expressing GTs from *Dianthus caryophyllus* (carnation) or *Scutellaria baicalensis* have been successfully applied as whole-cell biocatalyst for the production of naringenin glucosides [177] and scutellarein 7-*O*-glucoside [164], respectively. Furthermore, yeast expressing an *Arabidopsis* GT was used for the production of zearalenone 4-*O*-glucoside [60], and *Schizosaccharomyces pompe* cells expressing human GTs were utilized for the generation of drug metabolites [180]. Additionally, the GT-mediated production of rebaudioside A from stevioside was catalyzed by an engineered *S. cerevisiae* strain [74].

Plant cells have also been used as biocatalysts in glycosylation processes. Glucose esters of cinnamic acid, 4-coumaric acid, caffeic acid and ferulic acid were produced by suspension-cultured cells of *Eucalyptus perriniana* [193], and biotransformation of raspberry ketone and zingerone to the correspondent glucosides was achieved by cultured suspension cells of *P. americana* [194]. β-Thujaplicin (hinokitiol), a tropolone derivative present in the heartwood of cupressaceous plants and used as medicine, food additive, and preservative, is transformed by cultured plant cells of *Nicotiana tabacum* to two glucosides and two gentiobiosides [195].

9.6.3 Process Optimization

Although living cells as biocatalysts provide advantages over *in vitro* cell-free systems, there are also limitations. The harvested product suspension may contain cell debris, secretion products of the host metabolism, and other by-products, which result in more intensive downstream purification. Some organisms also express their own glycosyltransferases potentially interfering by drawing co-factors and UDP-sugars, and produce unwanted side-products. Depending on the choice of host organism, only specific metabolic pathways employing specific carbon sources are available for glucoside production. However, decisive progress has been made to overcome these hindrances. Nowadays, many properties can be added to or removed from the employed host or enzyme using strategies such as those shown in Fig. 9.5.

9.6.3.1 Vector Conveyed Process Optimization

Many studies engaged in optimizing the product formation use vector construct transformation strategies (Fig. 9.5a). Glycosides produced by *E. coli* strains are mostly conjugates of glucose or galactose. However, more and more UDP-sugar modifying enzymes are being identified and investigated. Microorganisms containing plasmids that harbor one or more of these enzymes can produce a broader range of different activated sugars and corresponding glycosides. Bioactive flavonol 3-*O*-rhamnosides have been successfully synthesized via co-transformation of two vectors in *E. coli* containing a flavonol rhamnosyltransferase and a rhamnose synthase both from *A. thaliana* [172]. The heterologously expressed rhamnosyltransferase produced rhamnosides and the rhamnose synthase ensured the conversion of UDP-glucose into UDP-rhamnose, thereby increasing the product formation rate. The NDP-sugar biosynthesis circuit of *E. coli* was also shifted towards the production of flavonoid glucosides and rhamnosides by employing a multi-monocistronic synthetic vector containing multiple genes [183]. Another study employed GTs and a sucrose synthase in enzymatic cascade reactions to enable utilization of sucrose as a carbon source, and to facilitate *in-situ* recycling of the NTP-sugar donor from the NDP leaving group [162]. Correspondingly, shifting of the nucleotide sugar pathways of *E. coli* towards the production of UDP-xylose and -arabinose for GT-conveyed synthesis of flavonoid *O*-pentosides was demonstrated [86]. To produce flavonoids attached to sugars such as glucuronic acid and galactoside, *E. coli* was genetically modified to express GTs specific for UDP-glucuronic acid (AmUGT10 from *Antirrhinum majus* and VvUGT from *V. vinifera*) and UDP-galactose (PhUGT from *Petunia hybrida*), along with the appropriate nucleotide biosynthetic genes to enable simultaneous production of their substrates, UDP-glucuronic acid and UDP-galactose [169]. Using these strategies, luteolin-7-*O*-glucuronide, quercetin-3-*O*-glucuronide, and quercetin 3-*O*-galactoside were successfully synthesized.

Overall, the success of metabolic engineering depends on a balanced expression of the biosynthesis enzymes. Factors that influence protein levels include gene copy number, promotor strength, ribosomal binding site (RBS), inducer concentration and codon usage [191, 196]. Besides, coupling to a fusion partner like GST (glutathione S-transferase), NusA (N-utilization substance protein A), MBP (maltose-binding protein), Trx (thioredoxin) or SUMO (small ubiquitin-related modifier) can increase enzyme solubility and prevent accumulation of inactive protein aggregates, thus providing enhanced biocatalytic activity [191].

9.6.3.2 UGT Optimization

Another strategy to improve the product yield and diversity is to optimize enzyme function (Fig. 9.5b). Usually, the GT is chosen for a specific activity or for its promiscuity towards a spectrum of aglycon substrates. However, if an enzyme with the desired substrate specificity is not available, known GTs can be modified. For example, the protein sequence of the GT OleD from *Streptomyces antibioticus* was mutated to broaden the accepted substrate spectrum. A more than 180-fold higher activity towards the therapeutically important acceptor

7-hydroxycoumarin-4-acetic acid was achieved [197]. The most common approach to achieve improved or altered substrate specificity, usually involves mutating amino acids positioned in the active cleft of the enzyme. For example, the activity of GTs from *Panax ginseng* towards ginsenosides, bioactive natural glycosides from the ginseng plant, was altered by this approach [184]. However, other researchers demonstrated that also amino acid substitutions outside of the active cleft, but in its vicinity, can procure changed product conversion rates [53, 54], and alteration of sugar donor specificities of plant GTs [103]. The catalytic mechanism and basis for *O*- and *N*-glucosylation specificity was also probed by mutagenesis and domain shuffling [198]. Mutation of an *O*-specific GT at just two positions installed high levels of *N*-GT activity, whereas molecular modeling revealed the connectivity of these residues to the catalytically active histidine-19 on UGT72B1, with its mutagenesis exclusively defining *N*-GT activity in UGT72B1.

Furthermore, it is imperative to secure a balanced protein expression, as overexpressed enzymes are a metabolic burden for the cell. This can lead to protein aggregation or incorrect folding, in addition to subsequent formation of enzymatically inactive inclusion bodies. Accumulation of inactive protein aggregates can be prevented by co-expression of proteins that assist in folding. It has been demonstrated that co-expression of recombinant GT enzymes with chaperones resulted in an increased enzymatic activity and a lower aggregation rate [199].

9.6.3.3 Host Genome Optimization – Metabolic Engineering

Along with process optimization through extrachromosomal gene expression, modification of the host metabolism can also be achieved by direct insertion/deletion of sequences into/from the genome (Fig. 9.5c). Thereby, futile metabolic cycles can be shut down, and product utilization by the host and nonessential metabolic effluxes can be avoided. Although plasmids are a quick and easy way to test out new enzyme additions without having to design a whole new expression cassette, permanent changes by gene deletion/insertion secure a stable metabolism within the cell population. However, in most cases, a combined approach of extra- and intra-chromosomal modifications achieves the best results. Figure 9.6 illustrates how the metabolic carbon flow of *E. coli* Waksman was modified to exclusively use sucrose and fructose as cheap carbon sources for the production of phenolic glucose esters [186]. The vector conveyed introduction of the sucrose phosphorylase from *Bifidobacterium adolescentis*, and the permanent deletion of the endogenous phosphoglucomutase gene created a split in the usage of sucrose. In this manner, half of the carbon was utilized for the production of phenolic glucose esters, while the second half was spared for other cell metabolism. Another strategy to increase GT product formation is the deletion of UDP-glucose consuming pathways, e.g. the UDP-glucose hydrolase (*ushA*) in *E. coli*, to prevent metabolization of the co-substrate UDP-glucose and to promote glycosylation reactions [163, 175] (Fig. 9.7a). The production of 3-*O*-xylosyl quercetin was improved in mutant strains deficient in phosphoglucoisomerase (*pgi*), D-glucose-6-phosphate dehydrogenase (*zwf*) and ushA genes, paired with over-expressed

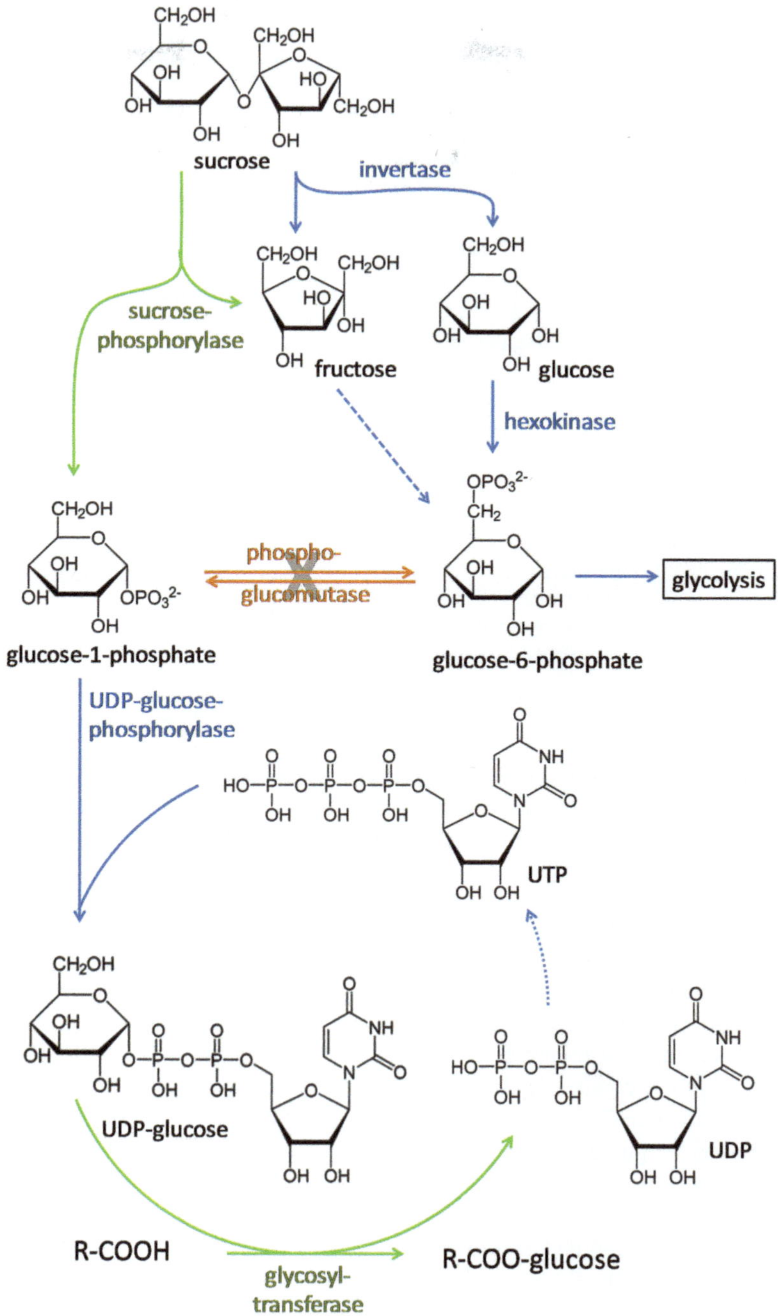

Fig. 9.6 Optimization strategy in *E. coli* Waksman for improved production of phenolic glucose esters. Introduction of sucrose phosphorylase facilitates direct formation of glucose-1-phosphate, and deletion of phosphoglucomutase prevents its degradation [186]. Blue: Endogenous *E. coli* Waksman pathway. Orange: Change procured by gene deletion. Green: Extrachromosomal gene additions

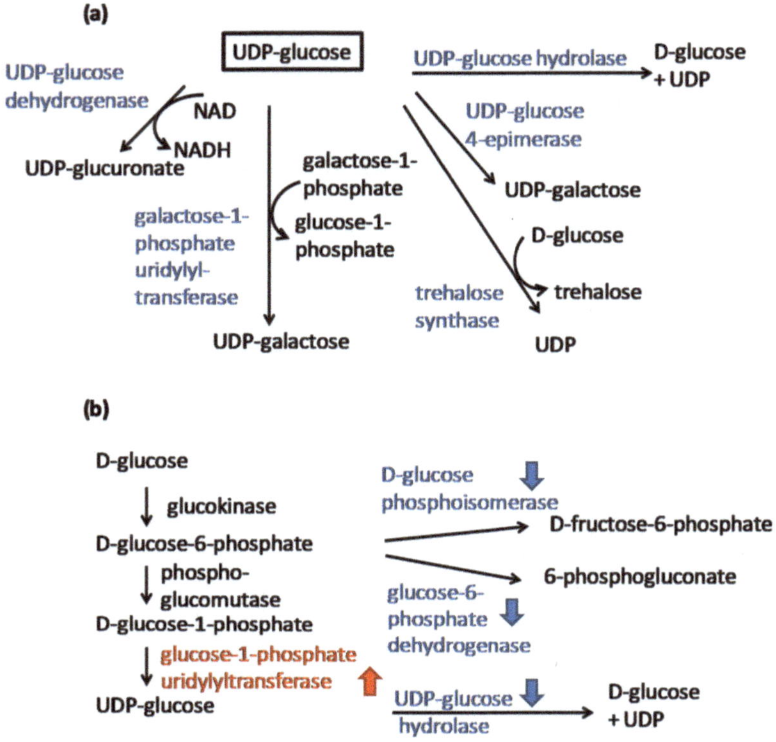

Fig. 9.7 UDP-glucose consuming enzymes of competing metabolic pathways (**a**). Schematic diagram of the UDP-glucose biosynthesis pathway showing the mutations of the engineered *E. coli* strain used for glycoside production (**b**) [170]. Up-regulated and deleted enzymes are shown in red and blue, respectively

UDP-xylose biosynthetic genes phosphoglucomutase (*pgm*), glucose 1-phosphate uridylyltransferase (*galU*), UDP-glucose dehydrogenase (*calS8*) and UDP-glucuronic acid decarboxylase (*calS9*) [170]. Thereby, the level of glucose-1-phosphate, UDP-glucose and consequently of UDP-xylose was increased (Fig. 9.7). A similar engineered *E. coli* mutant, lacking the over-expressed UDP-xylose biosynthetic genes, was further developed for the efficient whole-cell biocatalysis of flavone 7-*O*-*β*-D-glucopyranosides [168], phloretin glucosides [187] and isoflavonoid glucosides [188].

The deletion of galU can also be used to shift the metabolic pathway of *E. coli* towards the formation of alternative sugar donors, such as UDP-*N*-acetyl-D-glucosamine [174] and dTDP-6-deoxytalose [173], thereby increasing the production of novel quercetin glycosides employing extrachromosomal GTs.

Combinatorial intra- and extrachromosomal optimization approaches have recently been introduced as a successful strategy to increase product yields [167]. The authors made use of previously published advances, such as the optimization of the medium pH and the UDP-glucose supply, as well as introducing gene expression cassettes [176, 179, 200]. They combined these conditions with (i) optimized culture conditions and induction parameters, (ii) a bicistronic expression cassette for balanced co-expression of a GT and a precursor forming enzyme, and (iii) overexpression of transporter proteins for improved transportation of substrate and product across the host cell membrane. In doing so, a final titer of 350 mg/L of cyanidin 3-*O*-glucoside was achieved [167].

Aside from *E. coli*, other biocatalysts have been successfully engineered as well. For example, the metabolism of yeast strains was optimized by integration of multiple genes including GTs into the genome for *de novo* biosynthesis of ginsenosides and vanillin glucosides [14, 181, 184, 185] (**see also Chap. 1 of this book**).

9.7 Recent Applications of Glycosyltransferases for Production of Small Molecule Glycosides

Numerous remarkable characteristics arise from the glycosylation of natural products providing auspicious applications in new drug development, making it a hotspot in natural product biosynthesis and modification. Natural and biological approaches for glycosylation of aglycons to form glycosides have attracted substantial attention and interest, as it involves the formation of novel compounds under mild conditions. In comparison to 'traditional' chemical synthesis methods, biotransformation implements an environmentally friendly option for the synthesis of fine chemicals. GT enzymes that partake in the biosynthesis of natural products have demonstrated to be advantageous for the chemo-enzymatic synthesis or biosynthesis of functional compounds with new bioactivities. This enzymatic process influences their biological and chemical properties by the attachment or alteration of sugar moieties in natural or synthetic compounds. The common strategy utilizes the promiscuity feature of natural product GTs by transferring various sugar moieties to different aglycons *in-vivo* or *in-vitro*. Precise regio/stereo-selectivity and moderate reaction conditions lay the groundwork to manipulate and apply the enzymatic glycosylation of natural product GTs. In the recent years, the composed research and known features aid in the comprehensive understanding and has shed light on many GT sequences, which have been correlated with novel glycosylation reactions and perspective applications of GTs.

9.7.1 Glycosyltransferases and Glycosylation in Product for Consumer Consumption

In the 1960s, the discovery and extraction of glycol-conjugated forms of monoterpene alcohols has unraveled a new expanse of nonvolatile aroma precursors where researchers can indulge in the study of scents and flavor [4]. In general, nonvolatile, glycosylated aroma precursors produce an aroma when the glycosidic bond is removed and the molecule becomes volatile, and is then able to interact with the olfactory receptors [4, 67, 201–203]. The applications of fragrance and flavor chemical compounds have been vastly applied in food, medicine, tobacco, textiles, leather, papermaking, cosmetics, and further products for consumer consumption

[4]. Among the many aromatic compounds, vanillin, phenylethanol, benzyl acetate, linalool, menthol, and geraniol are economically important aroma chemicals, which exceed annual consumption rates of 5000 tonnes [4, 137] (**see also Chap. 1 of this book**).

Terpenes are volatile, unsaturated hydrocarbons that serve as essential oil constituents of many plants and are often used as natural flavor additives for food, fragrances in perfumery, as well as medicine, and aromatherapy [4, 204]. A number of volatile hemiterpenoids (C5), monoterpenoids (C10), and sesquiterpenoids (C15) greatly contribute to the odor of a product. At the same time, higher terpenoids can contribute to the perception of taste. As an example, steviosides are sweet tasting glycosides of the diterpenoid steviol extracted from *Stevia rebaudiana*, a herb from the *Asteraceae* family [205]. These glycosylated secondary metabolites are bioactive constituents of the commonly used sweetener, *Stevia*. Aside from other metabolites found and glycosylated in the fruit tissue, the steviol glycosides are biosynthesized in the leaf tissue [206]. Steviol glycoside synthesis commences with steviol, which is a product from the MEP pathway (Fig. 9.2). The subsequent four steps include the addition of sugar molecules to the carbon backbone catalyzed by GTs. The final product of this multi-step process is the glycosylated rebaudioside A, which is transported to the vacuole for storage [207]. Previous studies of the various enzymes involved in steviol glucosylation led to the isolation of two protein fractions from *S. rebaudiana* leaves revealing significant GT activity [207, 208]. Each enzyme possessed a unique activity; one that exhibited high specificity and ability to catalyze the transfer of UDP-glucose to steviol and subsequent glycosides, and a second that exhibited low specificity and activity. Interestingly, both enzymes were found to also accept flavonol substrates, such as kaempferol, quercetin, and hydroquinone [206]. The different properties of the individual glycosides can vary significantly, and their characteristics are determined by the type of sugar and pattern of glycosylation. Among various tested steviolglycosides, rebaudioside A was ascertained as the sweetest tasting and least bitter glycoside [73]. Thus, several biotechnological processes involving GTs have already been suggested to manipulate the glycosylation status of steviosides [70–74].

Similarly, a plant GT partially purified from pomelo fruits has been immobilized and used in bioprocessing for de-bittering citrus juice by converting the astringent triterpene limonoid into glycosides [209].

Monoterpenes (linalool, menthol, and geraniol) are volatile and predominantly poorly water-soluble flavor and fragrance compounds, thus limiting their utility for industrial applications. The glycosylation of these monoterpenes enhances their water-solubility, thus rendering them odorless. Therefore, they are of interest for the flavor and fragrance industry due to the possibility of a controlled release of the bound aroma compound upon cleavage of the glycosidic bond. Moreover, monoterpene glycosides have been proposed to inhibit unpleasant odor for various products, enhance the quality of cigarette smoke, augment the aroma of freshly cut flowers, and improve longer-lasting deodorants. Besides, they can be utilized in personal hygiene products for continuous release of scents and as air fresheners [4]. Utilizing odorless volatiles bound via a glycosidic bond as fragrance materials was exemplified by incubating skin microflora with numerous glycosides. The majority of the glycosidically bound volatiles, in particular β-D-glucosides of monoterpenes, are hydrolyzed by glucosidases excreted by the skin microbiome, thereby releasing the aglycones as fragrance ingredients [210, 211].

GTs are mainly involved in natural product glycosylations via *O*-glycosidic bonds. In contrast, glycosylation at the carbon atom (*C*-glycosylations) is quite rare, and only a few *C*-GT enzymes are therefore available [155]. Aryl glycosides are derived from aglycons resembling flavonoids, which denote either an *O*- or *C*-glycosidic bond [212]. These glycosides have been proposed to show significant biological activities, including antioxidant properties, antiviral, and cytotoxic impacts [213, 214]. Commonly, direct isolation from plants is impractical and chemical methodologies involve a series of steps resulting in poor yield and toxic byproducts. Therefore, a single-step reaction catalyzed by GTs *in-vitro* represents a powerful tool for the synthesis of aryl *C*-glycosides [215]. Nothofagin (3-*C*-glucoside of phloretin) is a natural secondary metabolite found in redbush herbal tea with applications in the food industry [216]. For a proficient synthesis of nothofagin, a *C*-GT reaction was coupled to an enzymatic supply of glucosyl donor substrate *in-situ.* This step-by-step reaction results in efficient and high-yielding bio-catalytic production of nothofagin [215].

9.7.2 Glycorandomization

GT enzymes with broad substrate specificity have been utilized in the development of powerful tools for glycorandomization [217]. This has led to the development of an *E. coli* platform for the combinational biosynthesis of antibiotics where numerous new glycoderivates were generated, some of which may become valuable drug candidates [218]. Another '*in-vitro* glycorandomization' experiment is based on the flexibility of glycosyltransferases fD and fE (GtfD, GtfE). These two GTs were used on NDP-sugar libraries to generate glycorandomized natural products and then the applied method of chemo-selective ligation produced monoglycosylated vancomycins. The obtained products' bioactivity varied significantly and at the same time possessed notably improved antibacterial properties [219].

9.7.3 Glycosyltransferases and Their Role in Cancer Therapies

In the 1970s, mitoxantrole (MXT), a synthetic anthracenedione, was discovered and developed for the treatment of various human cancers [220, 221]. Additionally, MXT has been proposed to exhibit *in-vivo* activity against rheumatoid arthritis in several animal models [222]. Moreover, MXT supposedly acts as anti-tuberculosis agent, functioning through the inhibition of a specific mycobacterial kinase (PknB) controlling pathogen growth and evolvement [223]. However, MXT treatments are associated with various serious side effects such as irreversible cardiomyopathy. Therefore, stimulating research to generate analogues with diverse therapeutic effects, such as increased potency and reduced cardiotoxicity, has been conducted [224–226]. The recently identified GT OleD produced by *S. antibioticus* catalyzes the transfer of a glucose moiety from UDP-Glucose to various macrolide antibiotics [227]. Protein engineering of OleD and subsequent screening for variants capable of executing extended glycosylation reactions has led to the discovery of a promiscuous triple mutant, OleD-ASP [228]. Initial liquid chromatography mass-spectrometry (LC-MS) evaluation of the OleD-ASP acceptor flexibility revealed its ability to glycosylate a range of pharmaceuticals including anthraquinones, indolocarbozoles, polyenes, cardenolides, steroids, macrolides, beta-lactams, and enediyenes [217]. As engineered GT with unique promiscuity, OleD-ASP, was shown to regio- and stereo-selectively modify and glycosylate also MXT, thus providing an asymmetric MXT 4′-β-D-glucoside [228]. Interestingly, OleD-ASP is the first engineered GT that is able to asymmetrically glycosylate an anticancer drug whilst retaining its activity. At the same time, this is a single-step reaction requiring no protecting groups or sugar activation guidance. The single glucoside of MXT may potentially offer a beneficial toxicity profile leading to a reduction of side effects with the potential to be further optimized [228].

9.8 Conclusions and Future Prospects

Synthetic biology has been driven by the development of new powerful tools for DNA synthesis, sequencing and genome editing, which enable microbial engineering for the production of pharmaceuticals and other high-value chemicals. There is great diversity in the type and number of sugar units that could be added to naturally occurring aglycones, and the biological relevance remains elusive [229]. It is anticipated, that the attachment of sugars to natural products, or altering an existing sugar moiety, can improve pharmacological properties and specificity at multiple levels.

Glycotechnologies have been used for the generation of novel glycosylated compounds either in *in-vitro* or *in-vivo* experiments. Several GTs are suitable for altering glycosylation patterns, but strict substrate specificity remains a limiting factor in natural product diversification. Engineering GTs is the most promising way to discover and develop new GTs with clearly defined specificities. Suitable high-throughput screening systems will support glycodiversification technologies. Further structure elucidation of GTs will help to understand the mode of action of these enzymes [229]. Modifying the specificity of current natural product GTs and enhancing the biosynthetic technologies in the discovery of new GTs to improve the yields of applicable natural products, is increasingly becoming of utmost importance for researchers.

References

1. André I, Potocki-Véronèse G, Barbe S, Moulis C, Remaud-Siméon M. CAZyme discovery and design for sweet dreams. Curr Opin Chem Biol. 2014;19:17–24. https://doi.org/10.1016/j.cbpa.2013.11.014.
2. Huber GW, Chheda JN, Barrett CJ, Dumesic JA. Production of liquid alkanes by aqueous-phase processing of biomass-derived carbohydrates. Science. 2005;308:1446–50. https://doi.org/10.1126/science.1111166.
3. Singh S, Phillips GN Jr, Thorson JS. The structural biology of enzymes involved in natural product glycosylation. Nat Prod Rep. 2012;29:1201–37. https://doi.org/10.1039/c2np20039b.
4. Schwab W, Fischer TC, Giri A, Wüst M. Potential applications of glucosyltransferases in terpene glucoside production: impacts on the use of aroma and fragrance. Appl Microbiol Biotechnol. 2014;99:165–74. https://doi.org/10.1007/s00253-014-6229-y.
5. Desmet T, Soetaert W, Bojarová P, Křen V, Dijkhuizen L, Eastwick-Field V, et al. Enzymatic glycosylation of small molecules: challenging substrates require tailored catalysts. Chem Eur J. 2012;18:10786–801. https://doi.org/10.1002/chem.201103069.
6. Wang J, Ma X, Kojima M, Sakakibara H, Hou B. *N*-Glucosyltransferase UGT76C2 is involved in cytokinin homeostasis and cytokinin response in *Arabidopsis thaliana*. Plant Cell Physiol. 2011;52:2200–13. https://doi.org/10.1093/pcp/pcr152.
7. Kopycki J, Wieduwild E, Kohlschmidt J, Brandt W, Stepanova AN, Alonso JM, et al. Kinetic analysis of *Arabidopsis* glucosyltransferase UGT74B1 illustrates a general mechanism by which enzymes can escape product inhibition. Biochem J. 2013;450:37. https://doi.org/10.1042/BJ20121403.
8. Hirade Y, Kotoku N, Terasaka K, Saijo-Hamano Y, Fukumoto A, Mizukami H. Identification and functional analysis of 2-hydroxyflavanone *C*-glucosyltransferase in soybean (*Glycine max*). FEBS Lett. 2015;589:1778–86. https://doi.org/10.1016/j.febslet.2015.05.010.
9. Schwab W, Fischer T, Wüst M. Terpene glucoside production: improved biocatalytic processes using glycosyltransferases. Eng Life Sci. 2015;15:376–86. https://doi.org/10.1002/elsc.201400156.
10. Lim E, Higgins GS, Li Y, Bowles DJ. Regioselectivity of glucosylation of caffeic acid by a UDP-glucose:glucosyltransferase is maintained in planta. Biochem J. 2003;373:987–92. https://doi.org/10.1042/BJ20021453.
11. Jackson R, Knisley D, McIntosh C, Pfeiffer P. Predicting flavonoid UGT regioselectivity. Adv Bioinforma. 2011:506583. https://doi.org/10.1155/2011/506583.
12. Graefe EU, Wittig J, Mueller S, Riethling A, Uehleke B, Drewelow B, et al. Pharmacokinetics and bioavailability of quercetin glycosides in humans. J Clin Pharmacol. 2001;41:492–9. https://doi.org/10.1177/00912700122010366.
13. Makino T, Shimizu R, Kanemaru M, Suzuki Y, Moriwaki M, Mizukami H. Enzymatically modified isoquercitrin, α-oligoglucosyl quercetin 3-*O*-glucoside, is absorbed more easily than other quercetin glycosides or aglycone after oral administration in rats. Biol Pharm Bull. 2009;32:2034–40. https://doi.org/10.1248/bpb.32.2034.
14. Brochado AR, Matos C, Møller BL, Hansen J, Mortensen UH, Patil KR. Improved vanillin production in baker's yeast through in silico design. Microb Cell Factories. 2010;9:84. https://doi.org/10.1186/1475-2859-9-84.
15. Yang C, Tanaka O. Advances in plant glycosides, chemistry and biology: proceedings of the international symposium on plant glycosides, Aug 12–15, 1997 Kunming, China. 1st edn. Amsterdam/New York: Elsevier; 1999.
16. Woo H, Kang H, Nguyen TTH, Kim G, Kim Y, Park J, et al. Synthesis and characterization of ampelopsin glucosides using dextransucrase from *Leuconostoc mesenteroides* B-1299CB4: glucosylation enhancing physicochemical properties. Enzym Microb Technol. 2012;51:311–8. https://doi.org/10.1016/j.enzmictec.2012.07.014.
17. Hofer B. Recent developments in the enzymatic *O*-glycosylation of flavonoids. Appl Microbiol Biotechnol. 2016;100:4269–81. https://doi.org/10.1007/s00253-016-7465-0.

18. Grotewold E. The science of flavonoids. New York: Springer; 2006.
19. Lepak A, Gutmann A, Kulmer ST, Nidetzky B. Creating a water-soluble resveratrol-based antioxidant by site-selective enzymatic glucosylation. Chembiochem. 2015;16:1870–4. https://doi.org/10.1002/cbic.201500284.
20. Fuggetta M, Mattivi F. The immunomodulating activities of resveratrol glucosides in humans. Recent Pat Food Nutr Agric. 2011;3:81–90.
21. Pandey RP, Parajuli P, Shin JY, Lee J, Lee S, Hong Y, et al. Enzymatic biosynthesis of novel resveratrol glucoside and glycoside derivatives. Appl Environ Microbiol. 2014;80:7235–43. https://doi.org/10.1128/AEM.02076-14.
22. Chung MJ, Kang A, Lee KM, Oh E, Jun H, Kim S, et al. Water-soluble genistin glycoside isoflavones up-regulate antioxidant metallothionein expression and scavenge free radicals. J Agric Food Chem. 2006;54:3819–26. https://doi.org/10.1021/jf060510y.
23. Haskins AH, Su C, Engen A, Salinas VA, Maeda J, Uesaka M, et al. Data for induction of cytotoxic response by natural and novel quercetin glycosides. Data Brief. 2016;6:262–6. https://doi.org/10.1016/j.dib.2015.11.066.
24. Yonekura-Sakakibara K, Nakayama T, Yamazaki M, Saito K. Modification and stabilization of anthocyanins. In: Winefield C, Davies K, Gould K, editors. Anthocyanins: biosynthesis, functions, and applications. New York: Springer; 2009. p. 169–90. https://doi.org/10.1007/978-0-387-77335-3_6.
25. Sasaki N, Nishizaki Y, Ozeki Y, Miyahara T. The role of acyl-glucose in anthocyanin modifications. Molecules. 2014;19:18747–66. https://doi.org/10.3390/molecules191118747.
26. Yoshida K, Mori M, Kondo T. Blue flower color development by anthocyanins: from chemical structure to cell physiology. Nat Prod Rep. 2009;26:884–915. https://doi.org/10.1039/b800165k.
27. Mäkilä L, Laaksonen O, Alanne A, Kortesniemi M, Kallio H, Yang B. Stability of hydroxycinnamic acid derivatives, flavonol glycosides, and anthocyanins in black currant juice. J Agric Food Chem. 2016;64:4584–98. https://doi.org/10.1021/acs.jafc.6b01005.
28. Roscher R, Schwab W, Schreier P. Stability of naturally occurring 2,5-dimethyl-4-hydroxy-3[2H]-furanone derivatives. Z Lebensm Unters F A. 1997;204:438–41. https://doi.org/10.1007/s002170050109.
29. Saslowsky D, Winkel-Shirley B. Localization of flavonoid enzymes in *Arabidopsis* roots. Plant J. 2001;27:37–48. https://doi.org/10.1046/j.1365-313x.2001.01073.x.
30. Grotewold E, Davies K. Trafficking and sequestration of anthocyanins. Nat Prod Commun. 2008;3:1251–8.
31. Petrussa E, Braidot E, Zancani M, Peresson C, Bertolini A, Patui S, et al. Plant flavonoids – biosynthesis, transport and involvement in stress responses. Int J Mol Sci. 2013;14:14950–73. https://doi.org/10.3390/ijms140714950.
32. Kitamura S, Shikazono N, Tanaka A. TRANSPARENT TESTA 19 is involved in the accumulation of both anthocyanins and proanthocyanidins in *Arabidopsis*. Plant J. 2004;37:104–14. https://doi.org/10.1046/j.1365-313X.2003.01943.x.
33. Conn S, Curtin C, Bézier A, Franco C, Zhang W. Purification, molecular cloning, and characterization of glutathione S-transferases (GSTs) from pigmented *Vitis vinifera* L. cell suspension cultures as putative anthocyanin transport proteins. J Exp Bot. 2008;59:3621–34. https://doi.org/10.1093/jxb/ern217.
34. Sun Y, Li H, Huang J. *Arabidopsis* TT19 functions as a carrier to transport anthocyanin from the cytosol to tonoplasts. Mol Plant. 2012;5:387–400. https://doi.org/10.1093/mp/ssr110.
35. Härtl K, Denton A, Franz-Oberdorf K, Hoffmann T, Spornraft M, Usadel B, et al. Early metabolic and transcriptional variations in fruit of natural white-fruited *Fragaria vesca* genotypes. Sci Rep. 2017;7:45113. https://doi.org/10.1038/srep45113.
36. Gomez C, Conejero G, Torregrosa L, Cheynier V, Terrier N, Ageorges A. *In vivo* grapevine anthocyanin transport involves vesicle-mediated trafficking and the contribution of anthoMATE transporters and GST. Plant J. 2011;67:960–70. https://doi.org/10.1111/j.1365-313X.2011.04648.x.

37. Poustka F, Irani NG, Feller A, Lu Y, Pourcel L, Frame K, et al. A trafficking pathway for anthocyanins overlaps with the endoplasmic reticulum-to-vacuole protein-sorting route in *Arabidopsis* and contributes to the formation of vacuolar inclusions. Plant Physiol. 2007;145:1323–35. https://doi.org/10.1104/pp.107.105064.
38. Goodman CD, Casati P, Walbot V. A multidrug resistance–associated protein involved in anthocyanin transport in *Zea mays*. Plant Cell. 2004;16:1812–26. https://doi.org/10.1105/tpc.022574.
39. Francisco RM, Regalado A, Ageorges A, Burla BJ, Bassin B, Eisenach C, et al. ABCC1, an ATP binding cassette protein from grape berry, transports anthocyanidin 3-*O*-glucosides. Plant Cell. 2013;25:1840–54. https://doi.org/10.1105/tpc.112.102152.
40. Marinova K, Pourcel L, Weder B, Schwarz M, Barron D, Routaboul J, et al. The *Arabidopsis* MATE transporter TT12 acts as a vacuolar flavonoid/H+-antiporter active in proanthocyanidin-accumulating cells of the seed coat. Plant Cell. 2007;19:2023–38. https://doi.org/10.1105/tpc.106.046029.
41. Gomez C, Terrier N, Torregrosa L, Vialet S, Fournier-Level A, Verriès C, et al. Grapevine MATE-type proteins act as vacuolar H+-dependent acylated anthocyanin transporters. Plant Physiol. 2009;150:402–15. https://doi.org/10.1104/pp.109.135624.
42. Zhao J, Huhman D, Shadle G, He X, Sumner LW, Tang Y, et al. MATE2 mediates vacuolar sequestration of flavonoid glycosides and glycoside malonates in *Medicago truncatula*. Plant Cell. 2011;23:1536–55. https://doi.org/10.1105/tpc.110.080804.
43. Burla B, Pfrunder S, Nagy R, Francisco RM, Lee Y, Martinoia E. Vacuolar transport of abscisic acid glucosyl ester is mediated by ATP-binding cassette and proton-antiport mechanisms in *Arabidopsis*. Plant Physiol. 2013;163:1446–58. https://doi.org/10.1104/pp.113.222547.
44. Chanoca A, Kovinich N, Burkel B, Stecha S, Bohorquez-Restrepo A, Ueda T, et al. Anthocyanin vacuolar inclusions form by a microautophagy mechanism. Plant Cell. 2015;27:2545–59. https://doi.org/10.1105/tpc.15.00589.
45. Passamonti S, Cocolo A, Braidot E, Petrussa E, Peresson C, Medic N, et al. Characterization of electrogenic bromosulfophthalein transport in carnation petal microsomes and its inhibition by antibodies against bilitranslocase. FEBS J. 2005;272:3282–96. https://doi.org/10.1111/j.1742-4658.2005.04751.x.
46. Cutler SR, Rodriguez PL, Finkelstein RR, Abrams SR. Abscisic acid: emergence of a core signaling network. Annu Rev Plant Biol. 2010;61:651–79.
47. Nambara E, Marion-Poll A. Abscisic acid biosynthesis and catabolism. Annu Rev Plant Biol. 2005;56:165–85. https://doi.org/10.1146/annurev.arplant.56.032604.144046.
48. Xu Z, Kim DH, Hwang I. ABA homeostasis and signaling involving multiple subcellular compartments and multiple receptors. Plant Cell Rep. 2013;32:807–13. https://doi.org/10.1007/s00299-013-1396-3.
49. Kushiro T, Okamoto M, Nakabayashi K, Yamagishi K, Kitamura S, Asami T, et al. The *Arabidopsis* cytochrome P450 CYP707A encodes ABA 8′-hydroxylases: key enzymes in ABA catabolism. EMBO J. 2004;23:1647–56. https://doi.org/10.1038/sj.emboj.7600121.
50. Wang P, Liu H, Hua H, Wang L, Song C. A vacuole localized β-glucosidase contributes to drought tolerance in *Arabidopsis*. Chin Sci Bull. 2011;56:3538–46. https://doi.org/10.1007/s11434-011-4802-7.
51. Xu ZY, Lee KH, Dong T, Jeong JC, Jin JB, Kanno Y, et al. A vacuolar beta-glucosidase homolog that possesses glucose-conjugated abscisic acid hydrolyzing activity plays an important role in osmotic stress responses in *Arabidopsis*. Plant Cell. 2012;24:2184–99. https://doi.org/10.1105/tpc.112.095935.
52. Lätari K, Wust F, Hubner M, Schaub P, Beisel KG, Matsubara S, et al. Tissue-specific apocarotenoid glycosylation contributes to carotenoid homeostasis in *Arabidopsis* leaves. Plant Physiol. 2015;168:1550–62. https://doi.org/10.1104/pp.15.00243.
53. Bönisch F, Frotscher J, Stanitzek S, Ruhl E, Wust M, Bitz O, et al. A UDP-glucose:monoterpenol glucosyltransferase adds to the chemical diversity of the grapevine metabolome. Plant Physiol. 2014;165:561–81. https://doi.org/10.1104/pp.113.232470.

54. Schulenburg K, Feller A, Hoffmann T, Schecker JH, Martens S, Schwab W. Formation of β-glucogallin, the precursor of ellagic acid in strawberry and raspberry. J Exp Bot. 2016;67:2299–308. https://doi.org/10.1093/jxb/erw036.
55. Khater F, Fournand D, Vialet S, Meudec E, Cheynier V, Terrier N. Identification and functional characterization of cDNAs coding for hydroxybenzoate/hydroxycinnamate glucosyltransferases co-expressed with genes related to proanthocyanidin biosynthesis. J Exp Bot. 2012;63:1201–14. https://doi.org/10.1093/jxb/err340.
56. Mittasch J, Böttcher C, Frolova N, Bonn M, Milkowski C. Identification of UGT84A13 as a candidate enzyme for the first committed step of gallotannin biosynthesis in pedunculate oak (*Quercus robur*). Phytochemistry. 2014;99:44–51. https://doi.org/10.1016/j.phytochem.2013.11.023.
57. Ono NN, Qin X, Wilson AE, Li G, Tian L. Two UGT84 family glycosyltransferases catalyze a critical reaction of hydrolyzable tannin biosynthesis in pomegranate (*Punica granatum*). PLoS One. 2016;11:e0156319. https://doi.org/10.1371/journal.pone.0156319.
58. Park HS, Stierle A, Strobel GA. Metabolism of maculosin, a host-specific phytotoxin produced by *Alternaria alternata* on spotted knapweed (*Centaurea maculosa*). Phytochemistry. 1993;35:101–6. https://doi.org/10.1016/S0031-9422(00)90516-8.
59. Poppenberger B, Berthiller F, Lucyshyn D, Sieberer T, Schuhmacher R, Krska R, et al. Detoxification of the Fusarium mycotoxin deoxynivalenol by a UDP-glucosyltransferase from *Arabidopsis thaliana*. J Biol Chem. 2003;278:47905–14. https://doi.org/10.1074/jbc.M307552200.
60. Poppenberger B, Berthiller F, Bachmann H, Lucyshyn D, Peterbauer C, Mitterbauer R, et al. Heterologous expression of *Arabidopsis* UDP-glucosyltransferases in *Saccharomyces cerevisiae* for production of zearalenone-4-*O*-glucoside. Appl Environ Microbiol. 2006;72:4404–10. https://doi.org/10.1128/AEM.02544-05.
61. Berthiller F, Crews C, Dall'Asta C, Saeger SD, Haesaert G, Karlovsky P, et al. Masked mycotoxins: a review. Mol Nutr Food Res. 2013;57:165–86. https://doi.org/10.1002/mnfr.201100764.
62. Høj P, Pretorius I, Blair R. The Australian wine research institute, Annual Report. Eds; The Australian wine research institute: Adelaide, Australia 2003:7–38.
63. Kennison KR, Wilkinson KL, Williams HG, Smith JH, Gibberd MR. Smoke-derived taint in wine: effect of postharvest smoke exposure of grapes on the chemical composition and sensory characteristics of wine. J Agric Food Chem. 2007;55:10897–901. https://doi.org/10.1021/jf072509k.
64. Hayasaka Y, Baldock GA, Pardon KH, Jeffery DW, Herderich MJ. Investigation into the formation of guaiacol conjugates in berries and leaves of grapevine *Vitis vinifera* L. Cv. cabernet sauvignon using stable isotope tracers combined with HPLC-MS and MS/MS analysis. J Agric Food Chem. 2010;58:2076–81. https://doi.org/10.1021/jf903732p.
65. Hayasaka Y, Dungey KA, Baldock GA, Kennison KR, Wilkinson KL. Identification of a β-D-glucopyranoside precursor to guaiacol in grape juice following grapevine exposure to smoke. Anal Chim Acta. 2010;660:143–8. https://doi.org/10.1016/j.aca.2009.10.039.
66. Kelly D, Zerihun A, Hayasaka Y, Gibberd M. Winemaking practice affects the extraction of smoke-borne phenols from grapes into wines. Aust J Grape Wine Res. 2014;20:386–93.
67. Mayr CM, Parker M, Baldock GA, Black CA, Pardon KH, Williamson PO, et al. Determination of the importance of in-mouth release of volatile phenol glycoconjugates to the flavor of smoke-tainted wines. J Agric Food Chem. 2014;62:2327–36. https://doi.org/10.1021/jf405327s.
68. Parker M, Osidacz P, Baldock GA, Hayasaka Y, Black CA, Pardon KH, et al. Contribution of several volatile phenols and their glycoconjugates to smoke-related sensory properties of red wine. J Agric Food Chem. 2012;60:2629–37. https://doi.org/10.1021/jf2040548.
69. Hellfritsch C, Brockhoff A, Stähler F, Meyerhof W, Hofmann T. Human psychometric and taste receptor responses to steviol glycosides. J Agric Food Chem. 2012;60:6782–93. https://doi.org/10.1021/jf301297n.

70. Wang Y, Chen L, Li Y, Li Y, Yan M, Chen K, et al. Efficient enzymatic production of rebaudioside A from stevioside. Biosci Biotechnol Biochem. 2016;80:67–73. https://doi.org/10.1080/09168451.2015.1072457.
71. Olsson K, Carlsen S, Semmler A, Simón E, Mikkelsen MD, Møller BL. Microbial production of next-generation stevia sweeteners. Microb Cell Factories. 2016;15:207. https://doi.org/10.1186/s12934-016-0609-1.
72. Yang Y, Huang S, Han Y, Yuan H, Gu C, Zhao Y. Base substitution mutations in uridinediphosphate-dependent glycosyltransferase 76G1 gene of *Stevia rebaudiana* causes the low levels of rebaudioside A: mutations in UGT76G1, a key gene of steviol glycosides synthesis. Plant Physiol Biochem. 2014;80:220–5. https://doi.org/10.1016/j.plaphy.2014.04.005.
73. Madhav H, Bhasker S, Chinnamma M. Functional and structural variation of uridine diphosphate glycosyltransferase (UGT) gene of *Stevia rebaudiana*–UGTSr involved in the synthesis of rebaudioside A. Plant Physiol Biochem. 2013;63:245–53. https://doi.org/10.1016/j.plaphy.2012.11.029.
74. Li Y, Li Y, Wang Y, Chen L, Yan M, Chen K, et al. Production of rebaudioside A from stevioside catalyzed by the engineered *Saccharomyces cerevisiae*. Appl Biochem Biotechnol. 2016;178:1586–98. https://doi.org/10.1007/s12010-015-1969-4.
75. Demchenko AV. General aspects of the glycosidic bond formation. In: Demchenko AV, editor. Handbook of chemical glycosylation: advances in stereoselectivity and therapeutic relevance. Weinheim: Wiley-VCH Verlag GmbH & Co. KGaA; 2008. p. 1–27.
76. Koenigs W, Knorr E. Ueber einige Derivate des Traubenzuckers und der Galactose. Ber Dtsch Chem Ges. 1901;34:957–81. https://doi.org/10.1002/cber.190103401162.
77. Křen V, Thiem J. Glycosylation employing bio-systems: from enzymes to whole cells. Chem Soc Rev. 1997;26:463–73. https://doi.org/10.1039/CS9972600463.
78. Lim E. Plant glycosyltransferases: their potential as novel biocatalysts. Chem Eur J. 2005;11:5486–94. https://doi.org/10.1002/chem.200500115.
79. Sinnott ML. Catalytic mechanism of enzymic glycosyl transfer. Chem Rev. 1990;90:1171–202. https://doi.org/10.1021/cr00105a006.
80. Campbell JA, Davies GJ, Bulone V, Henrissat B. A classification of nucleotide-diphospho-sugar glycosyltransferases based on amino acid sequence similarities. Biochem J. 1997;326:929–39.
81. Coutinho PM, Deleury E, Davies GJ, Henrissat B. An evolving hierarchical family classification for glycosyltransferases. J Mol Biol. 2003;328:307–17. https://doi.org/10.1016/S0022-2836(03)00307-3.
82. Lairson LL, Henrissat B, Davies GJ, Withers SG. Glycosyltransferases: structures, functions, and mechanisms. Annu Rev Biochem. 2008;77:521–55. https://doi.org/10.1146/annurev.biochem.76.061005.092322.
83. Ekstrom A, Taujale R, McGinn N, Yin Y. PlantCAZyme: a database for plant carbohydrate-active enzymes. Database (Oxford). 2014. https://doi.org/10.1093/database/bau079.
84. Mackenzie PI, Rogers A, Treloar J, Jorgensen BR, Miners JO, Meech R. Identification of UDP glycosyltransferase 3A1 as a UDP *N*-acetylglucosaminyltransferase. J Biol Chem. 2008;283:36205–10. https://doi.org/10.1074/jbc.M807961200.
85. Shibuya M, Nishimura K, Yasuyama N, Ebizuka Y. Identification and characterization of glycosyltransferases involved in the biosynthesis of soyasaponin I in *Glycine max*. FEBS Lett. 2010;584:2258–64. https://doi.org/10.1016/j.febslet.2010.03.037.
86. Han SH, Kim BG, Yoon JA, Chong Y, Ahn J. Synthesis of flavonoid *O*-pentosides by *Escherichia coli* through engineering of nucleotide sugar pathways and glycosyltransferase. Appl Environ Microbiol. 2014;80:2754–62. https://doi.org/10.1128/AEM.03797-13.
87. Montefiori M, Espley RV, Stevenson D, Cooney J, Datson PM, Saiz A, et al. Identification and characterisation of F3GT1 and F3GGT1, two glycosyltransferases responsible for anthocyanin biosynthesis in red-fleshed kiwifruit (*Actinidia chinensis*). Plant J. 2011;65:106–18. https://doi.org/10.1111/j.1365-313X.2010.04409.x.

88. Unligil UM, Zhou S, Yuwaraj S, Sarkar M, Schachter H, Rini JM. X-ray crystal structure of rabbit *N*-acetylglucosaminyltransferase I: catalytic mechanism and a new protein superfamily. EMBO J. 2000;19:5269–80. https://doi.org/10.1093/emboj/19.20.5269.
89. Hiromoto T, Honjo E, Tamada T, Noda N, Kazuma K, Suzuki M, et al. Crystal structure of UDP-glucose:anthocyanidin 3-*O*-glucosyltransferase from *Clitoria ternatea*. J Synchrotron Radiat. 2013;20:894–8. https://doi.org/10.1107/S0909049513020712.
90. Lizak C, Gerber S, Numao S, Aebi M, Locher KP. X-ray structure of a bacterial oligosaccharyltransferase. Nature. 2011;474:350–5. https://doi.org/10.1038/nature10151.
91. Zhang H, Zhu F, Yang T, Ding L, Zhou M, Li J, et al. The highly conserved domain of unknown function 1792 has a distinct glycosyltransferase fold. Nat Commun. 2014;5:4339. https://doi.org/10.1038/ncomms5339.
92. Liang D, Liu J, Wu H, Wang B, Zhu H, Qiao J. Glycosyltransferases: mechanisms and applications in natural product development. Chem Soc Rev. 2015;44:8350–74. https://doi.org/10.1039/c5cs00600g.
93. Busch C, Hofmann F, Selzer J, Munro S, Jeckel D, Aktories K. A common motif of eukaryotic glycosyltransferases is essential for the enzyme activity of large clostridial cytotoxins. J Biol Chem. 1998;273:19566–72. https://doi.org/10.1074/jbc.273.31.19566.
94. Charnock SJ, Davies GJ. Structure of the nucleotide-diphospho-sugar transferase, SpsA from *Bacillus subtilis*, in native and nucleotide-complexed forms. Biochemistry. 1999;38:6380–5. https://doi.org/10.1021/bi990270y.
95. Vrielink A, Rüger W, Driessen HP, Freemont PS. Crystal structure of the DNA modifying enzyme beta-glucosyltransferase in the presence and absence of the substrate uridine diphosphoglucose. EMBO J. 1994;13:3413–22.
96. Morera S, Lariviere L, Kurzeck J, Aschke-Sonnenborn U, Freemont PS, Janin J, et al. High resolution crystal structures of T4 phage β-glucosyltransferase: induced fit and effect of substrate and metal binding. J Mol Biol. 2001;311:569–77. https://doi.org/10.1006/jmbi.2001.4905.
97. Lovering AL, de Castro LH, Liza H, Lim D, Strynadka NCJ. Structural insight into the transglycosylation step of bacterial cell-wall biosynthesis. Science. 2007;315:1402–5. https://doi.org/10.1126/science.1136611.
98. Yuan Y, Barrett D, Zhang Y, Kahne D, Sliz P, Walker S. Crystal structure of a peptidoglycan glycosyltransferase suggests a model for processive glycan chain synthesis. Proc Natl Acad Sci U S A. 2007;104:5348–53. https://doi.org/10.1073/pnas.0701160104.
99. Igura M, Maita N, Kamishikiryo J, Yamada M, Obita T, Maenaka K, et al. Structure-guided identification of a new catalytic motif of oligosaccharyltransferase. EMBO J. 2008;27:234–43. https://doi.org/10.1038/sj.emboj.7601940.
100. Bowles D, Lim E, Poppenberger B, Vaistij FE. Glycosyltransferases of lipophilic small molecules. Annu Rev Plant Biol. 2006;57:567–97. https://doi.org/10.1146/annurev.arplant.57.032905.105429.
101. Gachon CMM, Langlois-Meurinne M, Saindrenan P. Plant secondary metabolism glycosyltransferases: the emerging functional analysis. Trends Plant Sci. 2005;10:542–9. https://doi.org/10.1016/j.tplants.2005.09.007.
102. Hans J, Brandt W, Vogt T. Site-directed mutagenesis and protein 3D-homology modelling suggest a catalytic mechanism for UDP-glucose-dependent betanidin 5-*O*-glucosyltransferase from *Dorotheanthus bellidiformis*. Plant J. 2004;39:319–33. https://doi.org/10.1111/j.1365-313X.2004.02133.x.
103. Kubo A, Arai Y, Nagashima S, Yoshikawa T. Alteration of sugar donor specificities of plant glycosyltransferases by a single point mutation. Arch Biochem Biophys. 2004;429:198–203. https://doi.org/10.1016/j.abb.2004.06.021.
104. Dixon RA, Achnine L, Kota P, Liu C, Reddy MSS, Wang L. The phenylpropanoid pathway and plant defence-a genomics perspective. Mol Plant Pathol. 2002;3:371–90. https://doi.org/10.1046/j.1364-3703.2002.00131.x.
105. Umezawa T. The cinnamate/monolignol pathway. Phytochem Rev. 2010;9:1–17. https://doi.org/10.1007/s11101-009-9155-3.

106. Mandal SM, Chakraborty D, Dey S. Phenolic acids act as signaling molecules in plant-microbe symbioses. Plant Signal Behav. 2010;5:359–68. https://doi.org/10.4161/psb.5.4.10871.
107. Ferrer J, Austin MB, Stewart C Jr, Noel JP. Structure and function of enzymes involved in the biosynthesis of phenylpropanoids. Plant Physiol Biochem. 2008;46:356–70. https://doi.org/10.1016/j.plaphy.2007.12.009.
108. Castelluccio C, Paganga G, Melikian N, Paul Bolwell G, Pridham J, Sampson J, et al. Antioxidant potential of intermediates in phenylpropanoid metabolism in higher plants. FEBS Lett. 1995;368:188–92. https://doi.org/10.1016/0014-5793(95)00639-Q.
109. Barber MS, McConnell VS, DeCaux BS. Antimicrobial intermediates of the general phenylpropanoid and lignin specific pathways. Phytochemistry. 2000;54:53–6. https://doi.org/10.1016/S0031-9422(00)00038-8.
110. Bruyere C, Genovese S, Lallemand B, Ionescu-Motatu A, Curini M, Kiss R, et al. Growth inhibitory activities of oxyprenylated and non-prenylated naturally occurring phenylpropanoids in cancer cell lines. Bioorg Med Chem Lett. 2011;21:4174–9. https://doi.org/10.1016/j.bmcl.2011.05.089.
111. Lunkenbein S, Bellido M, Aharoni A, Salentijn EMJ, Kaldenhoff R, Coiner HA, et al. Cinnamate metabolism in ripening fruit. Characterization of a UDP-glucose:cinnamate glucosyltransferase from strawberry. Plant Physiol. 2006;140:1047–58. https://doi.org/10.1104/pp.105.074955.
112. Cui L, Yao S, Dai X, Yin Q, Liu Y, Jiang X, et al. Identification of UDP-glycosyltransferases involved in the biosynthesis of astringent taste compounds in tea (*Camellia sinensis*). J Exp Bot. 2016;67:2285–97. https://doi.org/10.1093/jxb/erw053.
113. Lin J, Huang X, Li Q, Cao Y, Bao Y, Meng X, et al. UDP-glycosyltransferase 72B1 catalyzes the glucose conjugation of monolignols and is essential for the normal cell wall lignification in *Arabidopsis thaliana*. Plant J. 2016;n/a. https://doi.org/10.1111/tpj.13229.
114. Wang Y, Wang W, Jin S, Wang J, Wang B, Hou B. Over-expression of a putative poplar glycosyltransferase gene, PtGT1, in tobacco increases lignin content and causes early flowering. J Exp Bot. 2012;63:2799–808. https://doi.org/10.1093/jxb/ers001.
115. Le Roy J, Huss B, Creach A, Hawkins S, Neutelings G. Glycosylation is a major regulator of phenylpropanoid availability and biological activity in plants. Front Plant Sci. 2016;7:132. https://doi.org/10.3389/fpls.2016.00735.
116. Chong J, Baltz R, Schmitt C, Beffa R, Fritig B, Saindrenan P. Downregulation of a pathogen-responsive tobacco UDP-glc:phenylpropanoid glucosyltransferase reduces scopoletin glucoside accumulation, enhances oxidative stress, and weakens virus resistance. Plant Cell. 2002;14:1093–107. https://doi.org/10.1105/tpc.010436.
117. Kanoh H, Kawauchi M, Kuroyanagi M, Arima T. Molecular cloning and characterization of coumarin glucosyltransferase in hairy roots of *Pharbitis nil* (*Ipomoea nil*). Plant Biotechnol. 2014;31:21–8. https://doi.org/10.5511/plantbiotechnology.13.1203a.
118. Griesser M, Hoffmann T, Bellido ML, Rosati C, Fink B, Kurtzer R, et al. Redirection of flavonoid biosynthesis through the down-regulation of an anthocyanidin glucosyltransferase in ripening strawberry fruit. Plant Physiol. 2008;146:1528–39. https://doi.org/10.1104/pp.107.114280.
119. He F, Chen W, Yu K, Ji X, Duan C, Reeves MJ, et al. Molecular and biochemical characterization of the UDP-glucose: anthocyanin 5-*O*-glucosyltransferase from *Vitis amurensis*. Phytochemistry. 2015;117:363–72. https://doi.org/10.1016/j.phytochem.2015.06.023.
120. Jánváry L, Hoffmann T, Pfeiffer J, Hausmann L, Töpfer R, Fischer TC, et al. A double mutation in the anthocyanin 5-*O*-glucosyltransferase gene disrupts enzymatic activity in *Vitis vinifera* L. J Agric Food Chem. 2009;57:3512–8. https://doi.org/10.1021/jf900146a.
121. Wang L, Han W, Xie C, Hou J, Fang Q, Gu J, et al. Comparing the acceptor promiscuity of a *Rosa hybrida* glucosyltransferase RhGT1 and an engineered microbial glucosyltransferase OleDPSA toward a small flavonoid library. Carbohydr Res. 2013;368:73–7. https://doi.org/10.1016/j.carres.2012.12.012.

122. Gosch C, Halbwirth H, Stich K. Phloridzin: biosynthesis, distribution and physiological relevance in plants. Phytochemistry. 2010;71:838–43. https://doi.org/10.1016/j.phytochem.2010.03.003.
123. Gosch C, Halbwirth H, Schneider B, Hölscher D, Stich K. Cloning and heterologous expression of glycosyltransferases from *Malus x domestica* and *Pyrus communis*, which convert phloretin to phloretin 2′-*O*-glucoside (phloridzin). Plant Sci. 2010;178:299–306. https://doi.org/10.1016/j.plantsci.2009.12.009.
124. Zhang T, Liang J, Wang P, Xu Y, Wang Y, Wei X, et al. Purification and characterization of a novel phloretin-2′-*O*-glycosyltransferase favoring phloridzin biosynthesis. Sci Rep. 2016;6:35274. https://doi.org/10.1038/srep35274.
125. Jugdé H, Nguy D, Moller I, Cooney JM, Atkinson RG. Isolation and characterization of a novel glycosyltransferase that converts phloretin to phlorizin, a potent antioxidant in apple. FEBS J. 2008;275:3804–14. https://doi.org/10.1111/j.1742-4658.2008.06526.x.
126. Gutmann A, Bungaruang L, Weber H, Leypold M, Breinbauer R, Nidetzky B. Towards the synthesis of glycosylated dihydrochalcone natural products using glycosyltransferase-catalysed cascade reactions. Green Chem. 2014;16:4417–25. https://doi.org/10.1039/C4GC00960F.
127. Yahyaa M, Davidovich-Rikanati R, Eyal Y, Sheachter A, Marzouk S, Lewinsohn E, et al. Identification and characterization of UDP-glucose: phloretin 4′-*O*-glycosyltransferase from *Malus x domestica* Borkh. Phytochemistry. 2016;130:47–55. https://doi.org/10.1016/j.phytochem.2016.06.004.
128. Song C, Ring L, Hoffmann T, Huang F, Slovin JP, Schwab W. Acylphloroglucinol biosynthesis in strawberry fruit. Plant Physiol. 2015. https://doi.org/10.1104/pp.15.00794.
129. Song C, Zhao S, Hong X, Liu J, Schulenburg K, Schwab W. A UDP-glucosyltransferase functions in both acylphloroglucinol glucoside and anthocyanin biosynthesis in strawberry (*Fragaria* × *ananassa*). Plant J. 2016;85:730–42. https://doi.org/10.1111/tpj.13140.
130. Song C, Hong X, Zhao S, Liu J, Schulenburg K, Huang F, et al. Glucosylation of 4-hydroxy-2,5-dimethyl-3(2H)-furanone, the key strawberry flavor compound in strawberry fruit. Plant Physiol. 2016;171:139–51. https://doi.org/10.1104/pp.16.00226.
131. Kiselev KV, Aleynova OA, Grigorchuk VP, Dubrovina AS. Stilbene accumulation and expression of stilbene biosynthesis pathway genes in wild grapevine *Vitis amurensis* Rupr. Planta. 2016:1–9. https://doi.org/10.1007/s00425-016-2598-z.
132. Mozafari M, Nekooeian AA, Mashghoolozekr E, Panjeshahin MR. The cardioprotective effects of resveratrol in rats with simultaneous type 2 diabetes and renal hypertension. Nat Prod Commun. 2015;10:335–8.
133. Ozaki S, Imai H, Iwakiri T, Sato T, Shimoda K, Nakayama T, et al. Regioselective glucosidation of trans-resveratrol in *Escherichia coli* expressing glucosyltransferase from *Phytolacca americana*. Biotechnol Lett. 2012;34:475–81. https://doi.org/10.1007/s10529-011-0784-4.
134. Hall D, de Luca V. Mesocarp localization of a bi-functional resveratrol/hydroxycinnamic acid glucosyltransferase of concord grape (*Vitis labrusca*). Plant J. 2007;49:579–91. https://doi.org/10.1111/j.1365-313X.2006.02987.x.
135. Kaminaga Y, Sahin FP, Mizukami H. Molecular cloning and characterization of a glucosyltransferase catalyzing glucosylation of curcumin in cultured *Catharanthus roseus* cells. FEBS Lett. 2004;567:197–202. https://doi.org/10.1016/j.febslet.2004.04.056.
136. Oguchi Y, Masada S, Kondo T, Terasaka K, Mizukami H. Purification and characterization of UDP-glucose: curcumin glucoside 1,6-glucosyltransferase from *Catharanthus roseus* cell suspension cultures. Plant Cell Physiol. 2007;48:1635–43. https://doi.org/10.1093/pcp/pcm138.
137. Schwab W, Davidovich-Rikanati R, Lewinsohn E. Biosynthesis of plant-derived flavor compounds. Plant J. 2008;54:712–32. https://doi.org/10.1111/j.1365-313X.2008.03446.x.
138. Gershenzon J, Dudareva N. The function of terpene natural products in the natural world. Nat Chem Biol. 2007;3:408–14. https://doi.org/10.1038/nchembio.2007.5.
139. Chen W, Viljoen AM. Geraniol — a review of a commercially important fragrance material. S Afr J Bot. 2010;76:643–51. https://doi.org/10.1016/j.sajb.2010.05.008.

140. Bönisch F, Frotscher J, Stanitzek S, Ruhl E, Wust M, Bitz O, et al. Activity-based profiling of a physiologic aglycone library reveals sugar acceptor promiscuity of family 1 UDP-glucosyltransferases from grape. Plant Physiol. 2014;166:23–39. https://doi.org/10.1104/pp.114.242578.
141. Yauk Y, Ged C, Wang MY, Matich AJ, Tessarotto L, Cooney JM, et al. Manipulation of flavour and aroma compound sequestration and release using a glycosyltransferase with specificity for terpene alcohols. Plant J. 2014;80:317–30. https://doi.org/10.1111/tpj.12634.
142. Ohgami S, Ono E, Horikawa M, Murata J, Totsuka K, Toyonaga H, et al. Volatile glycosylation in tea plants: sequential glycosylations for the biosynthesis of aroma β-primeverosides are catalyzed by two *Camellia sinensis* glycosyltransferases. Plant Physiol. 2015;168:464–77. https://doi.org/10.1104/pp.15.00403.
143. Dong T, Hwang I. Contribution of ABA UDP-glucosyltransferases in coordination of ABA biosynthesis and catabolism for ABA homeostasis. Plant Signal Behav. 2014;9:e28888. https://doi.org/10.4161/psb.28888.
144. Poppenberger B, Fujioka S, Soeno K, George GL, Vaistij FE, Hiranuma S, et al. The UGT73C5 of *Arabidopsis thaliana* glucosylates brassinosteroids. Proc Natl Acad Sci U S A. 2005;102:15253–8. https://doi.org/10.1073/pnas.0504279102.
145. Husar S, Berthiller F, Fujioka S, Rozhon W, Khan M, Kalaivanan F, et al. Overexpression of the UGT73C6 alters brassinosteroid glucoside formation in *Arabidopsis thaliana*. BMC Plant Biol. 2011;11:51. https://doi.org/10.1186/1471-2229-11-51.
146. Kudo T, Makita N, Kojima M, Tokunaga H, Sakakibara H. Cytokinin activity of cis-zeatin and phenotypic alterations induced by overexpression of putative cis-zeatin-*O*-glucosyltransferase in rice. Plant Physiol. 2012;160:319–31. https://doi.org/10.1104/pp.112.196733.
147. Tognetti VB, van Aken O, Morreel K, Vandenbroucke K, van de Cotte B, de Clercq I, et al. Perturbation of indole-3-butyric acid homeostasis by the UDP-glucosyltransferase UGT74E2 modulates *Arabidopsis* architecture and water stress tolerance. Plant Cell. 2010;22:2660–79. https://doi.org/10.1105/tpc.109.071316.
148. Liu Z, Yan J, Li D, Luo Q, Yan Q, Liu Z, et al. UDP-glucosyltransferase71C5, a major glucosyltransferase, mediates abscisic acid homeostasis in *Arabidopsis*. Plant Physiol. 2015;167:1659–70. https://doi.org/10.1104/pp.15.00053.
149. Suzuki H, Hayase H, Nakayama A, Yamaguchi I, Asami T, Nakajima M. Identification and characterization of an *Ipomoea nil* glucosyltransferase which metabolizes some phytohormones. Biochem Biophys Res Commun. 2007;361:980–6. https://doi.org/10.1016/j.bbrc.2007.07.147.
150. Noguchi A, Kunikane S, Homma H, Liu W, Sekiya T, Hosoya M, et al. Identification of an inducible glucosyltransferase from *Phytolacca americana* L. cells that are capable of glucosylating capsaicin. Plant Biotechnol. 2009;26:285–92. https://doi.org/10.5511/plantbiotechnology.26.285.
151. Dick R, Rattei T, Haslbeck M, Schwab W, Gierl A, Frey M. Comparative analysis of benzoxazinoid biosynthesis in monocots and dicots: independent recruitment of stabilization and activation functions. Plant Cell. 2012;24:915–28. https://doi.org/10.1105/tpc.112.096461.
152. Sue M, Nakamura C, Nomura T. Dispersed benzoxazinone gene cluster: molecular characterization and chromosomal localization of glucosyltransferase and glucosidase genes in wheat and rye. Plant Physiol. 2011;157:985–97. https://doi.org/10.1104/pp.111.182378.
153. Schwab W. Natural 4-hydroxy-2,5-dimethyl-3(2H)-furanone (Furaneol(R)). Molecules. 2013;18:6936–51. https://doi.org/10.3390/molecules18066936.
154. Gandia-Herrero F, Lorenz A, Larson T, Graham IA, Bowles DJ, Rylott EL, et al. Detoxification of the explosive 2,4,6-trinitrotoluene in *Arabidopsis*: discovery of bifunctional *O*- and *C*-glucosyltransferases. Plant J. 2008;56:963–74. https://doi.org/10.1111/j.1365-313X.2008.03653.x.
155. Brazier-Hicks M, Evans KM, Gershater MC, Puschmann H, Steel PG, Edwards R. The *C*-glycosylation of flavonoids in cereals. J Biol Chem. 2009;284:17926–34. https://doi.org/10.1074/jbc.M109.009258.

156. Landmann C, Fink B, Schwab W. FaGT2: a multifunctional enzyme from strawberry (*Fragaria* × *ananassa*) fruits involved in the metabolism of natural and xenobiotic compounds. Planta. 2007;226:417–28. https://doi.org/10.1007/s00425-007-0492-4.
157. Yu B, Sun J, Yang X. Assembly of naturally occurring glycosides, evolved tactics, and glycosylation methods. Acc Chem Res. 2012;45:1227–36. https://doi.org/10.1021/ar200296m.
158. Grynkiewicz G, Szeja W. Synthetic glycosides and glycoconjugates of low molecular weight natural products. Curr Pharm Des. 2016;22:1592–627.
159. Thibodeaux CJ, Melançon CE, Liu H. Natural-product sugar biosynthesis and enzymatic glycodiversification. Angew Chem Int Ed. 2008;47:9814–59. https://doi.org/10.1002/anie.200801204.
160. Masada S, Kawase Y, Nagatoshi M, Oguchi Y, Terasaka K, Mizukami H. An efficient chemoenzymatic production of small molecule glucosides with in situ UDP-glucose recycling. FEBS Lett. 2007;581:2562–6. https://doi.org/10.1016/j.febslet.2007.04.074.
161. Terasaka K, Mizutani Y, Nagatsu A, Mizukami H. In situ UDP-glucose regeneration unravels diverse functions of plant secondary product glycosyltransferases. FEBS Lett. 2012;586:4344–50. https://doi.org/10.1016/j.febslet.2012.10.045.
162. Huang FC, Hinkelmann J, Hermenau A, Schwab W. Enhanced production of β-glucosides by *in-situ* UDP-glucose regeneration. J Biotechnol. 2016;224:35–44. https://doi.org/10.1016/j.jbiotec.2016.02.022.
163. de Bruyn F, van Brempt M, Maertens J, van Bellegem W, Duchi D, de Mey M. Metabolic engineering of *Escherichia coli* into a versatile glycosylation platform: production of bioactive quercetin glycosides. Microb Cell Factories. 2015;14:138. https://doi.org/10.1186/s12934-015-0326-1.
164. Wang H, Yang Y, Lin L, Zhou W, Liu M, Cheng K, et al. Engineering *Saccharomyces cerevisiae* with the deletion of endogenous glucosidases for the production of flavonoid glucosides. Microb Cell Factories. 2016;15:134. https://doi.org/10.1186/s12934-016-0535-2.
165. Rivas F, Parra A, Martinez A, Garcia-Granados A. Enzymatic glycosylation of terpenoids. Phytochem Rev. 2013;12:327–39. https://doi.org/10.1007/s11101-013-9301-9.
166. Caputi L, Lim E, Bowles DJ. Discovery of new biocatalysts for the glycosylation of terpenoid scaffolds. Chem Eur J. 2008;14:6656–62. https://doi.org/10.1002/chem.200800548.
167. Lim CG, Wong L, Bhan N, Dvora H, Xu P, Venkiteswaran S, et al. Development of a recombinant *Escherichia coli* strain for overproduction of the plant pigment anthocyanin. Appl Environ Microbiol. 2015;81:6276–84. https://doi.org/10.1128/AEM.01448-15.
168. Thuan NH, Park JW, Sohng JK. Toward the production of flavone-7-*O*-β-D-glucopyranosides using *Arabidopsis* glycosyltransferase in *Escherichia coli*. Process Biochem. 2013;48:1744–8. https://doi.org/10.1016/j.procbio.2013.07.005.
169. Kim SY, Lee HR, Park K, Kim B, Ahn J. Metabolic engineering of *Escherichia coli* for the biosynthesis of flavonoid-*O*-glucuronides and flavonoid-*O*-galactoside. Appl Microbiol Biotechnol. 2015;99:2233–42. https://doi.org/10.1007/s00253-014-6282-6.
170. Pandey RP, Malla S, Simkhada D, Kim B, Sohng JK. Production of 3-*O*-xylosyl quercetin in *Escherichia coli*. Appl Microbiol Biotechnol. 2013;97:1889–901. https://doi.org/10.1007/s00253-012-4438-9.
171. Pei J, Dong P, Wu T, Zhao L, Fang X, Cao F, et al. Metabolic engineering of *Escherichia coli* for astragalin biosynthesis. J Agric Food Chem. 2016;64:7966–72. https://doi.org/10.1021/acs.jafc.6b03447.
172. Kim B, Kim HJ, Ahn J. Production of bioactive flavonol rhamnosides by expression of plant genes in *Escherichia coli*. J Agric Food Chem. 2012;60:11143–8. https://doi.org/10.1021/jf302123c.
173. Yoon J, Kim B, Lee WJ, Lim Y, Chong Y, Ahn J. Production of a novel quercetin glycoside through metabolic engineering of *Escherichia coli*. Appl Environ Microbiol. 2012;78:4256–62. https://doi.org/10.1128/AEM.00275-12.
174. Kim B, Sung SH, Ahn J. Biological synthesis of quercetin 3-*O*-*N*-acetylglucosamine conjugate using engineered *Escherichia coli* expressing *UGT78D2*. Appl Microbiol Biotechnol. 2012;93:2447–53. https://doi.org/10.1007/s00253-011-3747-8.

175. Malla S, Pandey RP, Kim B, Sohng JK. Regiospecific modifications of naringenin for astragalin production in *Escherichia coli*. Biotechnol Bioeng. 2013;110:2525–35. https://doi.org/10.1002/bit.24919.
176. Yan Y, Li Z, Koffas MAG. High-yield anthocyanin biosynthesis in engineered *Escherichia coli*. Biotechnol Bioeng. 2008;100:126–40. https://doi.org/10.1002/bit.21721.
177. Werner SR, Morgan JA. Expression of a *Dianthus* flavonoid glucosyltransferase in *Saccharomyces cerevisiae* for whole-cell biocatalysis. J Biotechnol. 2009;142:233–41. https://doi.org/10.1016/j.jbiotec.2009.05.008.
178. Ito T, Fujimoto S, Shimosaka M, Taguchi G. Production of *C*-glucosides of flavonoids and related compounds by *Escherichia coli* expressing buckwheat *C*-glucosyltransferase. Plant Biotechnol. 2014;advpub. doi:https://doi.org/10.5511/plantbiotechnology.14.1016a.
179. Yan Y, Chemler J, Huang L, Martens S, Koffas MAG. Metabolic engineering of anthocyanin biosynthesis in *Escherichia coli*. Appl Environ Microbiol. 2005;71:3617–23. https://doi.org/10.1128/AEM.71.7.3617-3623.2005.
180. Drăgan C, Buchheit D, Bischoff D, Ebner T, Bureik M. Glucuronide production by whole-cell biotransformation using genetically engineered fission yeast *Schizosaccharomyces pombe*. Drug Metab Dispos. 2010;38:509. https://doi.org/10.1124/dmd.109.030965.
181. Hansen EH, Møller BL, Kock GR, Bünner CM, Kristensen C, Jensen OR, et al. De novo biosynthesis of vanillin in fission yeast (*Schizosaccharomyces pombe*) and baker's yeast (*Saccharomyces cerevisiae*). Appl Environ Microbiol. 2009;75:2765–74. https://doi.org/10.1128/AEM.02681-08.
182. Xue F, Guo H, Hu Y, Liu R, Huang L, Lv H, et al. Expression of codon-optimized plant glycosyltransferase UGT72B14 in *Escherichia coli* enhances salidroside production. Biomed Res Int. 2016:9845927. https://doi.org/10.1155/2016/9845927.
183. Parajuli P, Pandey RP, Trang NTH, Chaudhary AK, Sohng JK. Synthetic sugar cassettes for the efficient production of flavonol glycosides in *Escherichia coli*. Microb Cell Factories. 2015;14:76. https://doi.org/10.1186/s12934-015-0261-1.
184. Wei W, Wang P, Wei Y, Liu Q, Yang C, Zhao G, et al. Characterization of *Panax ginseng* UDP-glycosyltransferases catalyzing protopanaxatriol and biosyntheses of bioactive ginsenosides F1 and Rh1 in metabolically engineered yeasts. Mol Plant. 2015;8:1412–24. https://doi.org/10.1016/j.molp.2015.05.010.
185. Wang P, Wei Y, Fan Y, Liu Q, Wei W, Yang C, et al. Production of bioactive ginsenosides Rh2 and Rg3 by metabolically engineered yeasts. Metab Eng. 2015;29:97–105. https://doi.org/10.1016/j.ymben.2015.03.003.
186. de Bruyn F, de Paepe B, Maertens J, Beauprez J, de Cocker P, Mincke S, et al. Development of an in vivo glucosylation platform by coupling production to growth: production of phenolic glucosides by a glycosyltransferase of *Vitis vinifera*. Biotechnol Bioeng. 2015;112:1594–603. https://doi.org/10.1002/bit.25570.
187. Pandey RP, Li TF, Kim E, Yamaguchi T, Park YI, Kim JS, et al. Enzymatic synthesis of novel phloretin glucosides. Appl Environ Microbiol. 2013;79:3516–21. https://doi.org/10.1128/AEM.00409-13.
188. Pandey RP, Parajuli P, Koirala N, Lee JH, Park YI, Sohng JK. Glucosylation of isoflavonoids in engineered *Escherichia coli*. Mol Cells. 2014;37:172–7. 10.14348/molcells.2014.2348.
189. Straathof AJJ, Panke S, Schmid A. The production of fine chemicals by biotransformations. Curr Opin Biotechnol. 2002;13:548–56. https://doi.org/10.1016/S0958-1669(02)00360-9.
190. Takors R. Scale-up of microbial processes: impacts, tools and open questions. J Biotechnol. 2012;160:3–9. https://doi.org/10.1016/j.jbiotec.2011.12.010.
191. Rosano GL, Ceccarelli EA. Recombinant protein expression in *Escherichia coli*: advances and challenges. Front Microbiol. 2014;5:172. https://doi.org/10.3389/fmicb.2014.00172.
192. Kim BG, Yang SM, Kim SY, Cha MN, Ahn J. Biosynthesis and production of glycosylated flavonoids in *Escherichia coli*: current state and perspectives. Appl Microbiol Biotechnol. 2015;99:2979–88. https://doi.org/10.1007/s00253-015-6504-6.
193. Katsuragi H, Shimoda K, Kubota N, Nakajima N, Hamada H, Hamada H. Biotransformation of cinnamic acid, *p*-coumaric acid, caffeic acid, and ferulic acid by plant cell cultures

of *Eucalyptus perriniana*. Biosci Biotechnol Biochem. 2010;74:1920–4. https://doi.org/10.1271/bbb.100335.
194. Shimoda K, Harada T, Hamada H, Nakajima N, Hamada H. Biotransformation of raspberry ketone and zingerone by cultured cells of *Phytolacca americana*. Phytochemistry. 2007;68:487–92. https://doi.org/10.1016/j.phytochem.2006.11.030.
195. Kwon S, Shimoda K, Hamada H, Ishihara K, Masuoka N. High production of β-thujaplicin glycosides by immobilized plant cells of *Nicotiana tabacum*. Acta Biol Hung. 2008;59:347–55. https://doi.org/10.1556/ABiol.59.2008.3.8.
196. Jones JA, Koffas, M. A. G. Optimizing metabolic pathways for the improved production of natural products, chap. 8. In: Sarah E. O'Connor, editor. Methods in enzymology: synthetic biology and metabolic engineering in plants and microbes part A: metabolism in microbes. Academic; 2016. p. 179–193. doi:https://doi.org/10.1016/bs.mie.2016.02.010.
197. Williams GJ, Zhang C, Thorson JS. Expanding the promiscuity of a natural-product glycosyltransferase by directed evolution. Nat Chem Biol. 2007;3:657–62. https://doi.org/10.1038/nchembio.2007.28.
198. Brazier-Hicks M, Offen WA, Gershater MC, Revett TJ, Lim E, Bowles DJ, et al. Characterization and engineering of the bifunctional *N*- and *O*-glucosyltransferase involved in xenobiotic metabolism in plants. Proc Natl Acad Sci U S A. 2007;104:20238–43. https://doi.org/10.1073/pnas.0706421104.
199. Choi SH, Ryu M, Yoon YJ, Kim D, Lee EY. Glycosylation of various flavonoids by recombinant oleandomycin glycosyltransferase from *Streptomyces antibioticus* in batch and repeated batch modes. Biotechnol Lett. 2012;34:499–505. https://doi.org/10.1007/s10529-011-0789-z.
200. Leonard E, Yan Y, Fowler ZL, Li Z, Lim C, Lim K, et al. Strain improvement of recombinant *Escherichia coli* for efficient production of plant flavonoids. Mol Pharm. 2008;5:257–65. https://doi.org/10.1021/mp7001472.
201. Rowan DD. Volatile metabolites. Meta. 2011;1:41–63.
202. Marks LE, Veldhuizen MG, Shepard TG, Shavit AY. Detecting gustatory–olfactory flavor mixtures: models of probability summation. Chem Senses. 2012;37:263–77. https://doi.org/10.1093/chemse/bjr103.
203. Dunkel A, Steinhaus M, Kotthoff M, Nowak B, Krautwurst D, Schieberle P, et al. Nature's chemical signatures in human olfaction: a foodborne perspective for future biotechnology. Angew Chem Int Ed. 2014;53:7124–43. https://doi.org/10.1002/anie.201309508.
204. Caputi L, Aprea E. Use of terpenoids as natural flavouring compounds in food industry. Recent Pat Food Nutr Agric. 2011;3:9–16. https://doi.org/10.1016/j.tibtech.2005.02.003.
205. Geuns JMC. Stevioside. Phytochemistry. 2003;64:913–21. https://doi.org/10.1016/S0031-9422(03)00426-6.
206. Richman A, Swanson A, Humphrey T, Chapman R, McGarvey B, Pocs R, et al. Functional genomics uncovers three glucosyltransferases involved in the synthesis of the major sweet glucosides of *Stevia rebaudiana*. Plant J. 2005;41:56–67. https://doi.org/10.1111/j.1365-313X.2004.02275.x.
207. Shibata H, Sonoke S, Ochiai H, Nishihashi H, Yamada M. Glucosylation of steviol and steviol-glucosides in extracts from *Stevia rebaudiana* Bertoni. Plant Physiol. 1991;95:152–6. https://doi.org/10.1104/pp.95.1.152.
208. Shibata H, Sawa Y, Oka T, Sonoke S, Kim KK, Yoshioka M. Steviol and steviol-glycoside: glucosyltransferase activities in *Stevia rebaudiana* Bertoni – purification and partial characterization. Arch Biochem Biophys. 1995;321:390–6. https://doi.org/10.1006/abbi.1995.1409.
209. Karim MR, Hashinaga F. Preparation and properties of immobilized pummelo limonoid glucosyltransferase. Process Biochem. 2002;38:809–14. https://doi.org/10.1016/S0032-9592(02)00233-9.
210. Ikemoto T, Okabe B, Mimura K, Kitahara T. Formation of fragrance materials from odourless glycosidically-bound volatiles by skin microflora (part 1). Flavour Fragr J. 2002;17:452–5. https://doi.org/10.1002/ffj.1127.

211. Ikemoto T, Mimura K, Kitahara T. Formation of fragrant materials from odourless glycosidically-bound volatiles on skin microflora (part 2). Flavour Fragr J. 2003;18:45–7. https://doi.org/10.1002/ffj.1150.
212. Veitch NC, Grayer RJ. Flavonoids and their glycosides, including anthocyanins. Nat Prod Rep. 2011;28:1626–95.
213. Harborne JB, Williams CA. Advances in flavonoid research since 1992. Phytochemistry. 2000;55:481–504. https://doi.org/10.1016/S0031-9422(00)00235-1.
214. Pietta P. Flavonoids as antioxidants. J Nat Prod. 2000;63:1035–42. https://doi.org/10.1021/np9904509.
215. Bungaruang L, Gutmann A, Nidetzky B. Leloir glycosyltransferases and natural product glycosylation: biocatalytic synthesis of the *C*-glucoside nothofagin, a major antioxidant of redbush herbal tea. Adv Synth Catal. 2013;355:2757–63. https://doi.org/10.1002/adsc.201300251.
216. Krafczyk N, Glomb MA. Characterization of phenolic compounds in rooibos tea. J Agric Food Chem. 2008;56:3368–76. https://doi.org/10.1021/jf703701n.
217. Gantt RW, Goff RD, Williams GJ, Thorson JS. Probing the aglycon promiscuity of an engineered glycosyltransferase. Angew Chem Int Ed. 2008;47:8889–92. https://doi.org/10.1002/anie.200803508.
218. Jiang M, Zhang H, Park S, Li Y, Pfeifer BA. Deoxysugar pathway interchange for erythromycin analogues heterologously produced through *Escherichia coli*. Metab Eng. 2013;20:92–100. https://doi.org/10.1016/j.ymben.2013.09.005.
219. Fu X, Albermann C, Jiang J, Liao J, Zhang C, Thorson JS. Antibiotic optimization via in vitro glycorandomization. Nat Biotech. 2003;21:1467–9. https://doi.org/10.1038/nbt909.
220. Weiss RB. Mitoxantrone: its development and role in clinical practice. Oncology (Williston Park). 1989;3:135–41. discussion 141–3, 147–8
221. van der Graaf WTA, de Vries EG. Mitoxantrone: bluebeard for malignancies. Anti-Cancer Drugs. 1990;1(2):109–25.
222. Verdrengh M, Isaksson O, Tarkowski A. Topoisomerase II inhibitors, irrespective of their chemical composition, ameliorate experimental arthritis. Rheumatology. 2005;44:183–6. https://doi.org/10.1093/rheumatology/keh444.
223. Wehenkel A, Fernandez P, Bellinzoni M, Catherinot V, Barilone N, Labesse G, et al. The structure of PknB in complex with mitoxantrone, an ATP-competitive inhibitor, suggests a mode of protein kinase regulation in mycobacteria. FEBS Lett. 2006;580:3018–22. https://doi.org/10.1016/j.febslet.2006.04.046.
224. Cavalletti E, Crippa L, Mainardi P, Oggioni N, Cavagnoli R, Bellini O, et al. Pixantrone (BBR 2778) has reduced cardiotoxic potential in mice pretreated with doxorubicin: comparative studies against doxorubicin and mitoxantrone. Investig New Drugs. 2007;25:187–95. https://doi.org/10.1007/s10637-007-9037-8.
225. Evison BJ, Mansour OC, Menta E, Phillips DR, Cutts SM. Pixantrone can be activated by formaldehyde to generate a potent DNA adduct forming agent. Nucleic Acids Res. 2007;35:3581–9. https://doi.org/10.1093/nar/gkm285.
226. Fulbright JM, Huh W, Anderson P, Chandra J. Can anthracycline therapy for pediatric malignancies be less cardiotoxic? Curr Oncol Rep. 2010;12:411–9. https://doi.org/10.1007/s11912-010-0129-9.
227. Quirós LM, Carbajo RJ, Salas JA. Inversion of the anomeric configuration of the transferred sugar during inactivation of the macrolide antibiotic oleandomycin catalyzed by a macrolide glycosyltransferase. FEBS Lett. 2000;476:186–9. https://doi.org/10.1016/S0014-5793(00)01721-X.
228. Zhou M, Thorson JS. Asymmetric enzymatic glycosylation of mitoxantrone. Org Lett. 2011;13:2786–8. https://doi.org/10.1021/ol200977u.
229. Härle J, Bechthold A. The power of glycosyltransferases to generate bioactive natural compounds, chap. 12. In: Hopwood DA, editor. Methods in enzymology : complex enzymes in microbial natural product biosynthesis, part A: overview articles and peptides, vol. 458. Amsterdam: Academic; 2009. p. 309–33. https://doi.org/10.1016/S0076-6879(09)04812-5.

Part III

Analytical and Legal Aspects

10 Authenticity Control of Natural Products by Stable Isotope Ratio Analysis

Matthias Wüst

Abbreviations

BSIA	Bulk stable isotope analysis
CAM	Crassulacean acid metabolism
CSIA	Compound-specific isotope ratio analysis
EA	Elemental analyzer
GC	Gas chromatography
HT-RPLC	High-temperature reversed-phase liquid chromatography
IRMS	Isotope ratio mass spectrometry
LC	Liquid chromatography
PSIA	Position specific isotope analysis
SIRA	Stable isotope ratio analysis
SNIF-NMR	Site-specific natural abundance isotope fractionation-nuclear magnetic resonance spectroscopy
TMU	Tetramethyl urea

10.1 Introduction

There has been an ever increasing interest in the chemical industry to produce natural products either chemically or biotechnologically in order to be independent of unsecure natural resources. However, while the chemical structure of a synthesized product is identical to its natural counterpart, it is not of equal value for the consumer. From a psychological perspective this phenomenon has been described as

M. Wüst (✉)
Bioanalytics, Institute of Nutritional and Food Sciences (IEL),
University of Bonn, Endenicher Allee 11-13, 53115 Bonn, Germany
e-mail: matthias.wuest@uni-bonn.de

W. Schwab et al. (eds.), *Biotechnology of Natural Products*,
https://doi.org/10.1007/978-3-319-67903-7_10

Table 10.1 Elements and their stable isotopes that frequently occur in natural products (bioelements); for oxygen and sulfur additional stable isotopes exist that are not shown because their ratios are usually not used for authenticity control purposes

Element	Isotope Symbol	F [atom-%]	Standard	R (Standard)
Hydrogen	^{1}H	99.9855	Standard mean Ocean water (V-SMOW)[a]	0.00015576
	$^{2}H = D$	0.0145		
Carbon	^{12}C	98.892	Pee-Dee belemnite (V-PDB)[a]	0.0111802
	^{13}C	1.108		
Nitrogen	^{14}N	99.6337	Air Nitrogen (Air)	
	^{15}N			0.0036782
Oxygen	^{16}O	99.7587	Standard mean Ocean water (V-SMOW)[a]	
	^{17}O	0.0375		
	^{18}O	0.2039		0.00200520
Sulfur	^{32}S	95.018	Canyon diablo Troilite (CDT)	
	^{33}S	0.750		
	^{34}S	4.215		0.0441509
	^{36}S	0.02		

F mean relative abundance = [Isotope]/[Σ Isotopes], *R* isotope ratio of the two most abundant isotopes for the given standard = [Heavy isotope]/[Light isotope]
[a]V indicates the location of the international atomic energy agency IAEA in Vienna, Austria

"natural preference" and is documented particularly for the domain of food related natural products [30, 31]. Humans may have an innate desire ("biophilia") for the experience of their ancestral environment. Natural products that are obtained from natural sources or synthesized biotechnologically from natural precursors are representing thus an additional value to consumers and are perceived of being closer to nature. New terms like "all natural" or "minimally processed" are discussed in this context. In the EU, for example, three guiding principles for natural flavoring substances are published in the Regulation (EC) No. 1334/2008 on flavorings: (i) occurrence of a natural flavoring substance in nature, (ii) natural source materials and (iii) the use of permitted natural processes, their sequence and conditions thereof applied during manufacture [10] (**see also Chap. 11 of this book**). In order to safeguard the positioning and positive value recognition of natural products manufacturers and control authorities continuously need to develop and implement measures regarding the authenticity and the quality assurance of natural products in the EU and increasingly also for other regions [32]. This chapter will therefore provide an overview about new and well established strategies regarding the authenticity control of natural products by stable isotope ratio analysis (SIRA), which has become the most important method during the last years. The method is extremely powerful when stable isotope ratios of multiple elements are determined for a given target compound. Table 10.1 provides an overview of the most important elements in natural products and their stable isotopes. Characteristic deviations from their

Table 10.2 Overview of parameters influencing isotopic composition in plants for the bioelements measured by stable isotope ratio analysis

Element	Primary pool	Primary influencing parameter on isotope composition	Secondary influencing parameter on isotope composition
C	Atmospheric CO_2, geogenic bicarbonate, soil organic matter	Photosynthetic pathway including mode of CO_2-fixation (C3, C4 or CAM)	Climate (humidity, temperature, sunlight), cultivation practice, plant variety, ripening stage, secondary metabolism
H	Ocean water	Distance from coast, latitude, altitude	Temperature, transpiration, respiration
O	Ocean water, atmospheric oxygen	Distance from coast, latitude, altitude	Temperature, transpiration, respiration
N	Atmospheric N_2, soil	Agricultural practice, environmental and climatic parameters	

Adapted from [20]

relative mean abundances are detectable in natural products and can be explained by the differential behavior of isotopomeric and isotopologic molecules (isotopomers or isotopic isomers are isomers with isotopic atoms, having the same number of each isotope of each element but differing in their positions; isotopologues are molecules that differ only in their isotopic composition). The global and positional isotopic abundance of an element in a natural product is determined by thermodynamic and kinetic isotope effects during its biosynthesis and by climatic effects (Table 10.2). These deviations for a given element are reported as differences to the isotopic abundance of an international standard and can be interpreted as a fingerprint for a specific origin of a natural product. Because these deviations are in the ppm range, highly precise methods must be employed for their accurate measurement. To date, isotope ratio mass spectrometry (IRMS) and site-specific natural abundance isotope fractionation-nuclear magnetic resonance spectroscopy (SNIF-NMR) are methods that are employed for this purpose and are discussed below.

10.2 Isotope Ratio Mass Spectrometry – IRMS

10.2.1 Notations in IRMS

The molar ratios of the main isotopes of a given element are reported with respect to an international standard (Table 10.1) and are expressed as the so-called delta-value (δ) in ‰ [20]. For the $^{12}C/^{13}C$–ratio (R) the international standard is Vienna Pee Dee Belemnite (V-PDB) and the following notation is obtained:

$$\delta^{13}C[‰] = [(R_{sample} / R_{standard}) - 1] \times 1000$$

Negative δ-values indicate a depletion of the heavier isotope in the sample relative to the standard and positive values indicate an enrichment of the heavier isotope.

10.2.2 Elemental Analyzer-IRMS (EA-IRMS) and Gaschromatography-IRMS (GC-IRMS)

If the natural product to be characterized is available in pure form the isotope ratios of the elements can be measured directly by the so called elemental analyzer-IRMS (EA-IRMS). Here, mg quantities of the substance are converted to simple gases like hydrogen, nitrogen, carbon monoxide and carbon dioxide by combustion or pyrolysis in an elemental analyzer and separated by solid phase gas chromatography using molecular sieves as stationary phases. The gases are ionized by electron impact in an ion source and the ions are separated by mass spectrometry using a magnetic sector analyzer. The ion currents of the separated isotopic species are detected by collectors, the so called Faraday cylinders or cups. The use of triple collectors allows, for example, the determination of the isotope ratios of carbon ($\delta^{13}C$) by detecting ions of the mass/charge ratios of *m/z* 44, 45 and 46. This approach delivers the high analytical precision that is necessary to reveal the natural variations in the stable isotope ratios (Table 10.3) [19]. If the natural product to be characterized is not present in pure form, as for example vanillin in a vanilla extract, a GC allowing the separation of complex mixtures can be coupled online to a capillary reactor. This approach is called GC-IRMS or compound specific isotope analysis (CSIA). Figure 10.1 shows the instrumental setup for GC-IRMS employing an on-line combustion interface for the oxidation of compounds eluting from the GC for the production of carbon dioxide and nitrogen for subsequent IRMS measurement. Figure 10.2 shows the isotope ratio mass spectrometer in more detail with ion-source, magnet and the array of Faraday cups for simultaneous isotope detection. Table 10.3 summarizes the technical setups of stable isotope measurements for the most important bioelements and the analytical precision that can be achieved. It has to be kept in mind that the obtained value for a certain isotope ratio represents the mean for all atoms in a given molecule. Position-specific values can be obtained by a selective chemical degradation of the target molecule and subsequent detection of selected fragments [34, 36, 37]. New approaches rely on the on line-pyrolysis and separation of the generated fragments using GC followed by the IRMS measurement [5, 6, 8]. The determination of intramolecular isotope ratios of different

Table 10.3 Technical setups used in stable isotope analysis by mass spectrometry for the most important bio-elements

Isotope ratio	Measured species	Method	Mass detected	Temperature	Technical precision
$\delta^{13}C$	CO_2	Combustion	44, 45, 46	Up to 1000 °C	^{13}C 0.05‰ (0.56 ppm)
$\delta^{15}N$	N_2	Combustion, Reduction	28, 29	Up to 1000 °C	^{15}N 0.1‰ (0.27 ppm)
$\delta^{18}O$	CO	Pyrolysis	28, 30	1450 °C	^{18}O 0.1‰ (0.2 ppm)
$\delta^{2}H$	H_2	Pyrolysis	2, 3	1280 °C	^{2}H 1‰ (0.16 ppm)

Adapted from [19]

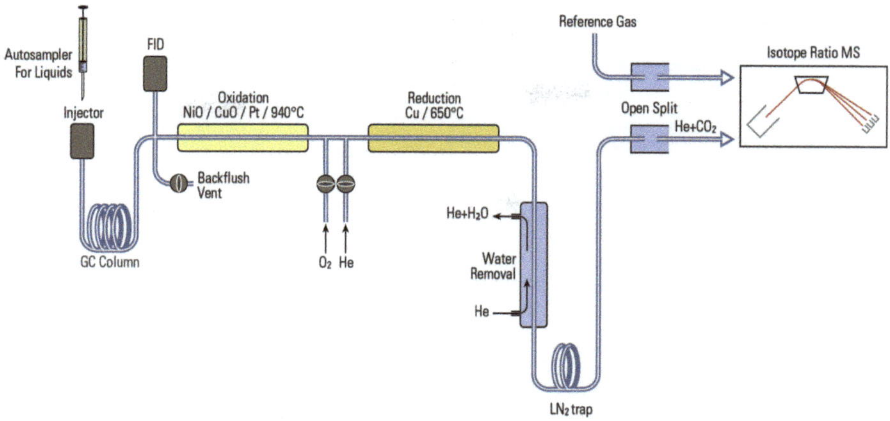

Fig. 10.1 On-line oxidation of compounds eluting from GC for the production of carbon dioxide and nitrogen for IRMS measurement (Copyright by Thermo Fisher Scientific Inc., reproduced with permission)

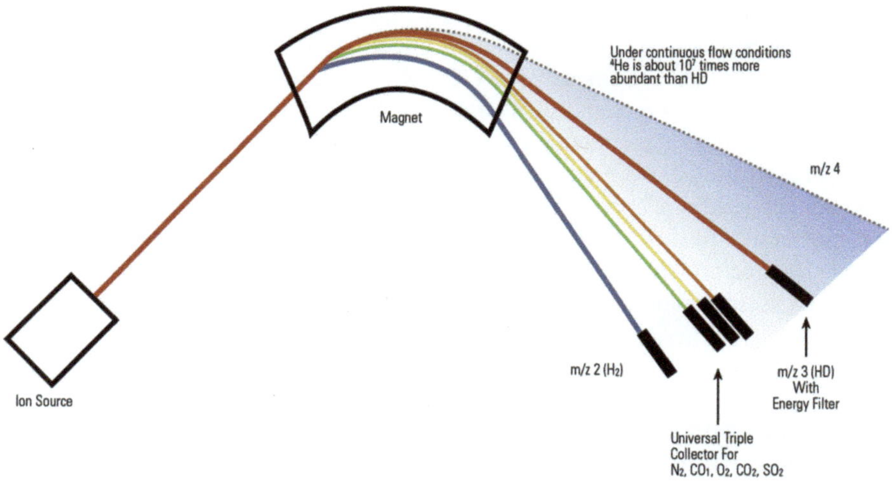

Fig. 10.2 Isotope ratio mass spectrometer with ion source, magnet and the array of Faraday cups for simultaneous ion detection (Copyright by Thermo Fisher Scientific Inc., reproduced with permission)

positions in a molecule has been termed position-specific isotope analysis (PSIA). However, PSIA by on line-pyrolysis and separation of the generated fragments is technically demanding and not yet widely employed in routine analysis. PSIA by selective chemical degradation of the target molecule can be quite tedious and time-consuming and carries the risk of undesirable isotope discriminations [24]. A more widely applied method to determine isotope ratios of pure substances is site-specific natural isotope fractionation-nuclear magnetic resonance spectroscopy (SNIF-NMR), which will be discussed below.

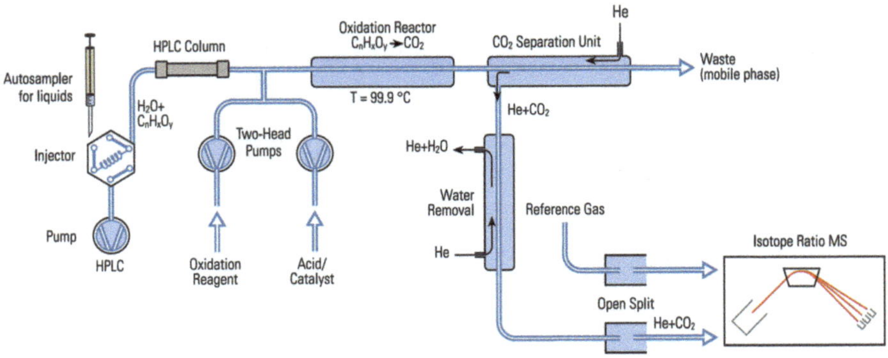

Fig. 10.3 Schematic representation of LC-IRMS hyphenation (LC IsoLink®) used for organic matter chemical oxidation and carbon dioxide extraction from aqueous eluent (Copyright by Thermo Fisher Scientific Inc., reproduced with permission)

10.2.3 Liquid Chromatography-IRMS (LC-IRMS)

LC-IRMS has gained growing interest during the last years as it is able to measure carbon isotope ratios of single compounds directly from complex mixtures that are not volatile and thus not amenable to GC [14]. The technical challenge is to completely oxidize the compound of interest in a liquid eluent that has to be free of interfering carbon i.e. an organic solvent. This is achieved using the powerful oxidant peroxodisulfate and phosphoric acid that are both mixed with the eluent in a post-column reactor at approx. 90 °C. The cooled solution passes through a separating membrane, flushed by helium on the outside, allowing the extraction of the carbon dioxide produced in the reactor (Fig. 10.3). The gas mixture is sent in the IRMS for isotope ratio analysis after crossing a Nafion membrane for gas drying [41].

10.2.4 Site-Specific Natural Isotope Fractionation-Nuclear Magnetic Resonance Spectroscopy (SNIF-NMR)

If the natural product to be characterized is available in pure form and in sufficient quantity (minimum 50–100 mg) the isotope ratios of the elements hydrogen and carbon can be measured by SNIF-NMR exploiting the joint structural and quantitative dimensions of NMR spectroscopy [24]. The method has proved to be particularly powerful for detecting the chaptalization of wines but can now also applied to provide an isotopic fingerprint of a wide variety of natural products. The notation usually used in SNIF-NMR is quite different from the one that is used in IRMS. Specific isotope ratios and molar fractions are widely used. For hydrogen, as an example, position specific isotope ratios $(D/H)_i$ in parts per million (ppm) or molar fractions f_i of the monodeuterated isotopomers are used. These values are obtained by internal referencing using a precisely known quantity of a certified

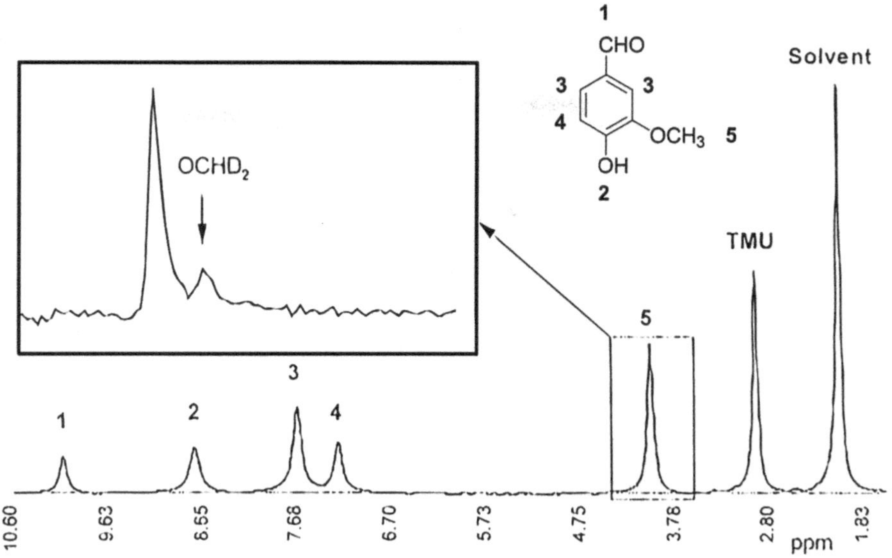

Fig. 10.4 Chemical structure and ^{2}H–NMR spectrum of a vanillin sample showing the presence of a small peak on the right side of the peak corresponding to the methoxy group (site 5). The small peak is due to molecules containing probably $OCHD_2$ (enlarged box in the left upper corner). This is an indication that an isotopic manipulation of vanillin was performed on that sample. The numbering of the different positions of the hydrogen/deuterium atoms is consistent with earlier publications and follows peak order in the ^{2}H–NMR spectra but is not based on IUPAC references [29] (Copyright by American Chemical Society, reproduced with permission)

working standard, which is frequently tetramethyl urea (TMU) [24]. The values for f_i and $(D/H)_i$ can be directly calculated from the signal area of the individual NMR signals S_i. Figure 10.4 shows a typical ^{2}H–NMR spectrum of vanillin. Signals are relatively broad due to the electric quadrupole moment of deuterium. However, only singlets are observed because of the low abundance of deuterium which does not give rise to spin-spin coupling that is typical for conventional ^{1}H–NMR spectra.

10.3 Practical Applications of Stable Isotope Analysis on Miscellaneous Natural Products

10.3.1 Phenylpropanoids (Vanillin)

Vanillin is one of the most important flavoring substances (**see also Chap. 1 of this book**). Beside natural vanillin isolated from vanilla pods, it can be produced by chemical synthesis and by biotechnological production processes [12]. Because natural vanillin (in the sense of the legal definition i.e. of natural or biotechnological origin) (**see also Chap. 11 of this book**) has a high consumer acceptance, it is often adulterated by addition of undeclared, inexpensive, chemically synthesized vanillin

Table 10.4 Stable carbon and hydrogen isotope values of vanillin and vanillin methoxyl groups

Origin or starting material of vanillin	Bulk $\delta^{13}C_{VPDB}$ (‰)	Bulk $\delta^{2}H_{VSMOW}$ (‰)	Methoxyl group $\delta^{13}C_{VPDB}$ (‰)	Methoxyl group $\delta^{2}H_{VSMOW}$ (‰)	Reference
Ex-bean (Madagascar)	−21.5 to −18.2	−57.7 to −10.4			[17, 33]
Ex-bean (Tahiti)	−19.7 to −15.5	−28.5 to −3.0			[17, 33]
Ex-bean			−7.08 to −24.08	−149.0 to −181.6	[15]
Ex-bean	−21.5 to −16.8	−115 to −52			[35]
Ex-guaiacol	−26.2 to −24.9	−23 to −27	−29.73	−121.4	[15, 35]
Ex-eugenol	−31.7 to −29.9				[35]
Ex-lignin	−28.7 to −26.5	−204 to −170	−37.15	−235.6	[15, 35]
Ex-feruclic acid (rice bran)	−36.4 to −36.1				[3, 39]

[29]. EA-IRMS, LC-IRMS and GC-IRMS have been proven to be efficient and reliable methods for the authenticity control of vanillin [1–4, 15, 17, 21, 29, 33, 35]. Table 10.4 provides an overview of the isotope ratios of carbon and hydrogen that have been obtained for authentic vanillin samples. The vanilla plant as a CAM-plant (crassulacean acid metabolism) biosynthesizes vanillin with less negative $\delta^{13}C$–values than vanillin, which is produced from precursors (guaiacol, eugenol, lignin and ferulic acid) obtained from C3-plants. By inspecting the values of the carbon isotope ratios it becomes clear that it is possible to distinguish between vanillin of different botanical origins and different production processes:

- Natural vanillin ex beans (*V. pompona, V. planifolia or V. tahitensis*) shows $\delta^{13}C$–values that are above (more positive) –22‰
- Vanillin that has been obtained by biotechnological processes ex ferulic acid isolated from rice bran shows characteristic values between −37 and −36‰
- Vanillin that has been chemically synthesized from lignin, eugenol or guaiacol shows values that are more negative than −25‰

However, measurement uncertainty of +/− 1‰, which has been determined by inter-laboratory trials, has to be taken into account for authenticity control purposes as recommended by the Society of German Chemists (GDCh) [1]. Moreover, in vanilla extracts prepared by an enzymatic curing process, $\delta^{13}C_{V\text{-}PDB}$ values for vanillin of −21.6 to −22.2‰ were found [13], which are significantly more negative than those of vanilla extracts from traditional curing. Data suggest that the less negative $\delta^{13}C_{V\text{-}PDB}$ values in traditional extracts are the result of an isotope discriminating

Table 10.5 Mean values of $(D/H)_i$ of vanillin from the main origins [29]. The numbering for $(D/H)_i$ of the different positions is consistent with earlier publications and follows peak order in the ^{2}H–NMR spectra but is not based on IUPAC references (see Fig. 10.4)

Origin	$(D/H)_1$ (ppm)	$(D/H)_3$ (ppm)	$(D/H)_4$ (ppm)	$(D/H)_5$ (ppm)
Ex-beans	130.8	157.3	196.4	126.6
Ex-lignin	119.9	132.1	168.8	105.9
Ex-guaiacol	315.2	138.8	143.8	139.1

degradation of vanillin in the traditional process [13]. Less negative $\delta^{13}C_{V\text{-}PDB}$ values were also reported for vanillin-flavored dairy products. An incomplete enzymatic oxidation of vanillin to vanillic acid was made likely to explain these unexpected analytical deviations [22]. Authenticity control of vanillin by $\delta^{13}C_{V\text{-}PDB}$ values can be supplemented by the determination of the $\delta^2H_{V\text{-}SMOW}$ values (Table 10.4) [3]. More sophisticated adulterations can be detected by degradation of the vanillin molecule and subsequent IRMS measurement of the generated product [15], by IRMS measurements of accompanying substances [21, 33] or by $\delta^{18}O_{V\text{-}SMOW}$ measurements of vanillin or its degradation products [3, 18].

If enough sample material is available in pure form, position-specific isotope analysis by SNIF-NMR is possible. ^{2}H–NMR and ^{13}C–NMR can be employed to distinguish between vanillin of natural, biotechnological and chemosynthetic origin [2, 24, 28, 29, 39]. Table 10.5 shows the $(D/H)_i$ values of vanillin samples of different origin. A relatively high value for $(D/H)_4$ is characteristic for vanillin ex-beans and can be explained by an isotope effect during the mono-oxygenase catalyzed hydroxylation reaction of cinnamic acid to para-coumaric acid which involves an NIH-shift [25]. Using additionally quantitative ^{13}C–NMR, the $^{13}C/^{12}C$ ratios at all eight carbon positions of the vanillin molecule can be exploited. Improved discrimination using all eight sites is preferable to differentiate between different methods of production from natural ferulic acid or between natural and lignin-derived vanillin on the basis of the $^{13}C/^{12}C$ ratios characteristic of different origins (Fig. 10.5) [39].

10.3.2 Terpenes (Monoterpenes and Tetraterpenes)

The monoterpene linalool, together with its esters like linalyl acetate, is one of the most frequently used fragrance substances and is produced in large quantities. In the 1950s, linalool was isolated from essential oils but currently this method plays no longer a commercial role. Since linalool is an important intermediate in the chemical synthesis of vitamin E, several large-scale processes are available today for its chemical synthesis [38]. α- and β-pinene or 6-methylhept-5-en-2-one can be used as starting materials. SNIF-NMR combined with IRMS can be used to characterize linalool and linalyl acetate obtained from chemical synthesis or extracted from essential oils of well-defined botanical and geographical origins [16, 26]. Other acyclic monoterpenes, citral and citronellal, hold key positions as fragrance and

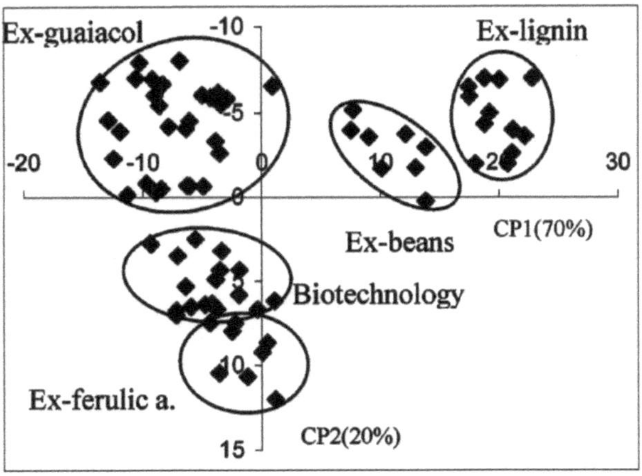

Fig. 10.5 Bidimensional representations of the PCA performed on partial reduced molar fractions f_i/F_i of sites (carbons) 1 and 5–8 of vanillin calculated from ^{13}C NMR spectra. The samples are represented in the plane of the two main axes (CP1 and CP2), and the relative weights are indicated in parentheses [39] (Copyright by American Chemical Society, reproduced with permission)

flavor chemicals and serve as starting materials for the synthesis of other flavor compounds [38]. Natural citral is nearly always a mixture of the two isomers geranial and neral. Since citral is used in bulk as a starting material for the synthesis of vitamin A, it is produced on a large scale by various chemical syntheses. Isotope data obtained from GC-IRMS measurements ($\delta^{13}C_{V\text{-}PDB}$ and $\delta^{2}H_{V\text{-}SMOW}$) allowed to determine the origin of these terpenes and to detect adulterations [27].

Carotenoids like beta-carotene, lycopene and lutein are of increasing commercial importance in the nutraceutical industry. There is thus an interest in developing a reliable method for authenticity assessment of these compounds. Applying EA-IRMS, the $\delta^{13}C_{V\text{-}PDB}$ and $\delta^{2}H_{V\text{-}SMOW}$ values of these carotenoids and carotene-based commercial dietary supplements were determined in comparison to those of synthetic, biotechnological and natural references [23]. It could be shown that the natural stable isotopic composition of carotenoids is a powerful tool for determining their origin.

10.3.3 Polyphenols (Resveratrol)

Resveratrol is a long known plant secondary metabolite that received considerable attention due to its biological activity (**see also Chap. 3 of this book**). It is commercially available as a food additive and there are two sources for this material: the costly plant extraction and the chemical synthesis from two readily accessible C-6-C-1 precursors, that is, anisaldehyde and 3,5-dimethoxybenzaldehyde. The C6-C2-C6 framework of suitably derivatized resveratrol can be degraded with ozone to the C6-C1 aldehydes [11]. These derivatives of either synthetic or natural origin can be characterized and distinguished by SNIF-NMR. The positional

$\delta^{18}O_{V\text{-}SMOW}$ values of resveratrol samples were also determined following a selective deoxygenation procedure. These results demonstrate the utility of simple chemical degradations in the stable isotope characterization of structurally complex food components.

10.3.4 Alkaloids (Caffeine)

Purine alkaloid-containing drinks like tea and coffee are the most popular type of hot beverage in the world, and synthetic caffeine (a purine alkaloid) is also added to energy drinks and cola-type soft drinks (**see also Chap. 6 of this book**). In the case of caffeine, the synthetic product is much depleted in ^{13}C and ^{15}N, and its characterization by SNIF-NMR and EA-IRMS is straightforward [7]. The $\delta^{13}CV_{V\text{-}PDB}$, $\delta^{2}H_{V\text{-}SMOW}$ and $\delta^{18}O_{V\text{-}SMOW}$ values of caffeine isolated from Arabica green coffee beans of different geographical origin have been determined by isotope ratio mass spectrometry (IRMS) using elemental analysis (EA) in the "combustion" (C) and "pyrolysis" (P) modes (EA-C/P-IRMS) [40]. Within the natural species, the climatic dependency enables caffeine from Africa and South America to be distinguished. A new method for compound-specific isotope analysis (CSIA) by coupling high-temperature reversed-phase liquid chromatography to isotope ratio mass spectrometry (HT-RPLC/IRMS) was developed for discrimination of natural and synthetic caffeine contained in all types of drinks [41]. More recently, a methodology has been developed that exploits the power of isotopic quantitative ^{13}C nuclear magnetic resonance (NMR) spectrometry combined with chemical modification of the xanthines (caffeine, theobromine, and theophylline) to enable the determination of positional intramolecular $^{13}C/^{12}C$ ratios ($\delta^{13}C_i$) with high precision [9].

10.4 Conclusions

An overview of the established methods for SIRA as well as their fields of application in the authenticity control of various natural products has been given. For this methodology to be successfully employed in the future by control authorities and quality managers, the implementation of databases is highly recommended to monitor the stable isotope ratios of the natural products of commercial importance. This is of vital interest to safeguard the interests of consumers and honest manufacturers.

References

1. Arbeitsgruppe Aromastoffe. Lebensmittelchemische Gesellschaft der Gesellschaft Deutscher Chemiker. Grundlagenpapier 'Herkunft und Authentizität von Vanillearomen' [Internet]. 2016 [cited 2016 Oct 4]. Available from: https://www.gdch.de/fileadmin/downloads/Netzwerk_und_Strukturen/Fachgruppen/Lebensmittelchemiker/Arbeitsgruppen/aromastoffe/vanille_aroma.pdf

2. Bayle K, Grand M, Chaintreau A, Robins RJ, Fieber W, Sommer H, et al. Internal referencing for 13C position-specific isotope analysis measured by NMR spectrometry. Anal Chem. 2015;87(15):7550–4.
3. Bensaid FF, Wietzerbin K, Martin GJ. Authentication of natural vanilla flavorings: isotopic characterization using degradation of vanillin into guaiacol. J Agric Food Chem. 2002;50(22):6271–5.
4. Bononi M, Quaglia G, Tateo F. Easy extraction method to evaluate δ13C vanillin by liquid chromatography–isotopic ratio mass spectrometry in chocolate bars and chocolate snack foods. J Agric Food Chem. 2015;63(19):4777–81.
5. Brenna JT. Natural intramolecular isotope measurements in physiology: elements of the case for an effort toward high-precision position-specific isotope analysis. Rapid Commun Mass Spectrom. 2001;15(15):1252–62.
6. Corso TN, Brenna JT. High-precision position-specific isotope analysis. Proc Natl Acad Sci. 1997;94(4):1049–53.
7. Danho D, Naulet N, Martin GJ. Deuterium, carbon and nitrogen isotopic analysis of natural and synthetic caffeines. Authentication of coffees and coffee extracts. Analusis. 1992;20(3):179–84.
8. Dias RF, Freeman KH, Franks SG. Gas chromatography–pyrolysis–isotope ratio mass spectrometry: a new method for investigating intramolecular isotopic variation in low molecular weight organic acids. Org Geochem. 2002;33(2):161–8.
9. Diomande DG, Martineau E, Gilbert A, Nun P, Murata A, Yamada K, et al. Position-specific isotope analysis of Xanthines: a 13C nuclear magnetic resonance method to determine the 13C intramolecular composition at natural abundance. Anal Chem. 2015;87(13):6600–6.
10. European Union. Regulation (EC) No 1334/2008 of the European Parliament and of the Council of 16 December 2008 on flavourings and certain food ingredients with flavouring properties for use in and on foods and amending Council Regulation (EEC) No 1601/91, Regulations (EC) No 2232/96 and (EC) No 110/2008 and Directive 2000/13/EC. Off J Eur Union. 2008;51:34–50.
11. Fronza G, Fuganti C, Serra S, Cisero M, Koziet J. Stable isotope labeling pattern of resveratrol and related natural stilbenes. J Agric Food Chem. 2002;50(10):2748–54.
12. Gallage NJ, Møller BL. Vanillin–bioconversion and bioengineering of the most popular plant flavor and its de novo biosynthesis in the vanilla orchid. Mol Plant. 2015;8(1):40–57.
13. Gassenmeier K, Binggeli E, Kirsch T, Otiv S. Modulation of the 13C/12C ratio of vanillin from vanilla beans during curing. Flavour Fragr J. 2013;28(1):25–9.
14. Godin JP, Breuillé D, Obled C, Papet I, Schierbeek H, Hopfgartner G, Fay LB. Liquid and gas chromatography coupled to isotope ratio mass spectrometry for the determination of 13C-valine isotopic ratios in complex biological samples. J Mass Spectrom. 2008;43(10):1334–43.
15. Greule M, Tumino LD, Kronewald T, Hener U, Schleucher J, Mosandl A, et al. Improved rapid authentication of vanillin using δ13C and δ2H values. Eur Food Res Technol. 2010;231(6):933–41.
16. Hanneguelle S, Thibault JN, Naulet N, Martin GJ. Authentication of essential oils containing linalool and linalyl acetate by isotopic methods. J Agric Food Chem. 1992;40(1):81–7.
17. Hansen A-MS, Fromberg A, Frandsen HL. Authenticity and traceability of vanilla flavors by analysis of stable isotopes of carbon and hydrogen. J Agric Food Chem. 2014;62(42):10326–31.
18. Hener U, Brand WA, Hilkert AW, Juchelka D, Mosandl A, Podebrad F. Simultaneous on-line analysis of 18O/16O and 13C/12C ratios of organic compounds using GC-pyrolysis-IRMS. Z Für Leb Forsch A. 1998;206(3):230–2.
19. Hübschmann H-J. Handbook of GC-MS: fundamentals and applications. Weinheim: Wiley; 2015.
20. Jochmann MA, Schmidt TC. Compound-specific stable isotope analysis. Cambridge: Royal Society of Chemistry; 2012.
21. Kaunzinger A, Juchelka D, Mosandl A. Progress in the authenticity assessment of vanilla. 1. Initiation of authenticity profiles. J Agric Food Chem. 1997;45(5):1752–7.

22. Kempe K, Kohnen M. Deterioration of natural vanilla flavours in dairy products during processing. Adv Food Sci. 1999;21(1–2):48–53.
23. Kroll H, Friedrich J, Menzel M, Schreier P. Carbon and hydrogen stable isotope ratios of carotenoids and β-carotene-based dietary supplements. J Agric Food Chem. 2008;56(11):4198–204.
24. Martin G, Remaud G, Martin GJ. Isotopic methods for control of natural flavours authenticity. Flavour Fragr J. 1993;8(2):97–107.
25. Martin GJ, Heck G, Djamaris-Zainal R, Martin ML. Isotopic criteria in the characterization of aromatic molecules. 1. Hydrogen affiliation in natural benzenoid/phenylpropanoid molecules. J Agric Food Chem. 2006;54(26):10112–9.
26. Martin GJ, Lavoine-Hanneguelle S, Mabon F, Martin ML. The fellowship of natural abundance 2H-isotopomers of monoterpenes. Phytochemistry. 2004;65(20):2815–31.
27. Nhu-Trang T-T, Casabianca H, Grenier-Loustalot M-F. Authenticity control of essential oils containing citronellal and citral by chiral and stable-isotope gas-chromatographic analysis. Anal Bioanal Chem. 2006;386(7–8):2141–52.
28. Remaud GS, Akoka S. A review of flavors authentication by position-specific isotope analysis by nuclear magnetic resonance spectrometry: the example of vanillin. Flavour Fragr J. 2017;32(2):77–84.
29. Remaud GS, Martin Y-L, Martin GG, Martin GJ. Detection of sophisticated adulterations of natural vanilla flavors and extracts: application of the SNIF-NMR method to vanillin and p-hydroxybenzaldehyde. J Agric Food Chem. 1997;45(3):859–66.
30. Rozin P. The meaning of 'natural' process more important than content. Psychol Sci. 2005;16(8):652–8.
31. Rozin P, Spranca M, Krieger Z, Neuhaus R, Surillo D, Swerdlin A, et al. Preference for natural: instrumental and ideational/moral motivations, and the contrast between foods and medicines. Appetite. 2004;43(2):147–54.
32. Schäfer U, Kiefl J, Geißler T, Geißler K, Kassing M, Ley J, et al. Recent advances in the authenticity control of bioflavours. Biofavour 2015 – book Abstr. Frankfurt/Main: Dechema e.V; 2015. p. 53.
33. Scharrer A. Vanille : Neues zur Authentizität [Internet]. [Franfurt am Main]: Universität Frankfurt am Main; 2002 [cited 2016 Oct 4]. Available from: http://publikationen.ub.uni-frankfurt.de/frontdoor/index/index/docId/5451/
34. Schmidt H-L. Fundamentals and systematics of the non-statistical distributions of isotopes in natural compounds. Naturwissenschaften. 2003;90(12):537–52.
35. Schmidt H-L, Mosandl A, Schreier P, Roßmann A, Werner RA. Stable isotope ratio analysis in quality control of flavourings. In: Ziegler H, editor. Flavourings: production, composition, applications, regulations. Weinheim:Wiley-VCH; 2007.
36. Schmidt H-L, Werner RA, Eisenreich W. Systematics of 2H patterns in natural compounds and its importance for the elucidation of biosynthetic pathways. Phytochem Rev. 2003;2(1–2):61–85.
37. Schmidt H-L, Werner RA, Roßmann A. 18O pattern and biosynthesis of natural plant products. Phytochemistry. 2001;58(1):9–32.
38. Surburg H, Panten J. Common fragrance and flavor materials: preparation, properties and uses. Weinheim: Wiley; 2016.
39. Tenailleau EJ, Lancelin P, Robins RJ, Akoka S. Authentication of the origin of vanillin using quantitative natural abundance 13C NMR. J Agric Food Chem. 2004;52(26):7782–7.
40. Weckerle B, Richling E, Heinrich S, Schreier P. Origin assessment of green coffee (Coffea arabica) by multi-element stable isotope analysis of caffeine. Anal Bioanal Chem. 2002;374(5):886–90.
41. Zhang L, Kujawinski DM, Federherr E, Schmidt TC, Jochmann MA. Caffeine in your drink: natural or synthetic? Anal Chem. 2012;84(6):2805–10.

Natural or Synthetic? The Legal Framework in the EU for the Production of Natural Flavouring Ingredients

11

Jan C.R. Demyttenaere

11.1 Introduction

In the EU, the legal framework for the safe and correct use of flavourings is the European Flavouring Regulation (EC) No 1334/2008 [1]. This regulation does not only cover "flavourings" as such but also other *"food ingredients with flavouring properties for use in and on foods"*. According to Recital (5) of the Regulation, the purpose is to protect the human health. This Regulation was published on 31/12/2008 and entered into force on 20 January 2009 with a transition period of 2 years until its application from 20 January 2011. It repeals the former Flavour Directive 88/388/EC of 1988 [2], which applied until almost a decade ago.

With the introduction of a "Flavouring Regulation" in 2008 the implementation by the different EU Member States should be better harmonised as stipulated by the last line of the Regulation: *"This Regulation shall be binding in its entirety and directly applicable in all Member States."*

This Chapter will mainly focus on definitions of natural flavouring ingredients (natural flavouring substances and flavouring preparations) and the permissible processes for their preparation. However some insight will also be given in the correct labelling (Business-to-Business) of flavourings (in particular natural flavourings) and on the safety evaluation of flavourings in the EU.

Disclaimer: The views expressed in this chapter are purely those of the author and may not in any circumstances be regarded as stating an official position of the European Flavour Association (EFFA), although the EFFA Guidance Documents are often referred to in this Chapter and have been consulted as basis for this contribution.

J.C.R. Demyttenaere (✉)
European Flavour Association (EFFA), Avenue des Arts, 6, 1210, Brussels, Belgium
e-mail: jdemyttenaere@effa.eu

W. Schwab et al. (eds.), *Biotechnology of Natural Products*,
https://doi.org/10.1007/978-3-319-67903-7_11

11.2 General Conditions

The Flavouring Regulation is built up of six chapters and five Annexes. Chapter I (Subject Matter, Scope and Definitions) covers Articles 1–3 including definitions which will be discussed in more detail in this chapter. Chapter IV (Labelling) will be addressed in more detail in the section on the labelling rules. However, also Recitals provide important information and clarification on certain requirements.

For example Recital (7) describes some criteria to which flavourings should comply: "*Flavourings are used to improve or modify the odour and/or taste of foods for the benefit of the consumer. Flavourings and food ingredients with flavouring properties should only be used if they fulfil the criteria laid down in this Regulation. They must be safe when used, and certain flavourings should, therefore, undergo a risk assessment before they can be permitted in food. Where possible, attention should be focused on whether or not the use of certain flavourings could have any negative consequences on vulnerable groups. The use of flavourings must not mislead the consumer and their presence in food should, therefore, always be indicated by appropriate labelling. Flavourings should, in particular, not be used in a way as to mislead the consumer about issues related to, amongst other things, the nature, freshness, quality of ingredients used, the naturalness of a product or of the production process, or the nutritional quality of the product. [...]*".

Thus, flavourings are used to impart or modify the odour and/or taste of foods; according to Art. 3(4) of the Flavouring Regulation, flavourings may contain food additives as permitted by Regulation (EC) No 1333/2008 (the Additives Regulation [3]) and/or other food ingredients incorporated for technological purposes.

The general conditions for the use of flavourings or food ingredients with flavouring properties are laid down in Art. 4: they can only be used in or on food if they do not, on the basis of the scientific evidence available, pose a safety risk to the health of the consumer, and if their use does not mislead the consumer.

For more information the author refers to the Guidance Documents of EFFA (European Flavour Association) which are available and can be consulted on the open EFFA website: www.effa.eu namely: EFFA Guidance Document on the European Flavour Regulation [4] and EFFA Guidance Document for the Production of Natural Flavouring Substances and (Natural) Flavouring Preparations in the EU [5].

11.3 Scope of the Regulation

The scope of the Flavouring Regulation is clearly outlined in Art. 2 of the Regulation which applies to:

- flavourings which are used or intended to be used in or on foods;
- food ingredients with flavouring properties;
- food containing flavourings and/or food ingredients with flavouring properties;

- source materials for flavourings and/or source materials for food ingredients with flavouring properties

However, the regulation does not apply to:

- substances which have exclusively a sweet, sour or salty taste;
- raw foods;
- non-compound foods and mixtures such as, but not exclusively, fresh, dried or frozen spices and/or herbs, mixtures of tea and mixtures for infusion as such as long as they have not been used as food ingredients.

The first bullet point above is very important as it clearly defines that substances which only impart a sweet (e.g. sweeteners), sour (e.g. acidulents) or salty (e.g. mineral salts) taste are out of the scope of the flavouring regulation. But also spices and herbs (fresh or dried or frozen) and tea and infusions are out of the scope of this regulation, unless they are used exclusively as "food ingredients" to impart a flavouring to the final food.

11.4 Definitions

11.4.1 Definitions of Flavourings in General

Art. 3 of the Flavouring Regulation defines a '**flavouring**' as products *not intended to be consumed as such, which are added to food in order to impart or modify odour and/or taste.*

It is important to stress that flavourings are ingredients which are added to food to give it a certain taste or aroma but they should not be consumed as such.

The flavouring regulation further defines six different categories of flavourings: flavouring substances (either natural or not), flavouring preparations (per definition natural), thermal process flavourings, smoke flavourings, flavour precursors and other flavourings or mixtures thereof.

Although all of these categories (or mixtures thereof) can be used and added to food, this chapter will mainly focus on natural flavouring substances and flavouring preparations (which are per definition natural flavouring ingredients according to Art. 3(2)(d)). The following paragraphs will focus on the production of natural flavouring substances and flavouring preparations and their labelling from a Business-to-Business perspective (B2B).

According to Art. 3(2)(b) a 'flavouring substance' means a "*defined chemical substance with flavouring properties*". Hence, a flavouring substance is a chemically defined substance which further complies with the criteria defined in Art. 3(2)(a). Flavouring substances can be natural in which case they have to comply with Art. 3(2)(c) on 'natural flavouring substance' – in the other case they are simply referred to as a "flavouring substance". The old distinction (from the former Flavour Directive 88/388/EC) between "*nature-identical flavouring substances*" and "*artificial flavouring substances*" no longer applies. Hence what used to be referred to as nature-identical (NI) or artificial flavouring substances is today simply called "flavouring substances". Note that also the annotation "*synthetic flavouring substances*" is not used in the regulation and would be a terminology without legal basis. The same applies to the terminology "*artificial*" which is not legally defined for flavourings. Consequently labels or claims on the front of final food products such as "*without artificial flavourings*" have no legal basis whatsoever – they are misleading to the consumers and should therefore be avoided.

11.4.2 Definition of Natural Flavouring Substance

As mentioned above, "natural flavouring substances" (which constitute one subcategory of "flavouring substances") are defined by Art. 3(2)(c):

> *(c) 'natural flavouring substance' shall mean a flavouring substance obtained by appropriate physical, enzymatic or microbiological processes from material of vegetable, animal or microbiological origin either in the raw state or after processing for human consumption by one or more of the traditional food preparation processes listed in Annex II. Natural flavouring substances correspond to substances that are naturally present and have been identified in nature;*

Thus in order to call a flavouring substance "natural" three cumulative basic requirements have to be checked:

1. The source material must be from material of vegetable, animal or microbiological origin, either in the raw state or after processing for human consumption by traditional food preparation processes listed in Annex II (or combinations thereof);
2. The identification in nature of the manufactured flavouring substance must be valid and meeting the criteria as further explained below.
3. The natural processes, their sequence and conditions thereof applied during manufacture: as stipulated by Art. 3(2)(k) and Annex II.

Substances which do not meet the above-mentioned requirements shall not be considered as being natural.

(a) **Source material requirements**

According to the first criterion (and as defined in Art. 3(2)(c) by the wording "*either in the raw state or after processing for human consumption by one or more of the traditional food preparation processes listed in Annex II…*") the source material for the production of natural flavouring substances is not restricted to the unprocessed/raw source materials (material of vegetable, animal or microbiological origin): also processed source materials can be used as a basis for the production of natural flavouring substances. However, the "processing" should be done by strict

Table 11.1 List of traditional food preparation processes (Annex II to Flavouring Regulation (EC) No 1334/2008) [* footnote by EFFA added (EFFA Guidance document)]

Used in the kitchen at home and in industrial food preparation facilities	
Chopping	Coating
Heating, cooking, baking, frying (up to 240 °C at atmospheric pressure) and pressure cooking (up to 120 °C)	Cooling
Cutting	Distillation/rectification
Drying	Emulsification
Evaporation	Extraction, incl. solvent extraction in accordance with Directive 88/344/EEC
Fermentation	Filtration
Grinding	Maceration
Infusion	Mixing
Microbiological processes*	Percolation
Peeling	Refrigeration/Freezing
Pressing	Squeezing
Roasting/Grilling	Steeping

*The Flavour Industry considers these to include "enzymatic processes"

criteria, namely using typical "traditional food preparation processes" as listed in Annex II of the Regulation – these are the processes typically used in a kitchen (at home) or in industrial food preparation facilities.

Annex II of the Flavouring regulation gives a list of "traditional food preparation processes" by which natural flavouring substances and (*natural*) flavouring preparations can be obtained (**see** Table 11.1).

(b) **Identified in nature**

The second requirement (identification in nature) is a very important one (and was not part of the former Flavour Directive 88/388/EC) and in particular applies to natural flavouring substances.

It clearly requires that a substance has to be identified in nature before it can be regarded as 'natural', so it is not only sufficient to produce it by a permissible processing. It has to be identical to a substance which is present in nature. This is to avoid that a substance that has never been identified in nature before (and is not naturally occurring), such as ethylvanillin, would be labelled as a 'natural flavouring substance', even if today a (new) "natural process" has been discovered (e.g. through biotechnology) for its production that was not described before.

This could be the case when a new enzymatic or microbial process would be developed by which a flavouring substance can be produced "*by enzymatic or microbial processes from material of vegetable origin*" (i.e. natural source materials), which has so far never been identified in nature.

Since such substance has not been "identified in nature" before, it can never be claimed to be "natural" even if the new production process is technically speaking a "natural process".

According to EFFA's interpretation, "identified in nature means":

- it has been identified in materials of plant, animal, microbiological, or mineral origin, and/or
- it has been identified in food in the raw state or processed or partly processed for human consumption.

In addition, within the global Flavour Industry, there is a general principle and agreement that in order to decide that a flavouring substance has been 'identified in nature', any identification needs to meet the criteria for the validity of identifications in nature as described by IOFI (International Organization of the Flavor Industry) [6] and where necessary in specific IOFI guidelines for using the technique of LC-MS for identifications [7].

A very important note though is the following EFFA consideration: a process used to produce a natural flavouring substance or flavouring preparation may not in itself be used to qualify the resulting product as natural if the end product cannot be found in nature or in products traditionally used as foods by human beings (see example above on ethylvanillin).

(c) **Production processes acceptable as natural**

The third requirement relates to the types of processes that can be applied to produce / obtain a natural flavouring substance. Processes can either be the same "traditional food preparation processes" (listed in Annex II; Table 11.1) which can also be applied to the raw source materials but can also be "appropriate physical, enzymatic and microbiological processes".

The EU Flavouring Regulation defines "appropriate physical process" in Art. 3(2)(k).

This definition reads as follows:

> *(k) 'appropriate physical process' shall mean a physical process which does not intentionally modify the chemical nature of the components of the flavouring, without prejudice to the listing of traditional food preparation processes in Annex II, and does not involve,* inter alia, *the use of singlet oxygen, ozone, inorganic catalysts, metal catalysts, organometallic reagents and/or UV radiation.*

Strictly speaking, any process that "*intentionally modifies the chemical nature of the components of the flavouring*" would not be regarded as an 'appropriate physical process'. However, the wording "*without prejudice to the listing of traditional food preparation processes in Annex II*" in Art. 3(2)(k) indicates that all processes listed in Annex II also fall under the definition of "*appropriate physical processes*" although it is well-known that e.g. frying up to 240 °C and fermentation modify the chemical structure of the components of the flavouring.

A scheme showing the production of Natural Flavouring Substances according to the definition (Art. 3(2)(c) of the Flavouring Regulation) is depicted in Fig. 11.1. A "Natural Flavouring Substance" can be obtained either directly from material of

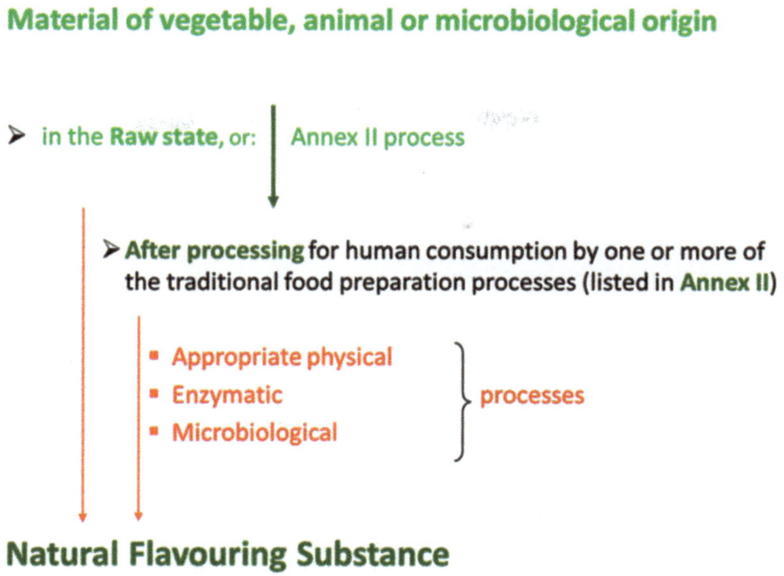

Fig. 11.1 Scheme depicting the production of Natural Flavouring Substances according to the definition (Art. 3(2)(c) of the Flavouring Regulation (EC) No 1334/2008)

vegetable, animal or microbiological origin "*in the raw state*" or "*after processing for human consumption*". In the first case (from the source material in the raw state), it can be obtained after various "*appropriate physical*" processes (as defined by Art. 3(2)(k) mentioned above) but also "*enzymatic or microbiological processes*" as listed in Annex II.

In the second case "*after processing for human consumption*" a two-step approach is involved, where the first step "*processing for human consumption*" should be one of the "*traditional food preparation processes*" listed in Annex II. The second step is again any other process (appropriate physical, enzymatic or microbiological process), which means not only processes listed in Annex II but also a process as defined by Art. 3(2)(k).

Some further considerations on the term "*for human consumption*" will be provided in the section on the safety evaluation of flavourings.

(d) **Consideration of geometric and optical isomers**

Many flavouring substances may exist as distinct geometric isomers (Z/E, more commonly referred to as *cis/trans*). It is EFFA's position that mixtures of such isomers do not have to be produced in the same ratio as they are found in a specific food or in a natural source. If all geometric isomers have been identified in nature, the production of a mixture of geometric isomers in any ratio should be allowed in order to call the mixture natural.

Flavouring substances may have one or more chiral centres and hence can exist as different optical isomers and stereoisomers (enantiomers and diastereoisomers, resp.). According to the EFFA Guidance, mixtures of optical isomers and stereoisomers shall be allowed in any ratio provided that all the isomers are identified in nature.

However, if only one out of two (or more) isomers has been identified in nature the mixture of the two (or more) isomers cannot be regarded as natural if the other isomer(s), not identified in nature is (are) present at levels that are higher than trace levels (i.e. unavoidable levels <1%).

In that respect EFFA considers that if one of the geometric or optical isomers has not (yet) been reported to have been identified in nature, it must be interpreted as an artefact of the natural process. Any flavouring substance thus produced would not qualify as natural unless the artefacts present are in small amounts (<1%: unavoidable traces) that do not contribute to the flavour of the natural ingredient.

11.4.3 Definitions of Flavouring Preparations

As already mentioned above, "flavouring preparations" (which constitute another category of flavourings) are always natural as defined by Art. 3(2)(d):

> (d) *'flavouring preparation' shall mean a product, other than a flavouring substance, obtained from:*
>
> (i) *food by appropriate physical, enzymatic or microbiological processes either in the raw state of the material or after processing for human consumption by one or more of the traditional food preparation processes listed in Annex II* (Table 11.1)*;*
>
> *and/or*
>
> *(ii) material of vegetable, animal or microbiological origin, other than food, by appropriate physical, enzymatic or microbiological processes, the material being taken as such or prepared by one or more of the traditional food preparation processes listed in Annex II;*

A very important element (which does not apply to flavouring substances) is the distinction between flavouring preparations obtained from food sources (Art. 3(2)(d)(i)) and those obtained from non-food sources (Art. 3(2)(d)(ii)). The reason behind is the potential need for a safety evaluation depending on the source (flavouring preparations from food source materials do not need to undergo a safety evaluation). This aspect will be addressed in the section on the safety evaluation.

Thus in order to call a flavouring a "flavouring preparation" (and thus natural) two cumulative basic requirements have to be checked:

1. The source material must be from food (raw or processed for human consumption, refer to Annex II, Table 11.1) or from material of vegetable, animal or microbiological origin, other than food (either taken as such or prepared by

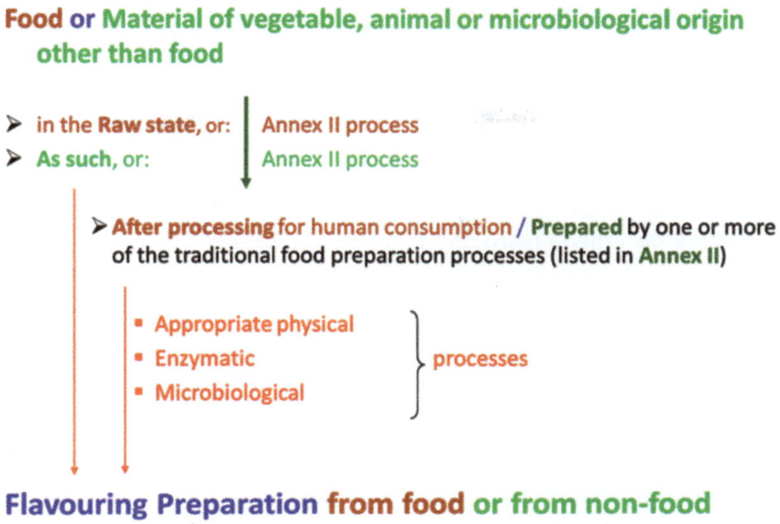

Fig. 11.2 Scheme depicting the production of (Natural) Flavouring Preparation from food sources [in red font] or from non-food sources [in green font] (according to the definition (Art. 3(2)(d)(i) & Art. 3(2)(d)(ii), resp. of the Flavouring Regulation (EC) No 1334/2008)

traditional food preparation processes listed in Annex II (Table 11.1, or combinations thereof));

2. The natural processes, their sequence and conditions thereof applied during manufacture are stipulated by Art. 3(2)(k) and Annex II (Table 11.1).

Flavourings which do not meet the above-mentioned requirements shall not be considered as "flavouring preparations" and thus not as natural. Depending on their sources and the production methods, such mixtures might still fall under the category 'thermal process flavourings' (Art. 3(2)(e)), 'smoke flavourings' (Art. 3(2)(f)) or 'other flavourings' (Art. 3(2)(h)).

(a) **Source material**

A scheme with the production of a (Natural) Flavouring Preparation from food sources according to the definition Art. 3(2)(d)(i) and from non-food sources (according to the definition Art. 3(2)(d)(ii)) is shown in Fig. 11.2 [red font is used for food sources, whereas green font is used for materials other than food].

In both cases (analogue to the production of natural flavouring substances), a preparation can be obtained either from the raw food [in red font] (or the non-food source material [in green font] taken as such – unprocessed) or from food [red font] (or a non-food source material [green font]) processed (or prepared, resp.) by one or more of the traditional food preparation processes listed in Annex II (Table 11.1). The second step involves again any other process (appropriate physical, enzymatic or microbiological process), which means not only a process listed in Annex II but also a process as defined by Art. 3(2)(k). The end result will be a flavouring preparation either from food sources [red font] according to Art. 3(2)(d)(i) or from non-food sources [green font] according to Art. 3(2)(d)(ii).

Typical examples of flavouring mixtures which are used in foods for their flavouring properties are essential oils and extracts and a key question is whether such mixtures would have the "natural" status and can be regarded as "flavouring preparations" according to Art. 3(2)(d).

Considering the criteria listed above, it can be concluded that according to this definition (Art. 3.2(d)) essential oils and extracts obtained from plant material (*material of vegetable origin*) prepared by distillation (which is a *traditional food preparation process* listed in Annex II, Table 11.1), followed by an *appropriate physical process* can be considered as a '*flavouring preparation*' and thus natural, as long as the chemical nature of the components is not intentionally modified during the physical process.

However, as it is EFFA's understanding (see above) that the definition of "appropriate physical processes (Art. 3(2)(k)) also covers all processes listed in Annex II (Table 11.1) an essential oil will still be regarded as a "flavouring preparation" and hence natural, even if the distillation (which is listed in Annex II, Table 11.1) and some of the following physical processes (e.g. extraction, drying, evaporation, concentration…) may modify the chemical nature of the components. The intention to apply those processes is to obtain the "flavouring preparations" for example in the expected purity or concentration range. It is not the intention to modify the chemical nature of the ingredients. Thus an occurring modification during a process like distillation (e.g. in the production of essential oils) is not intentionally and therefore does not prevent the material from being natural.

ISO Standard 9235 gives some examples of different flavouring preparations such as essential oils, extracts and tinctures [8].

With regard to the permitted source materials for the production of natural flavourings, EFFA in its Guidance Document makes a special note that the source materials which can be used are material of vegetable, animal or microbiological origin but not minerals.

EFFA further considers that these sources may be foods as well as non-foods. Source materials for the production of natural flavouring ingredients may also include less routinely consumed parts of plant material as well as co- and/or by-products of food production such as fibre, hulls, stems, shells etc. (see in this respect Recital (16) of the Regulation).

(b) Permissible processes for the production of (natural) flavouring preparations

The processes that can be used for the production of flavouring preparations are essentially the same as for the production of natural flavouring substances.

EFFA has developed a Guidance Document for the European Flavour Industry on the Permissible Processes to obtain natural flavouring ingredients (i.e. natural

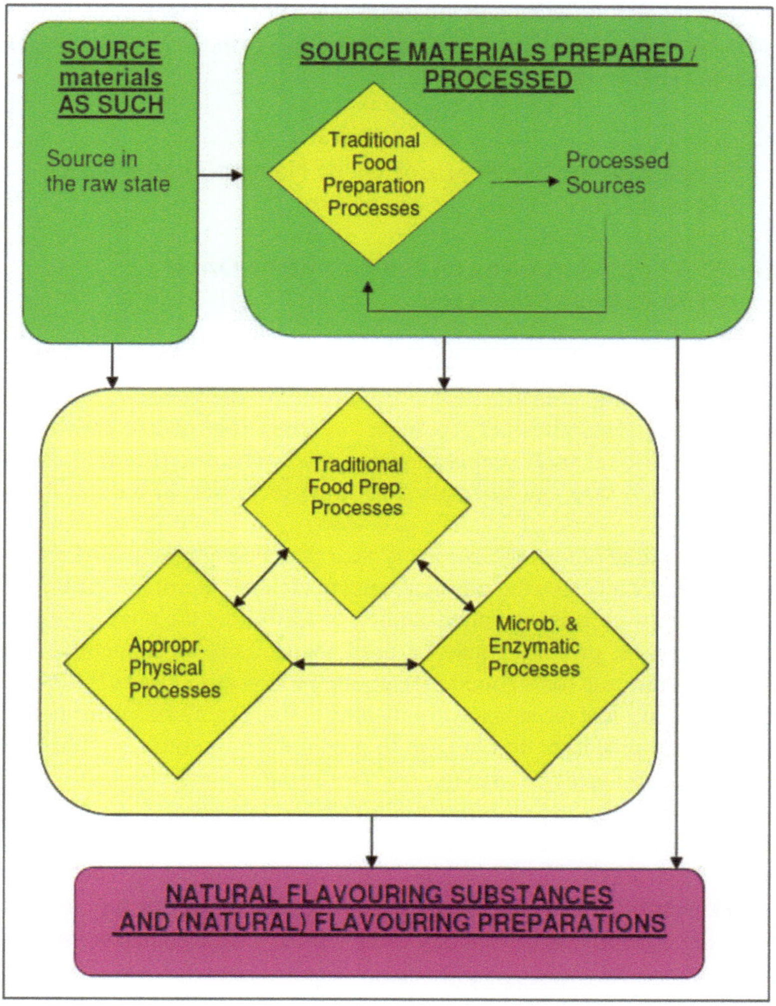

Fig. 11.3 Scheme on the permissible processes for the production of natural flavouring ingredients (Source: EFFA Guidance Document on natural flavouring ingredients [5])

flavouring substances and (natural) flavouring preparations) which is published on the EFFA-website [5].

A scheme on the permissible processes for the production of natural flavouring ingredients (according to the EFFA Guidance Document mentioned above) is shown in Fig. 11.3.

As can be seen from this scheme (yellow box in Fig. 11.3) all processes (including the appropriate physical processes (Art. 3(2)(k)) as well as the traditional food preparation processes (Annex II, Table 11.1) and microbial and enzymatic processes) are regarded as "permissible processes" to obtain natural flavouring ingredients.

This scheme also explains why essential oils are considered as natural even if some of the processes (like distillation) would be able to modify the chemical nature of the components.

11.5 Processes for the Production of Natural Flavouring Ingredients

11.5.1 EFFA Considerations on the Permitted Order of the Various Processes

In the EFFA Guidance document for the production of natural flavouring ingredients, some special considerations are given to the permissible processes (including appropriate physical processes, microbiological and enzymatic processes and traditional food preparation processes (Annex II, Table 11.1)) and the order in which they can be applied for the production of natural flavouring ingredients.

From the wording of the regulation it is clear that the traditional food preparation processes take place before the physical, enzymatic and microbiological processes are applied. However, it is the understanding of EFFA that the physical, enzymatic and microbiological processes may be used sequentially and repetitively in any order as is also the case for the food preparation processes in Annex II (Table 11.1). The regulation does not provide any limitations to this extent. EFFA further emphasises that natural processes do not have to mimic the route of formation by which the flavouring substances or preparations are formed in the vegetable, animal or microbiological source and/or during traditional food processing.

A natural flavouring substance or preparation may be produced by consecutive steps involving a series of intermediates. Each step must be recognised as a natural process. However, the intermediates themselves do not have to be recognised as flavouring ingredients (substances) nor as food intended for human consumption as such.

Several processes may be used for the production of natural flavouring ingredients. To further facilitate a transparent view of the consecutive steps used in natural flavouring production and how they are applied is covered in the Regulation, Chapter V of the EFFA Guidance Document which provides an open-ended list of such processes and includes concise descriptions and conditions of use.

A parameter which requires some special attention is the temperature. According to Annex II (Table 11.1) heating, cooking, baking and frying (the latter up to 240 °C at atmospheric pressure) as well as pressure cooking (up to 120 °C) are allowed traditional food preparation processes.

Although Annex II (Table 11.1) does not clearly specify a maximum temperature for heating and baking (in contrast to frying), EFFA considers that the temperature should be limited to 400 °C. The pressure should preferably be maintained below 400 bars (conditions which may be reached in a typical supercritical extraction of herbs and spices).

Also for roasting and grilling Annex II (Table 11.1) does not specify any maximum temperature. Although it will be higher than for frying, EFFA considers also a limit of 400 °C for these processes.

Certain grill-like flavourings, when obtained at higher temperature are subject to authorisation and evaluation (per Art. 9): see section on safety evaluation.

11.5.2 Biotechnology for the Production of Natural Flavouring Ingredients

Biotechnology is a very important technique in the production of natural flavouring ingredients. As biotechnology uses mainly fermentation and microbiological processes which are considered to include "enzymatic processes" it will lead to the formation of "natural flavourings" (in compliance with the legal requirements) when the other parameters (e.g. source materials requirements, other process conditions, identified in nature in the case of substances) are met.

In the EFFA Guidance Document on the production of natural flavouring ingredients [5] a specific chapter (Chapter IV) is devoted to microbiological and enzymatic processes. According to this Guidance Document, natural flavouring ingredients can be obtained by enzymatic or microbiological processes. EFFA considers the following biological entities as biosystems that are permitted for the production of natural flavourings: bacteria, yeasts and fungi, or higher organisms such as algae, plants or animals, used as such or in cell or tissue cultures, and enzymes derived thereof. These biosystems are grown and/or maintained during a fermentation process. Apart from those biosystems, also other enzymes can be used (food enzymes that can be typically used in the process of flavouring production are alpha-amylase, polygalacturonase, lipases and alpha-glucosidase amongst others).

The EFFA Guidance Document describes the conditions and general requirements with which the biosystems and enzymes have to comply. This covers the specific requirements for culture medium, nutrients, substrates, co-factors etc.

With regard to enzymes there is a requirement that the enzymes used for the production of natural flavourings are in compliance with the Enzyme Regulation (EC) No 1332/2008 [9].

One key question that is currently being addressed is what happens when the micro-organisms that perform the fermentation are genetically engineered or if the enzymes are produced by genetically modified organisms (GMO's)? Can these flavourings be labelled as 'natural'? Is there an obligation to mention the GMO nature of the process on the label of food products?

The EU-Commission is working on a Q&A-document for labelling of flavourings in food. The provisional answer to this question (pending the formal endorsement of the Q&A-document by all Member States) reads as follows:

> *The use of the term "natural" does not exclude the production of flavourings from GMO sources. For flavourings from GMO sources, the rules governing the labelling are the GMO labelling rules like for any other ingredient.*

> *If the flavouring is produced by fermentation using a genetically modified microorganism (GMM) which is kept under contained conditions and is not present in the final product, it is not included in the scope of Regulation (EC) No 1829/2003* [10]. *This food has to be considered as having been produced with the GMM, rather than from the GMM.*
>
> *In the case a GMM derived enzyme is used during the fermentation of a non-GM flavouring, the flavouring is also out of the scope of Regulation 1829/2003 since the GMM derived enzyme itself is out of the scope ("produced with" GMM).*
>
> *If, however, the flavouring is produced, in whole or in part, from a GMO, then the flavouring needs an authorisation according to Regulation (EC) No 1829/2003 and Regulation (EC) No 1334/2008 and should be labelled as "genetically modified".*

This answer is supported by Recital 16 of Regulation (EC) No 1829/2003 [10]:

> *This Regulation should cover food and feed produced 'from' a GMO but not food and feed 'with' a GMO.* ***The determining criterion is whether or not material derived from the genetically modified source material is present in the food or in the feed.*** *Processing aids which are only used during the food or feed production process are not covered by the definition of food or feed and, therefore, are not included in the scope of this Regulation. Nor are food and feed which are manufactured with the help of a genetically modified processing aid included in the scope of this Regulation. Thus, products obtained from animals fed with genetically modified feed or treated with genetically modified medicinal products will be subject neither to the authorisation requirements nor to the labelling requirements referred to in this Regulation.*

This is also supported by the EFFA position on the interpretation of GMO Legislation in the EU in relation to flavourings which states that food ingredients such as flavourings that are produced with genetically modified micro-organisms (GMMs) are in principle out of scope of the GMO-regulation (EC) no 1829/2003, as these GMMs are used as processing aids. The same applies to flavourings produced with enzymes obtained from GMM-sources.

In conclusion flavourings produced with enzymes obtained from GMMs or with GMMs as such, can be called "natural flavourings" if the other criteria are fulfilled (source materials requirements, other process conditions, identified in nature in the case of substances) and such flavourings are out of the scope of the GMO-regulation (EC) no 1829/2003, hence no GMO-labelling is required.

11.5.3 Further Considerations on the Production of Natural Flavouring Substances and Flavouring Preparations According to the EFFA Guidance Document

In its Guidance Document, EFFA has considered other elements that have to be taken into account when assessing the compliance of flavouring preparations with the EU regulation, esp. in relation to the presence of non-volatile constituents and other ingredients (e.g., solvents).

According to the EFFA interpretation, due to the way they are produced flavouring preparations are complex mixtures containing more than volatile flavouring molecules. Therefore the presence of constituents that are naturally occurring in the flavouring preparation due to their presence in the source materials, e.g. intrinsic fruit water, as well as foods / food ingredients used during the manufacturing process as processing aids, e.g. ethanol, edible oil, acetic acid, can be considered according to EFFA as part of the flavouring preparation.

Flavouring preparations shall be produced in line with appropriate processes as described in Article 3(2)(d) – incl. the traditional processes listed in Annex II (Table 11.1) – and Article 3(2)(k) taking into account the considerations expressed in Recital (25) and the requirements of Article 4 in order to ensure that consumers are not misled.

EFFA reminds its members that if solvents are used for extraction purposes to obtain flavouring preparations, only those listed in the EU Extraction Solvents Directive 2009/32/EC (as amended) shall be used and the applicable maximum residue levels should be observed [11]. The remaining amount of extraction solvents depending on the process, together with other intrinsic components from the respective source material e.g., fruit/plant sugars or cell water, are part of the entire flavouring preparation.

EFFA also notes that for other purposes than extraction (e.g. chromatography, crystallisation, azeotropic distillation) during the isolation / purification of the natural flavouring ingredients, the permitted solvents are not legally restricted to those listed in the Extraction Solvents Directive (as amended), however, solvents for these other uses are preferably also those solvents permitted by the Directive 2009/32/EC as amended.

EFFA further clearly stipulates in its Guidance Document that solvents or carriers, added for purposes other than extraction, e.g., for dilution, or standardization, or when used outside the conditions/limitations of EU Directive 2009/32/EC (as amended), will not become part of the entire flavouring preparation.

11.6 Analytical Methods to Assess Authenticity

Although the EFFA Guidance document [5] devotes a chapter on analytical methods to assess the authenticity of natural flavouring substances some clear warnings are given on the limitations of such analytical methods. For example EFFA cautions that the methods of analysis as listed in the EFFA Guidance Document are for the sole purpose of identifying compliance with the natural processes as described in the Guidance Document. These analytical techniques are not necessarily suitable for the identification of the natural source from which flavouring substances are produced (source authentication).

Some analytical methods are listed which can be used in the discrimination of the source of the flavouring substances. These comprise fingerprint analysis (such as the examination of impurities characteristic of a natural process or a non-natural process, including gas-chromatographic analysis), chiral analysis (even more effective when coupled with an isotopic method such as chiral GC/IRMS (isotopic ratio mass spectrometry)), site specific deuterium NMR, and more sophisticated techniques such as IRMS / SIRA (Stable Isotope Ratio Analysis) coupled with High

Resolution Gas Chromatography (HRGC) or Chiral Multidimensional Gas Chromatography (MDGC) (**see also Chap. 10 of this book**).

11.7 Labelling of Flavourings (B2B)

11.7.1 General Labelling Requirements

The 'business to business' (B2B) labelling requirements for flavourings are stipulated in Article 15 of the Flavouring Regulation. These requirements include the 'sales description' and are similar to those laid down in the former Directive 88/388/EEC with three additions:

- *date of minimum durability or use-by-date;*
- *allergen information according to the food labelling Directive 2000/13/EC* [12] *as amended* [which is now replaced by the Food Information to Consumers (Labelling) Regulation (EU) No 1169/2011 [13]*;*
- if necessary, *the special conditions for storage and/or use.*

This information has to be mentioned on a label on the packaging or container of the product. It is not sufficient that this information is only provided on the accompanying documents.

For general labelling purposes Art. 15(1)(a) is of particular importance as it allows to label a flavouring formula simply with the word 'flavouring' (which can be the labelling of choice for any flavouring) or in combination with a specific name or description of the flavouring. In particular for non-natural flavourings (since there are special provisions foreseen in Art. 16 for natural flavourings, see below) this is relevant. For example a non-natural flavouring which smells like apple can be labelled as "apple flavouring" even if the flavouring components are not coming from apple. Likewise, thermal processed flavourings which have a savoury smell and remind of barbecued or grilled meat (e.g., chicken) can be labelled as "barbecue flavouring" or "roasted chicken flavouring" if the flavouring has this smell or properties that are characteristic for the used descriptor(s). In those examples "apple" or "barbecue" or "roasted chicken" are a "more specific name or description" and the resulting sales description is in compliance with Art. 15(1)(a).

The EFFA Guidance Document further notes that where applicable (i.e. where the composition of the flavouring does allow) specific names such as 'orange oil', 'lemon oil', 'yeast extract', 'spice extracts' (i.e. mentioning the name of essential oils and extracts), and others, remain authorised.

It is important to note that for natural flavourings there is no obligation to apply exclusively the options stipulated in Art. 16, thus, a producer of a flavouring can always simply label a flavour as "flavouring" (according to Art. 15) with reference to a more specific name or description without referring to the natural status, even if it is a natural flavouring. For example a flavouring tasting/smelling like vanilla can be labelled as "vanilla flavour" irrespective of its source, production method and even if it is obtained via natural ways (**see also Chap. 1 of this book**). A reason could be if the flavouring is not exclusively (and not at least for 95%) obtained from vanilla beans (vanilla pods) but also from other natural sources and has an overall vanilla smell but would no longer comply with the rules stipulated in Art. 16(4).

11.7.2 Labelling of Natural Flavourings

Art. 16 sets out the specific requirements for the use of the term 'natural' in the B2B-labelling of natural flavourings and describes the rules. In order to use a reference to 'natural' in the labelling, the overall condition that needs to be met is that the flavouring component can only contain natural flavouring substances and/or flavouring preparations. In other words, the entire flavouring part has to be natural. There are four possible terms for the sales description of natural flavourings (see Article 16 paragraphs 3–6), namely:

- 'Natural flavouring substance(s)' (Art. 16(3));
- 'Natural <X > flavouring' (Art. 16(4));
- 'Natural <X> flavouring with other natural flavourings' (Art. 16(5));
- 'Natural flavouring' (Art. 16(6)),

where "<X>" stands for the food(s), food category or source(s) referred to in the labelling.

As mentioned above the use of these terms is optional, since 'flavouring' or 'a more specific name or description of the flavouring' remains possible (in line with Art. 15), without reference to the word "natural" also for natural flavourings.

(a) Natural flavouring substances

Art. 16(3) stipulates the labelling for natural flavouring substances as follows:

3. The term 'natural flavouring substance(s)' may only be used for flavourings in which the flavouring component contains exclusively natural flavouring substances.

This option can only be used if the flavouring contains exclusively chemically defined natural flavouring substances as defined by Art. 3(2)(c). However, the EFFA Guidance Document foresees the option, if preferred and if applicable, to use the term *'natural < X > flavourings'*, *'natural <X> flavouring with other natural flavourings'* or *'natural flavouring'* as an alternative.

(b) Natural <X > flavouring

Art. 16(4) defines the rules for the labelling of natural flavourings (natural flavouring substances and/or flavouring preparations) with a reference to the food source ("<X>") as follows:

4. The term 'natural' may only be used in combination with a reference to a food, food category or a vegetable or animal flavouring source if the flavouring component has been obtained exclusively or by at least 95 % by w/w from the source material referred to.

In order to be able to use this labelling option (Natural flavouring with reference to the food source from which the flavouring is obtained), three key criteria must be met: (1) the entire flavouring must be natural (not only the 95%-part); (2) the flavouring component should be obtained at least 95% by w/w from the source material referred to; and (3) the flavour perception of the named source needs to be easily recognized. Moreover, as indicated in Recital (26), there are also specific requirements for the other 5%-part: "*As the use of flavourings should not mislead the consumer, the other maximum 5 % can only be used for standardization or to give a, for example, more fresh, pungent, ripe or green note to the flavouring.*"

In addition, EFFA stipulates in its Guidance Document that the 5% part may not reproduce the total flavour profile of the 95% part from the source material referred to; otherwise the flavouring does not meet the provisions of Article 16(4).

In its Guidance Document, EFFA provides some clear rules and interpretation on the compositional assessment. The 95/5-ratio is examined on the basis of the formula composition. At least 95% by w/w of the flavouring component (i.e. flavouring preparations and/or natural flavouring substances as defined under Article 3 of the Regulation) have to be obtained from the source material(s) referred to. With regard to the compositional assessment, the wording of the Regulation indicates that when considering a "flavouring preparation" – as part of the "flavouring component" of a given "natural <X > flavouring" – in the quantitative determination pursuant to Art. 16(4), the entire "flavouring preparation" has to be taken into account and not only the volatile fraction (e.g. as determined by gas-chromatography).

(c) Natural <X > flavouring with other natural flavourings

For flavourings not obtained by at least 95% from the named source (and in combination with other natural flavourings) another labelling option is provided for by Art. 16(5):

5. The term 'natural "food(s) or food category or source(s)" flavouring with other natural flavourings' may only be used if the flavouring component is partially derived from the source material referred to, the flavour of which can easily be recognised.

To be able to use this labelling option ('*Natural <X> flavouring with other natural flavourings*' where <X> refers to the food(s) or food category or source(s)) the first requirement is of course that all flavouring components are natural, but it is also required that flavouring materials derived from all the named source(s) are present <u>and</u> that their flavour can easily be recognised in the final food. With regard to the requirement that "*the flavour can easily be recognised*" EFFA recommends that the qualification 'can easily be recognised' be based on expert opinion, e.g. by a flavourist or a sensory panel evaluating the final food product to which the flavouring has been added.

Table 11.2 Some examples of (natural) flavouring labelling (some as presented in the EFFA Guidance Document) (for more examples refer to Attachment IV to the EFFA Guidance Document [4])

Example of 'Flavouring' or a more specific name or description (Art. 15(1)(a))	**Labelling (Art. 15(1))**
80% w/w flavouring substances (natural or not) providing a banana flavour. 20% w/w flavouring substances (natural or not) used to introduce other notes [overall flavour is like banana]	'Flavouring' *or* 'Banana flavouring'
75% w/w flavouring materials providing a grilled chicken flavour 25% w/w flavouring substance (natural or not) providing spicy notes [overall flavour is like grilled chicken]	'Flavouring' *or* 'Grilled chicken flavouring'
Example of 'Natural flavouring substance(s)' (Art. 16(3))	**Natural Labelling (Art. 16(3)/(4))**
95% w/w natural flavouring substances derived from mint (e.g. menthol ex arvensis). 5% w/w natural flavouring substances derived from orange (e.g. limonene ex orange) which is used to introduce a special note	'Natural flavouring substances' *or* 'Natural mint flavouring' (Art. 16(4))
100% w/w menthol (natural)	'Natural flavouring substances' or 'Natural flavouring substances (menthol)'
Example of 'Natural <X> flavouring' (Art. 16(4))	**Natural Labelling (Art. 16(4))**
95% w/w flavouring preparations and/or natural flavouring substances derived from mint (e.g. mint oil and/or menthol ex arvensis). 5% w/w natural flavourings derived from other natural sources (e.g. orange oil or limonene ex orange) which is used to introduce a special note	'Natural mint flavouring'
97% w/w flavouring preparations and/or natural flavouring substances derived from raspberry (e.g. raspberry distillate and raspberry isolate). 3% w/w natural flavourings derived from other natural sources (e.g. natural flavouring substances used to adjust natural variations).	'Natural raspberry flavouring'
Example of 'Natural <X> flavouring with other natural flavourings' (Art. 16(5))	**Natural Labelling (Art. 16(5))**
85% w/w flavouring preparations and/or natural flavouring substances derived from mint (e.g. mint oil and/or menthol ex arvensis) [mint flavour easily recognized] 15% w/w natural flavourings from other natural sources (e.g. orange oil or limonene ex orange) which is used to introduce a special note	'Natural mint flavouring with other natural flavourings'

(continued)

Table 11.2 (continued)

80% w/w flavouring preparations and/or natural flavouring substances derived from banana [banana flavour easily recognized] 20% w/w natural flavourings from other natural sources (e.g. vanilla) which is used to introduce a round note or for standardisation	'Natural banana flavouring with other natural flavourings'
Example of 'Natural flavouring' (Art. 16(6))	**Natural Labelling (Art. 16(6))**
60% w/w flavouring preparations and/or natural flavouring substances derived from banana 40% w/w derived from other natural sources (e.g. mango, pear, lemon, etc). The overall flavour-profile is fruity but the source materials (banana, mango, pear, lemon) cannot easily be recognised.	'Natural flavouring'

(d) Natural flavouring

Only when the criteria for the option '*Natural <X>flavouring*' and '*Natural <X> flavouring with other natural flavourings*' are not met can the term '*Natural flavouring*' be used. This means that this term '*Natural flavouring*' is only possible to be used for flavourings when a clear relationship between the different source materials used in the flavouring component and the overall flavour-profile does not exist. Also in case of uncertainty about this relationship, it is recommended to use the term *'natural flavouring'*.

This can be understood from the wording of Art. 16(6):

6. The term 'natural flavouring' may only be used if the flavouring component is derived from different source materials and where a reference to the source materials would not reflect their flavour or taste.

Although Art. 16(6) refers to "*different source materials*" (in plural) it is EFFA's understanding that in case one source material is used the same principle can be adhered to.

For further information we refer to the EFFA Guidance Document on the Flavouring Regulation [4] and in particular the various examples on labelling in Attachment IV of the Guidance Document.

Some examples of natural flavouring labelling (as presented in the EFFA Guidance Document) are provided in Table 11.2 – for more examples the author refers to Attachment IV to the EFFA Guidance Document.

11.7.3 Considerations on Different Interpretations of the 95/5-Rule (Art. 16(4))

EFFA has been made aware of some different views of the application of the 95/5-rule (in relation to Art. 16(4)), in particular a different interpretation of the German Food Control Authorities. This has led to a decision of the OLG (Oberlandesgericht) in Düsseldorf (which is the Higher Regional Court of Düsseldorf) on 21/03/2012. This decision related to a final product (a strawberry yoghurt) and the declaration of the "natural strawberry flavour" (according to Art. 16(4)) that was challenged.

As a consequence of this Court Decision, EFFA has contacted the European Commission (DG SANTE) who has discussed the matter (EFFA interpretation versus the views of the German Food Control Authorities) with the EU Member States during a Standing Committee Meeting (Food Chain and Animal Health, Toxicology Section) on 31/07/2012.

The conclusion of this Standing Committee meeting (based on all the elements and background information provided by EFFA) was:

> *the Commission services consider that when assessing the "95/5-ratio" of a "flavouring component", it is necessary to take into account the amount of "flavouring preparations" and "natural flavouring substances" from the labelled source compared to the total of amount of flavouring component. For the purpose of this measurement the entire flavouring preparation from the labeled source should be included.*

This conclusion was formally published by the EU-Commission in a "Note to the Standing Committee on Food Chain and Animal Health, Toxicology Section" of 22/01/2013 [14].

Subsequent to this meeting and the publication of the note, EFFA received a letter from the EU-Commission (DG SANTE) stating the same conclusions and also noting that this interpretation was supported by the majority of the Member States in the Standing Committee meeting. In another (separate letter) from the EU-Commission (DG SANTE) to EFFA the Head of Unit stated that the Commission can agree with EFFA's view on the assessment of the 95/5-ratio of the flavouring component.

However, the EU-Commission also pointed out in their communications that ultimately it is for the Court of Justice of the European Union to provide legally binding interpretation of the provisions of the Regulation.

As a follow-up, EFFA has generated a pictorial representation (graphical depiction) of the correct interpretation of Art. 16(4) and the application of the 95/5-ratio with some examples of "Natural <X> Flavouring". In particular the labelling rules are illustrated through examples of natural strawberry flavouring and natural lemon flavourings: this graphical depiction has recently been added as Attachment XI to the EFFA Guidance Document which is available on EFFA's public website [4].

11.8 Safety Evaluation of Flavourings and Their Inclusion in the EU Union List

In relation to the discussion on the definition of flavouring preparations (see above) it is necessary to focus also on the difference between "from food sources" and "from non-food sources". Although all flavouring preparations are by definition natural (see Art. 3(2)(d)), only those obtained from food sources (covered by the Definition of Art. 3(2)(d)(i)) can be used without any need for safety evaluation as stipulated in Art. 8(1)(a). According to Art. 9(b), for flavouring preparations obtained from "material [...] other than food" (i.e. referred to in Art. 3(2)(d)(ii)) an evaluation and approval (by the European Food Safety Authority (EFSA)) is required.

For this reason, the definition according to Art. 3(2)(d)(i) specifies that any (traditional food preparation) process (i.e. Annex II, Table 11.1 process applied to the food) should be fit "for human consumption". For flavouring preparations from material other than food, defined by Art. 3(2)(d)(ii), there is no reference to the wording *"for human consumption"*. The definition only states that the material can be *"taken as such or prepared by one or more of the traditional food preparation processes"*. These are the preparations which need evaluation and approval.

In contrast to flavouring preparations, there is no distinction between natural flavouring substances from food sources and natural flavouring substances from non-food sources. Indeed, all flavouring substances (irrespective whether they are natural or not and regardless from which sources they are obtained) require an evaluation and approval, as stipulated by Art. 9(a).

All flavouring substances currently in use and legally introduced in the EU market are in the process of safety evaluation, carried out by the European Food Safety Authority (EFSA), which is in progress.

The basis for the evaluation program was the Regulation 2232/96/EC [15], a Community Procedure that established rules such as the need for drawing up a "List of Flavouring Substances". Initially this would become the EU-Register and later the Union List (see below). This Community Procedure also established a Notification procedure (procedure in which Member States had to notify the legal use and market introduction of certain flavouring substances in each EU-country to the EU-Commission), the adoption of the EU-Register (in 1999) [16] and finally the adoption of the "Evaluation Program" (in 2000). This Regulation further foresaw, after the completion of the evaluation program, for the adoption of "the list of flavouring substances" which is today known as the Union List of flavouring substances. The measures necessary for the adoption of the evaluation program were provided by the Regulation 1565/2000/EC [17].

All the flavouring substances which are approved for use in and on food in the EU are now listed on a positive list, the so-called *"EU Union List of flavouring substances"*. This list has been established and published with the Implementing Regulation (EU) No 872/2012 of 1 October 2012 [18]. The list currently contains more than 2500 chemically defined flavouring substances which can legally be used in or on foods in the EU (although for some substances certain restrictions (i.e. maximum use levels to certain food categories) apply, such as caffeine, quinine salts, ammonium chloride etc). This list does not differentiate between natural flavouring substances and other flavouring substances. The list with the chemically defined flavouring substances is in fact part (Part A) of a broader list, the so-called *"Union List of flavourings and food source materials"*, which is Annex I to the Flavouring Regulation. Parts B-F are currently empty and would in theory in the future constitute the following categories of flavourings:

- Part B: Flavouring preparations [from non-food sources]
- Part C: Thermal process flavourings
- Part D: Flavour precursors
- Part E: Other flavourings
- Part F: Source materials

It should be noted that only those flavourings/source materials for which an evaluation and approval is required (as listed in Art. 9) should ultimately be listed on Parts B-F.

More precisely these are the following categories (Art. 9):

- (b) flavouring preparations referred to in Article 3(2)(d)(ii), i.e. flavouring preparations from non-food sources;
- (c) thermal process flavourings obtained by heating ingredients which fall partially or totally within Article 3(2)(e)(ii) (i.e. from source material other than food) and/or for which the conditions for the production of thermal process flavourings and/or the maximum levels for certain undesirable substances set out in Annex V are not met;
- flavour precursors referred to in Article 3(2)(g)(ii) (i.e. from source material other than food);
- other flavourings referred to in Article 3(2)(h);
- source materials other than food referred to in Article 3(2)(j)(ii) (i.e. source material other than food).

In particular Part B would in the future only contain flavouring preparations from non-food sources. However, to the best of the author's knowledge no flavouring preparations from non-food sources are currently on the EU market for which an evaluation and approval is required and for which an application has been submitted by any company for its safety assessment. It is thus anticipated that in practice Part B of the Union List will remain empty since all flavouring preparations in use in EU are from food sources and do not require a safety evaluation.

Also with regard to Parts C, D and F, the author is not aware of applications which have up to now been submitted to the Risk Assessment and Risk Management Authorities in EU (EFSA and EU-Commission, resp.) by the flavour industry, although new applications might be submitted in future resulting from new research and innovation. However, in relation to Part E (Other flavourings) a few applications have been submitted by the flavour industry for the evaluation and authorisation of some "other flavourings" such as grill-like flavourings and a complex mixture called "Rum Ether". Some of these flavourings are currently still under evaluation by EFSA and may ultimately appear on Part E of the Union List, pending the completion of their evaluation.

For more information on the status of the evaluation by EFSA of the flavourings which are currently still under evaluation, the author refers to EFSA's Register of Questions: see http://registerofquestions.efsa.europa.eu/roqFrontend/ListOfQuestionsNoLogin?0.

Acknowledgements Although the views expressed in this chapter are purely those of the author and may not in any circumstances be regarded as stating an official position of the European Flavour Association, the author is referring to the EFFA Guidance Documents as basis for the clarifications. The author wishes to thank the resp. EFFA Working Groups who have elaborated the guidance documents and the members of the Working Groups who have reviewed the draft manuscript.

References

1. Regulation (EC) No 1334/2008 of the European Parliament and of the Council of 16 December 2008 on flavourings and certain food ingredients with flavouring properties for use in and on foods and amending Council Regulation (EEC) No 1601/91, Regulations (EC) No 2232/96 and (EC) No 110/2008 and Directive 2000/13/EC (OJ L 354/34, 31.12.2008).
2. Council Directive 88/388/EC of 22 June 1988 on the approximation of the laws of the Member States relating to flavourings for use in foodstuffs and to source materials for their production (OJ L 184, 15.7.1988, p. 61).
3. Regulation (EC) No 1333/2008 of the European Parliament and of the Council of 16 December 2008 on food additives (OJ L 354/16, 31.12.2008).
4. EFFA Guidance Document on the EC Regulation on Flavourings, 4th Revised version (V5.0), 29/02/16. The EFFA website: http://www.effa.eu ‡ Library ‡ Guidance Documents.
5. EFFA Guidance Document for the Production of Natural Flavouring Substances and (Natural) Flavouring Preparations in the EU, Revised version (V3.0), 29/09/16. The EFFA website: http://www.effa.eu ‡ Library ‡ Guidance Documents.
6. IOFI. Statement on the identification in nature of flavouring substances, made by the working group on methods of analysis of the International Organization of the Flavour Industry (IOFI). Flavour Fragr J. 2006;21:185.
7. IOFI (WGMA). Guidelines for LC–MS identifications of flavouring substances in nature, made by the working group on methods of analysis of the International Organization of the Flavor Industry (IOFI). Flavour Fragr J. 2010;25:2–3.
8. International Organization for Standardization (ISO): International Standard 9235 – Aromatic natural raw materials – Vocabulary. ISO 1997©.
9. Regulation (EC) No 1332/2008 of the European Parliament and of the Council of 16 December 2008 on food enzymes and amending Council Directive 83/417/EEC, Council Regulation (EC) No 1493/1999, Directive 2000/13/EC, Council Directive 2001/112/EC and Regulation (EC) No 258/97 (OJ L 354/7, 31.12.2008).
10. Regulation (EC) No 1829/2003 of the European Parliament and of the Council of 22 September 2003 on genetically modified food and feed (OJ L 268/1, 18.10.2003).
11. Directive 2009/32/EC of the European Parliament and of the Council of 23 April 2009 on the approximation of the laws of the Member States on extraction solvents used in the production of foodstuffs and food ingredients (OJ L 141/3, 6.6.2009).
12. Directive 2000/13/EC of the European Parliament and of the Council of 20 March 2000 on the approximation of the laws of the Member States relating to the labelling, presentation and advertising of foodstuffs (OJ L 109/29, 6.5.2000).
13. Regulation (EU) No 1169/2011 of the European Parliament and of the Council of 25 October 2011 on the provision of food information to consumers, amending Regulations (EC) No 1924/2006 and (EC) No 1925/2006 of the European Parliament and of the Council, and repealing Commission Directive 87/250/EEC, Council Directive 90/496/EEC, Commission Directive 1999/10/EC, Directive 2000/13/EC of the European Parliament and of the Council, Commission Directives 2002/67/EC and 2008/5/EC and Commission Regulation (EC) No 608/2004 (OJ L 304/19, 22.11.2011).

14. Note to the Standing Committee on Food Chain and Animal Health, Toxicology Section, European Commission, Brussels, 23 Jan 2013. http://ec.europa.eu/food/sites/food/files/safety/docs/reg-com_toxic_sum_31012013_flavourings_en.pdf.
15. Regulation (EC) No 2232/96 of the European Parliament and of the Council of 28 October 1996 laying down a Community procedure for flavouring substances used or intended for use in or on foodstuffs (OJ l 299/1, 23.11.96).
16. Commission Decision 1999/217/EC of 23 February 1999 adopting a register of flavouring substances used in or on foodstuffs drawn up in application of Regulation (EC) No 2232/96 of the European Parliament and of the Council of 28 October 1996. (OJ L 84/1, 27.3.1999).
17. Commission Regulation (EC) No 1565/2000 of 18 July 2000 laying down the measures necessary for the adoption of an evaluation programme in application of Regulation (EC) No 2232/96 of the European Parliament and of the Council (OJ L 180/8, 19.7.2000).
18. Commission Implementing Regulation (EU) No 872/2012 of 1 October 2012 adopting the list of flavouring substances provided for by Regulation (EC) No 2232/96 of the European Parliament and of the Council, introducing it in Annex I to Regulation (EC) No 1334/2008 of the European Parliament and of the Council and repealing Commission Regulation (EC) No 1565/2000 and Commission Decision 1999/217/EC (OJ L 267/1, 2.10.2012).

Index

W. Schwab et al. (eds.), *Biotechnology of Natural Products*,
https://doi.org/10.1007/978-3-319-67903-7

N

FSC
www.fsc.org
MIX
Papier | Fördert gute Waldnutzung
FSC® C083411

Kolibri 360 GmbH
Ferdinand-Jühlke-Straße 7
99095 Erfurt, Deutschland
produktsicherheit@kolibri360.de